Vehicle Design Processes

Vehicle Design Processes

Editors

Ralf Stetter
Udo Pulm
Markus Till

MDPI

Basel • Beijing • Wuhan • Barcelona • Belgrade • Novi Sad • Cluj • Manchester

Editors

Ralf Stetter
Mechanical Engineering
Ravensburg-Weingarten University
Weingarten
Germany

Udo Pulm
Mechanical Engineering and Production
Hamburg University of Applied Sciences
Hamburg
Germany

Markus Till
Mechanical Engineering
Ravensburg-Weingarten University
Weingarten
Germany

Editorial Office
MDPI
St. Alban-Anlage 66
4052 Basel, Switzerland

This is a reprint of articles from the Special Issue published online in the open access journal *Vehicles* (ISSN 2624-8921) (available at: www.mdpi.com/journal/vehicles/special_issues/design_processes).

For citation purposes, cite each article independently as indicated on the article page online and as indicated below:

Lastname, A.A.; Lastname, B.B. Article Title. *Journal Name* **Year**, *Volume Number*, Page Range.

ISBN 978-3-0365-8665-6 (Hbk)
ISBN 978-3-0365-8664-9 (PDF)
doi.org/10.3390/books978-3-0365-8664-9

Contents

Preface . . . . **vii**

Sven Schumacher, Stefan Schmid, Philipp Wieser, Ralf Stetter and Markus Till
Design, Simulation and Optimization of an Electrical Drive-Train
Reprinted from: *Vehicles* **2021**, *3*, 390-405, doi:10.3390/vehicles3030024 . . . . **1**

Kevin Holder, Sven Schumacher, Matthias Friedrich, Markus Till, Ralf Stetter and Walter Fichter et al.
Digital Development Process for the Drive System of a Balanced Two-Wheel Scooter
Reprinted from: *Vehicles* **2021**, *3*, 33-60, doi:10.3390/vehicles3010003 . . . . **17**

Raphael Mieth, Frank Gauterin, Felix Pauli and Harald Kraus
Transfer of Statistical Customer Data into Relevant Parameters for the Design of Vehicle Drive Systems
Reprinted from: *Vehicles* **2022**, *4*, 137-144, doi:10.3390/vehicles4010009 . . . . **45**

Raphael Mieth, Fabian Markschies, Ruixin Zhou, Frank Gauterin and Arnd Stephan
Enhanced Derivation of Customer-Specific Drive System Design Parameters with Time Frame-Based Maximum Load Analysis
Reprinted from: *Vehicles* **2023**, *5*, 306-320, doi:10.3390/vehicles5010017 . . . . **53**

Manh-Kien Tran, Mobaderin Akinsanya, Satyam Panchal, Roydon Fraser and Michael Fowler
Design of a Hybrid Electric Vehicle Powertrain for Performance Optimization Considering Various Powertrain Components and Configurations
Reprinted from: *Vehicles* **2020**, *3*, 20-32, doi:10.3390/vehicles3010002 . . . . **68**

Jan Schütte and Walter Sextro
Tire Wear Reduction Based on an Extended Multibody Rear Axle Model
Reprinted from: *Vehicles* **2021**, *3*, 233-256, doi:10.3390/vehicles3020015 . . . . **81**

Martin Dorynek, Lisa-Theres Derle, Martin Fleischer, Alex Thanos, Paul Weinmann and Michael Schreiber et al.
Potential Analysis for a New Vehicle Class in the Use Case of Ride-Pooling: How New Model Developments Could Satisfy Customers and Mobility Makers
Reprinted from: *Vehicles* **2022**, *4*, 199-218, doi:10.3390/vehicles4010013 . . . . **105**

Bruce W. Jo
Numerical Study of Longitudinal Inter-Distance and Operational Characteristics for High-Speed Capsular Train Systems
Reprinted from: *Vehicles* **2022**, *4*, 30-41, doi:10.3390/vehicles4010002 . . . . **125**

Jinhui Huang, Enrong Wang and Hailong Zhang
Analysis and Research on the Comprehensive Performance of Vehicle Magnetorheological Regenerative Suspension
Reprinted from: *Vehicles* **2020**, *2*, 576-588, doi:10.3390/vehicles2040033 . . . . **137**

Chun Lu, Jiliang Mo, Ruixue Sun, Yuanke Wu and Zhiyong Fan
Investigation into Multiaxial Character of Thermomechanical Fatigue Damage on High-Speed Railway Brake Disc
Reprinted from: *Vehicles* **2021**, *3*, 287-299, doi:10.3390/vehicles3020018 . . . . **150**

Casey J. Den Ouden, Paul S. Wills, Lucas Lopes, Joshua Sanderson and Bing Ouyang
Evolution of the Hybrid Aerial Underwater Robotic System (HAUCS) for Aquaculture: Sensor Payload and Extension Development
Reprinted from: *Vehicles* **2022**, *4*, 390-408, doi:10.3390/vehicles4020023 . . . . **163**

Cagri Un and Kadir Aydın
Modernization of Fire Vehicles with New Technologies and Chemicals
Reprinted from: *Vehicles* **2023**, *5*, 682-697, doi:10.3390/vehicles5020037 . . . . **182**

Preface

This reprint presents the current status of research activities concerning vehicle design processes. The design of vehicles is one of the most challenging tasks in engineering because of several reasons. The enormous consumer expectations as well as the intensive global competition aggravates vehicle design. Cost-driven design is a necessity and vehicles need to be economical in production, operation, and recycling; in fact, sustainable design is also imperative for ecological vehicles. The dynamics of vehicles have to be considered in the design of all components and light-weight design is of fundamental importance. Consumers expect convincing functional performance, high product quality, appealing appearance, high reliability, interconnected functionality as well as comprehensible and appealing user interfaces. More and more, additional services are connected to vehicles. These enormous requirements lead to complex multi-domain design processes of vehicles because most of the important decisions are made in the design phase. Production optimization and intelligent operation are important topics, but flaws and insufficiencies in the design stage lead to enormous expenditures in later stages and less-than-perfect products. The design processes of vehicles involve thousands of engineers are spread globally and need to consider multiple product versions and variants as well as multi-company product platforms. Very often, testing necessities and legal issues play an important role in these processes and the economic and ecological quality of the product has to be monitored throughout the processes. Even in the early stages, vehicle safety and ergonomic quality need to be considered. Needless to say, only digital support makes these processes feasible. For all domains, powerful computer tools for synthesis, analysis, evaluation, and optimization were created and numerous attempts try to sensibly link the data used in all these tools. However, many domain-specific and generic data formats as well as the sheer size of the data still create serious problems. It is important to note that design is also connected with scheduling and project management because certain design decisions can lead to long-term testing and production preparation processes. The listed challenges concerning the multi-domain design processes of vehicles lead to a prominent need for research activities aimed at supporting the designers in this endeavor; this reprint presents a collection of research activities in this area.

Ralf Stetter, Udo Pulm, and Markus Till
Editors

vehicles

MDPI

Article

Design, Simulation and Optimization of an Electrical Drive-Train

Sven Schumacher [1], Stefan Schmid [2], Philipp Wieser [2], Ralf Stetter [1,*] and Markus Till [1]

[1] Department of Mechanical Engineering, Ravensburg-Weingarten University (RWU), 88250 Weingarten, Germany; ss-191205@hs-weingarten.de (S.S.); markus.till@rwu.de (M.T.)

[2] AVL Deutschland GmbH, 70567 Stuttgart, Germany; stefan.schmid2@avl.com (S.S.); philipp.wieser@avl.com (P.W.)

* Correspondence: ralf.stetter@rwu.de

Abstract: Today, design engineers engaged in the development of a high-performance electrical drive-train are challenged by the multitude of possible topological choices and numerous mutually interconnected physical phenomena. Development teams around the globe struggle with this challenge; usually they employ several tools for simulation and topology optimization and transfer multiple versions of their product models in a mainly manual process. The research presented in this paper aims to explore a holistic possibility to realize a sensible analysis-synthesis cycle that takes into consideration current developments in design, simulation and optimization processes. This kind of process can enhance the transparency of design decisions, can reduce the risk of design and process flaws and can support the approach toward a holistic optimum. The investigation starts with the development of the topological concept of the drive-train and continues over the interconnected simulation of several decisive properties of the drive-train. Obviously, these properties concern several domains (mechanical, electrical, thermal and the control domain). The optimization of the drive-train takes into consideration the main requirement—in the investigated example, which is a formula student drive-train—the lap time. The result is a holistic concept for a design, simulation and optimization approach that considers topological variety, interconnected multi-domain simulation and a continuous connection to the decisive requirements.

Keywords: vehicle design processes; optimization; electrical drive-train; simulation

Citation: Schumacher, S.; Schmid, S.; Wieser, P.; Stetter, R.; Till, M. Design, Simulation and Optimization of an Electrical Drive-Train. *Vehicles* **2021**, 3, 390–405. https://doi.org/10.3390/vehicles3030024

Academic Editor: Adolfo Dannier

Received: 16 June 2021
Accepted: 16 July 2021
Published: 27 July 2021

1. Introduction

The automotive industry is currently in a major transition process focusing on e-mobility and the ongoing digitalization of the development process. According to recent research, the percentage of electric cars in new vehicle registrations in the EU will increase to 67% in the next 15 years [1]. This causes serious challenges for the automotive OEMs (original equipment manufacturers) regarding the choice of the concept, as electric power-trains allow very high flexibility in terms of the topology. In this research work, a holistic way was explored to realize a meaningful analysis-synthesis cycle that takes into account current developments in design, simulation and optimization methods.

To investigate this research question, a high-performance drive-train for a Formula Student race car was developed. The drive-train is intended to be used in the electric racing car of the Formula Student Team Weingarten. The team has developed previously internal combustion vehicles and will participate in the upcoming season with an electric racing car for the first time. At the beginning of the development, no vehicle data or previous experience with an electric race car were available. For this reason, this project is a good example to describe the complete development process starting from the requirements and ending with the system validation. The resulting drive-train of the Formula Student vehicle is shown in the following (Figure 1). The figure shows the wheel assemblies consisting of wheel hubs, planetary gearboxes, electric motors and cooling jackets. In the middle of the picture is the accumulator (colored yellow) with its housing. Above the accumulator are the inverters with cooling plates and in the rear the two air-water heat exchangers with the fans and expansion reservoirs are located.

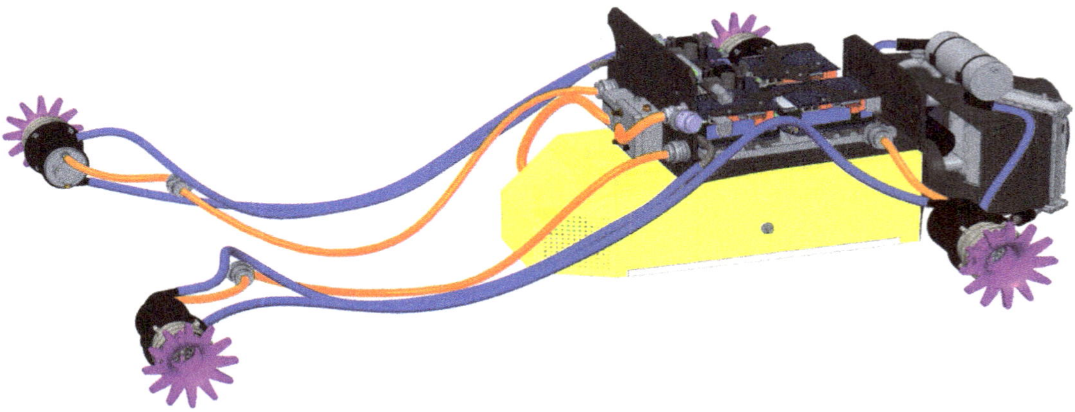

Figure 1. Drive-train of the electric race car with four motor packages, the accumulator, inverter and radiators.

In order to demonstrate the underlying product development process, the article is organized as follows. The general processes are explained in Section 2. Section 3 emphasizes the conception of the drive-train under development, Section 4 the synthesis of the simulation of this drive-train and Section 5 an expansion of this simulation. The focus of Section 6 is the validation of this simulation and of Section 7 the resulting optimization of the concept of the drive train. Section 8 describes the integration of different types of virtual, real-time prototype testing. Finally, conclusions are drawn and an outlook is given in Section 9.

2. Design, Simulation and Optimization Processes

For the development of the electrical drive-train, a continuous design process starting with an intensive analysis of the requirements was carried out. This strong emphasis on requirements is present in all current model-based systems engineering (MBSE) approaches. In September 2020, the well-known guideline VDI/VDE 2206 was republished as a draft under the new name "Development of cyber-physical mechatronic systems (CPMS)" [2]. This guideline basically offers the user support for all tasks in the development of cyber-physical mechatronic systems by presenting the main logical relationships. The core content of the guideline is an inherent flow logic for the development of mechatronic and cyber-physical systems; this is represented as an updated and extended V-model [3]. The central objective of the revision of the guideline is to adapt to greatly changed industrial environments; these changes result from the general trend towards cyber-physical systems, the increasing use of artificial intelligence and the analysis of large amounts of data, as well as further improved computer-aided synthesis and analysis methods. Current research in this topic area includes automated parameter selection [4], domain-specific modeling approaches [5], and the study of requirements change risk [6].

The research project continues to draw on the fundamentals of Model-Based System Engineering (MBSE) [7,8]. International research groups are currently working on the integration of verification processes into MBSE [9], tracking and communicating risks through MBSE tools [10], role models for the application of MBSE [11], and modular product architectures based on MBSE [12]. Currently, graph-based design methods (compare [13] are developed as a prominent means for realizing MBSE through knowledge representation in UML and compilation by means of a design compiler (DC43, IILS GmbH, Trochtelfingen, Germany [14]) leading to a machine-executable V-model. The approach includes representations of requirements [15], functions [16] and abstract physics [17] and was demonstrated on multiple examples such as the design of a balanced two-wheel scooter [18]. In the given case, the general procedure can be described applying a V-model which is included in the main V-model representing the development of the whole vehicle (Figure 2).

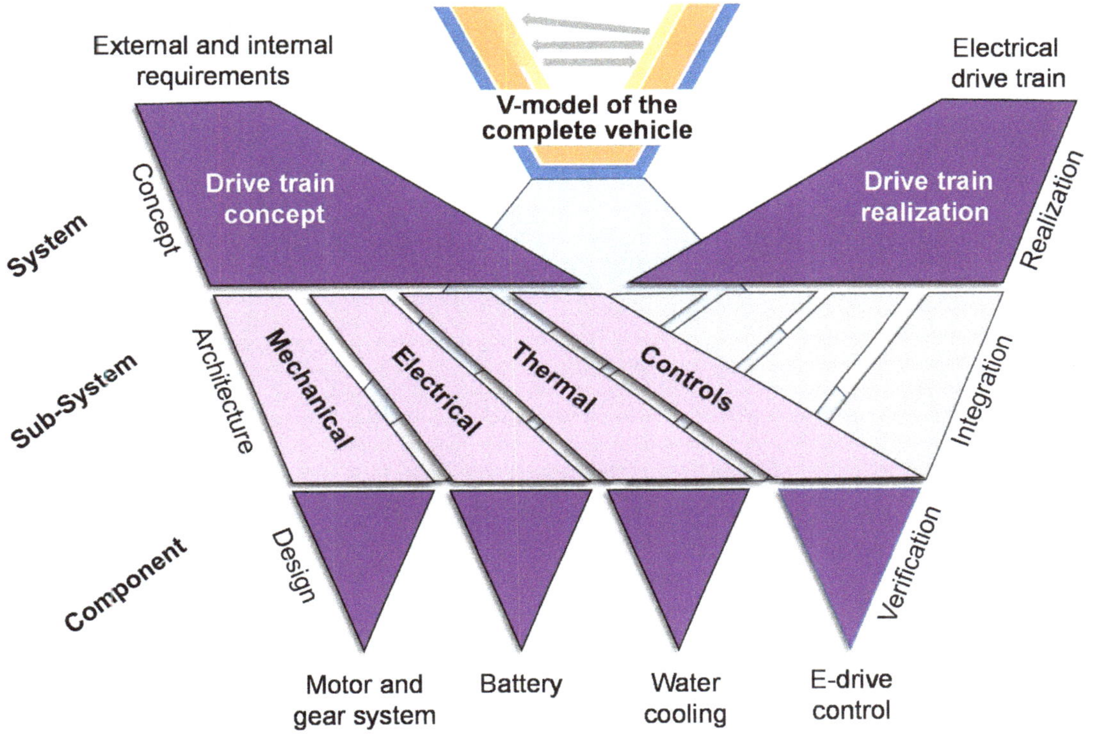

Figure 2. Process overview of design and simulation activities in the drive-train development.

The shown process addresses central challenges in the development of high-performance electrical drive trains which result from the multitude of possible topological choices and numerous mutually interconnected physical phenomena. In common industrial practice, several tools for simulation and topology optimization are applied. This requires the transfer of multiple versions of the product models; frequently many process steps have to be carried out manually. The shown processes offer a holistic possibility to realize a sensible analysis-synthesis cycle that takes into consideration current developments in design, simulation and optimization processes. The properties in these processes concern several domains (mechanical, electrical, thermal and the control domain). The process starts with internal and external requirements. On the system level, a domain spanning drive-train concept is being developed. On the sub-system level, the different domains find architectural solutions for this concept and allow the component design. A conscious verification and integration completes the design cycle including interconnected multi-domain simulation and a continuous connection to the decisive requirements. Consequently, from an initial phase concerning the general concept of the drive-train, several "action lines" are employed to realize important entities of the drive-train. These are verified by means of detailed simulations and integrated into a complete drive-train. The conception of this drive-train is discussed in the subsequent section.

3. Conception of the Drive-Train

For the development of the electric drive-train, the requirements are analyzed and transferred to the different sub-systems of the drive-train. The requirements are based on the technical regulations of the racing series (external requirements) and the team's demands regarding the vehicle performance (internal requirements). The requirements of the technical regulations include, for example, the maximum voltage for the high voltage system of 600 V, the maximum power of 80 kW or the maximum temperature of the accumulator cells, which is specified as 60 °C or the manufacturer's specifications. The lower temperature specification determines the maximum operating temperature [19]. The team's requirements are described more abstractly and demand the maximum possible acceleration in every driving situation, which is necessary to achieve the fastest possible lap time. Regarding the drive-train system, this goal can be achieved if the weight is minimal, high driving power is available, the mass moment of inertia is as low as possible and a low center of gravity of the vehicle is enabled. Furthermore, the driving characteristics should be as stable as possible [20]. The goal for developing the electric drive-train is to maximally exploit the external requirements to achieve the best possible internal requirements.

The concept of the electric drive-train is based on the perspective of the entire vehicle. For this purpose, the basics of the individual technical assemblies and their influence on the vehicle performance must be evaluated. Therefore, the following points are explained in this chapter:

1. Explanation of possible electric drive-train topologies and competitive analysis with subsequent choice of the drive-train topology;
2. Evaluation of the concept for the high-voltage accumulator (HV accumulator) and the choice of battery cells;
3. Selection of the gearbox and its conception with the associated boundary conditions;
4. Fundamentals of the vehicles thermal concept.

3.1. Electric Drive-Train Topologies

For the concept of the drive-train topology, there is a high degree of flexibility for electrically powered vehicles. Possible options are for example the variants with one motor and a conventional differential, the use of two motors on one axle or the use of one motor per wheel [21]. The Formula Student racing series is mainly dominated by teams using all-wheel-drive vehicles with wheel hub motors or internal-mounted motors which are connected to the wheel via drive shafts. Four main aspects were considered for the choice of the drive-train topology. The maximum possible acceleration, the efficiency of the drive system, the availability and the complexity.

The calculations and the literature show that the maximum acceleration can be achieved during accelerated circular driving if all four wheels are driven and the torque can be distributed fully variable [20]. This kind of drive control and drive topology enables a technique known as "torque vectoring", which is used for optimum traction exploitation. The second aspect concerning efficiency can also be fulfilled with an all-wheel-drive vehicle. Due to the effect of dynamic load transfer, which can be reduced but not eliminated by aerodynamics and suspension set-up, the rear axle can transfer increased force during acceleration and the front axle during braking [20]. Using an all-wheel-drive concept and four variably controllable motors, it is possible to maximize power transfer during acceleration and recover as much energy as possible during braking through "recuperation".

Finally, the drive concept will also be evaluated in terms of complexity and availability. The complexity of the all-wheel drive is significantly higher than the complexity of a single motor coupled to a conventional differential. For this reason, a motor-inverter package is selected, which was developed for electric vehicles with an all-wheel-drive concept used in the Formula Student racing series. On the one hand, this provides a competitive, reliable and rule-compliant drive package, and on the other hand, the use of the package significantly reduces the complexity, especially for the system implementation.

Furthermore, the drive package is quickly available for the team and allows early testing of the drive-train components.

3.2. Concept of the High-Voltage Accumulator

In order to recuperate as much energy as possible and to keep the transition losses as low as possible, the battery concept should be defined in order to minimize the amount of current that has to flow for the maximum drive and braking power supply. The concept for the battery is selected to ensure that it can discharge and charge with the maximum current (I). The current intensity is calculated by the maximum drive power (P) and the system voltage (U). To achieve the lowest current intensity by providing maximum power, the highest permissible voltage of 600 V has to be used. This is shown in Equation (1) and results in a maximum current of 133 A.

$$I = \frac{P}{U} = \frac{80,000\ W}{600\ V} \approx 133\ A \tag{1}$$

To achieve the lowest possible battery weight, as much energy as possible should be recuperated. For this reason, the cells must have a high charging rate. This requires high-performance cells with a high C-rate. Additionally important is a high energy density to achieve the lowest possible total weight of the accumulator. Besides the round cells, pouch cells and prismatic cells are available on the market. The pouch cell design is considered the most suitable for the race car application, as they have a very high energy density and a large surface area for the cooling of the cells [22]. Furthermore, these cells are available with high C-rates on the market and with the appropriate approvals (e.g., CE labeling). The initial concept used is a 144S2P configuration of a cell with a nominal voltage of 3.7 V and a capacity of 6.8 Ah. The review of this concept with an overall capacity of 7.2 kWh is presented in the following.

3.3. Concept of the Gearbox

The main task of vehicle transmission for electric drives is to achieve the best possible conversion of the tractive power offered by the electric motor into the tractive power of the vehicle. The goal is to achieve this over the widest possible speed range [23]. On the Formula Student circuits, a maximum driving speed (v_u) of about 120 km/h is possible. Combined with the maximum speed of the electric motor (n_1) of 20,000 rpm and a selected tire with a dynamic diameter (d) of 0.4 m, this results in a wheel speed (n_2) of about 1575 rpm and a gear ratio (i) of 12.7.

$$n_2 = \frac{v_u}{\pi * d} = \frac{33.\overline{3}\ \frac{m}{s}}{\pi * 0.4\ m} \approx 1575 \frac{1}{\text{min}} \tag{2}$$

$$i = \frac{n_1}{n_2} = \frac{20,000\ \frac{1}{\text{min}}}{1575\ \frac{1}{\text{min}}} \approx 12.7 \tag{3}$$

This estimation, from Equations (2) and (3), shows that a high transmission ratio is necessary for the conversion of the torque and speed. Therefore, a planetary gearbox with an in-between reduction on the planetary stage is chosen, due to space and weight reasons. The detailed design and optimization of the transmission concept will be explained in the following sections of this article.

3.4. Thermal Concept of the Vehicle

The requirements specification can be used to create an initial concept for the cooling system and the required heat sinks and material quantities for heat dissipation. For the drive package consisting of the motor and inverter, the supplier specifies constant cooling by using water as the cooling medium. Due to the required lower inlet temperature of the inverters, it can be derived from the concept that these should be flushed with the coldest fluid. Therefore, they are placed directly after the radiators. The inverters can be separated into two modules, which allows the cooling system to be divided into the right and left sides of the vehicle. The cooling medium, which flows through the left inverter is directed to the left motor on the front axle. Afterward, the motor placed on the left rear side is flowed through and the cooling medium is directed back into the left radiator (water–air heat exchanger). Afterward, the cooling flows of both vehicle sides are mixed and directed to both inverters again. For better comprehensibility, the concept is shown in the following concept sketch (Figure 3). The blue marked boxes with the letter M are the motors, P is the water pump and K are the radiators.

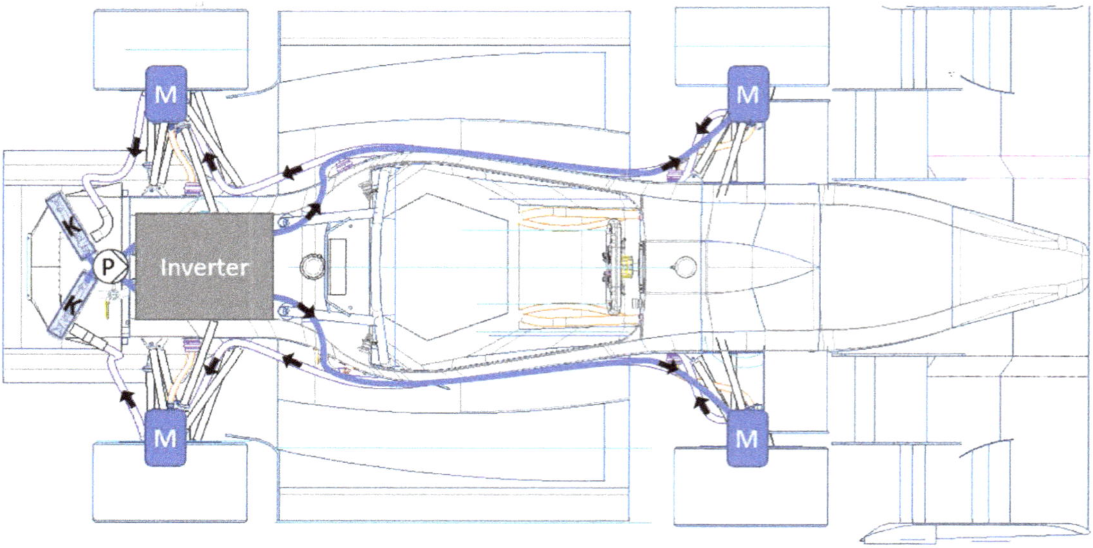

Figure 3. Concept sketch of the water-cooling system for the electric race car.

To keep the complexity of the cooling system to a minimum and to avoid possible failures due to leakages caused by the frequent removal and installation of the battery for service purposes, the concept of an air-cooling system for the accumulator is chosen. This also eliminates the additional risk of short-circuiting the battery cells in the event of undesired contact with the cooling medium. Due to the mentioned reasons and the additional weight saving, a water cooling or oil cooling system for the battery cells was not considered. The calculation of the air mass flow and the thermal cooling behavior of the battery and the water-cooling system are described in the following.

4. Synthesis of the Simulation Model

The development of a Formula Student racing car highlights the importance of pushing forward the learning curve. Especially since the e-drive is a completely new development and only nine months are available for developing and testing the vehicle, as many critical components as possible should be virtually analyzed in advance. In this context, simulations can have a key role in demonstrating behaviors of technical systems across all levels of the V-model and performing subsequent optimizations based on this information. One-dimensional simulation tools can provide important results for these multi-physics developments [24]. For this research, the AVL CRUISE^{TM} M simulation software (AVL CRUISE^{TM} M, AVL List GmbH, Graz, Austria) is used to model and simulate the electric drive-train. To evaluate the basic drive-train concept, a model is generated in the simulation environment to analyze the impact of the key components (Figure 4).

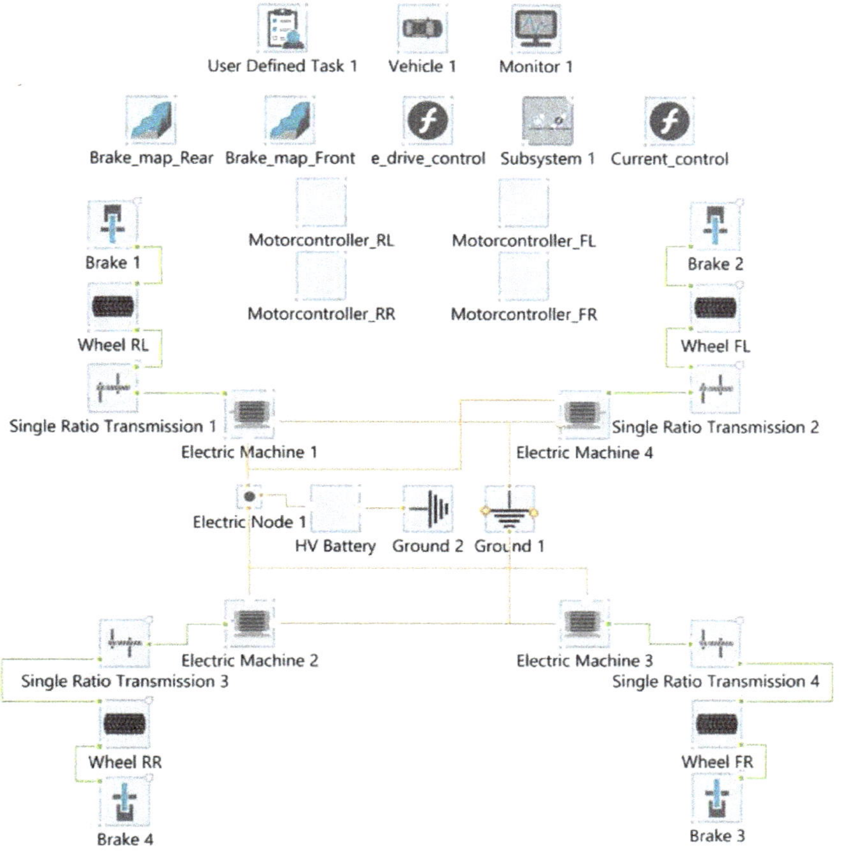

Figure 4. Initial model of the drive-train in the simulation environment.

The model consists of a high-voltage accumulator, the motors with the inverters, the gearboxes, the tires and the mechanical brakes. The remaining vehicle data (downforce, drag, weight, etc.), as well as the drive control, are integrated into the model via additional system blocks. Using a user-definable driving profile, the driving characteristics of the vehicle can be examined in terms of the relevant parameters, such as efficiency or acceleration capability.

The motor data of the motor "AMK DD5-14-10-POW 18600-B5" was derived from the supplier's specifications. This contains, for example, the torque, power losses, etc. for the inverter and the motors. For modeling the tire "Hoosier 16 $\times$ 7.5-10 LC0", measurement data from tests of the "Formula SAE Tire Testing Consortium" was used. The brake geometry was estimated according to the available installation space and the geometry of the brake pads. However, this factor should not be rated strongly, as the vehicle control system focuses on the best possible recuperation to reduce losses due to mechanical conventional brake use. The brake pressure through the travel on the brake pedal was determined according to the brake master cylinder in a map, which allows us to determine the resulting brake pressure. The vehicle center of gravity and weight data are derived from a concept model and data from previous combustion vehicles. For this purpose, the previous combustion cars were measured on an inertia measurement bench. The aerodynamic characteristics of the vehicle were determined based on the target values of the aerodynamics department. The coefficient of lift in the front is −2.576 and in the rear −3.025. The coefficient of drag is 2.18 due to the high aerodynamic load and the cross-sectional area of the racing car can be assumed to be one square meter.

For the boundary conditions of the environmental influences, the most critical race was selected due to the highest outside temperature. This is the Formula Student competition in Spain at the Circuit de Barcelona-Catalunya. The race is held under dry track conditions with an altitude of 128 m above sea level and an outside temperature of 40 °C. However, other racetracks were also simulated for comparison and to consider further critical load cases. For example, the race in Spielberg (Austria) was simulated for the evaluation of the cooling system, as high outside temperatures can also occur with significantly higher track altitude and thus lower air density. According to the regulations, a Formula Student racetrack always has a length of 22 km with a driver change halfway through the race.

For the load cycle, the speed profiles from log data from the previous combustion vehicles were used and new speed profiles were also generated with the help of a lap time simulation. The Formula Student race cars are redeveloped and rebuilt every year. In total, a maximum of 35 operating hours are achieved in the approximately three-month testing and racing phase under competitive conditions.

Using the initial model, it was possible, for example, to determine the energy consumption of the concept and to examine the accumulator dimensioning (Figure 5). In the picture, the energy dissipation of the accumulator over the endurance race with a length of 22 km can be seen. In the middle, a driver change is performed, which is why the discharge line stays constant. A total of approximately 6 kWh is used and the endurance race lasts about 1569 s.

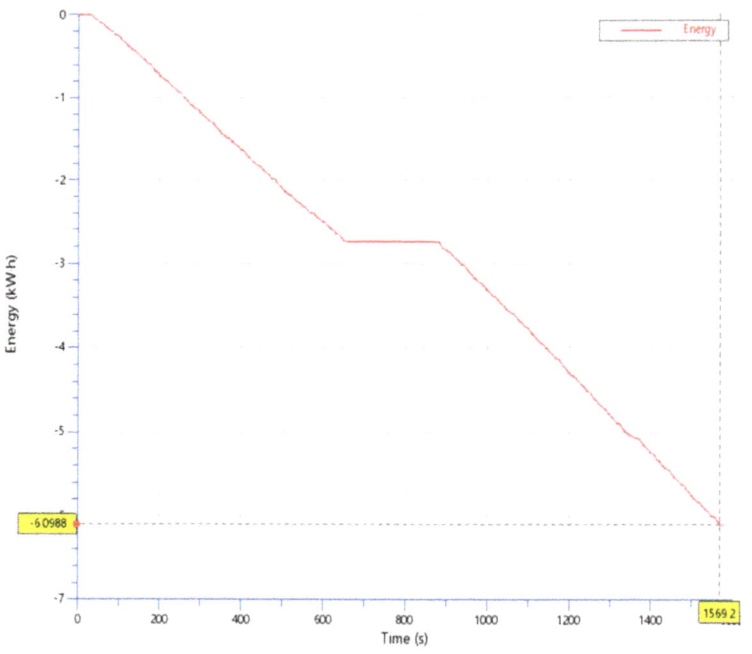

Figure 5. Energy consumption for an endurance run with the driver change in the middle.

5. Expansion of the Simulation Model

The basic vehicle model revealed the influence of the drive-train on many aspects of vehicle performance. To identify critical components and load cases, a failure mode and effects analysis (FMEA) was executed. Based on this FMEA, the simulation model was extended into several dimensions. To evaluate the cooling concept for the electric motors and the inverters, a new model is being built, based on the basic model. For this purpose, a new subsystem is created in which all components are integrated. Subsequently, the "water-air heat exchangers", are modeled and fed with data from analytical calculations and characteristic curves from the suppliers. After the model was created, results from the CFD software AVL FIRE™ (AVL FIRE™, AVL List GmbH, Graz, Austria) are transferred from the 3D CFD simulations to the 1D simulation with a significantly higher level of detail using the "Thermal Network Generator". This enables the integration of components into the system level as specified in the V-model. The complete product development process can be executed by using the AVL eSUITE™ (Figure 6).

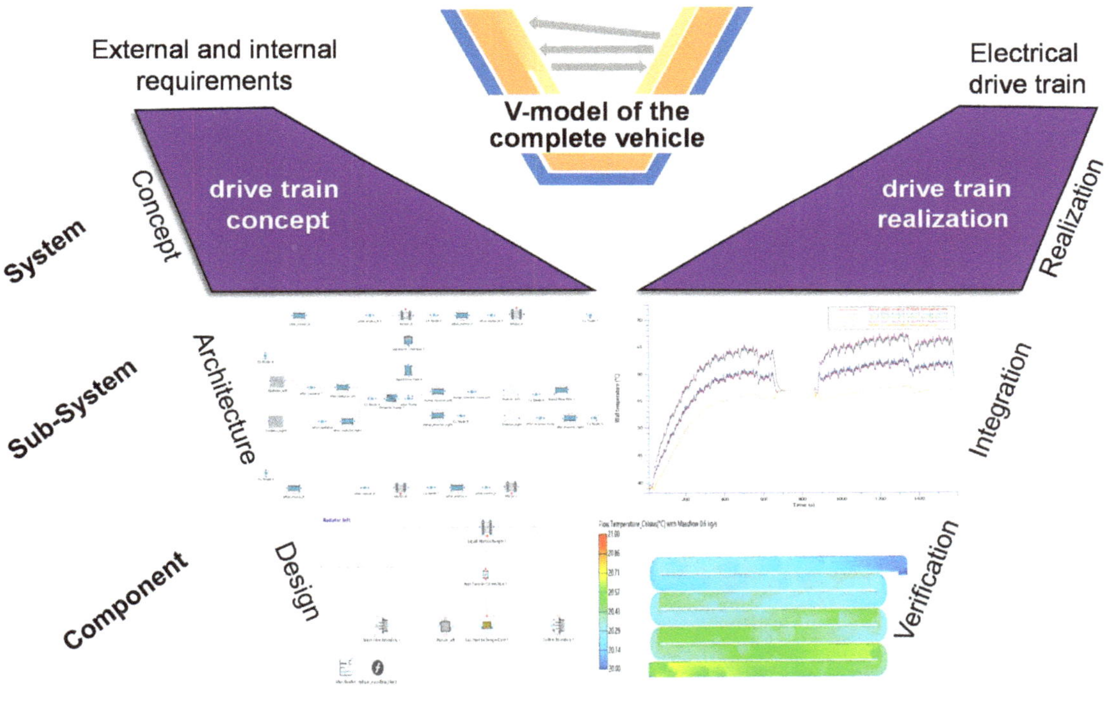

Figure 6. Realization of all levels of the V-model by using the simulation tools of the AVL eSUITE™ for the thermal system.

The figure shows the water-cooling circuit at the system and component level, followed by the integration and verification of the components and assemblies exemplarily. For electric vehicles, the thermal behavior of the battery and the individual cells is a significant aspect that can be investigated using simulation models [25]. For the investigation of the thermal behavior of the battery, a model was built in the simulation environment (Figure 7). The red lines represent the possible paths for the heat transfer and the blue lines symbolize the ducts of the cooling medium.

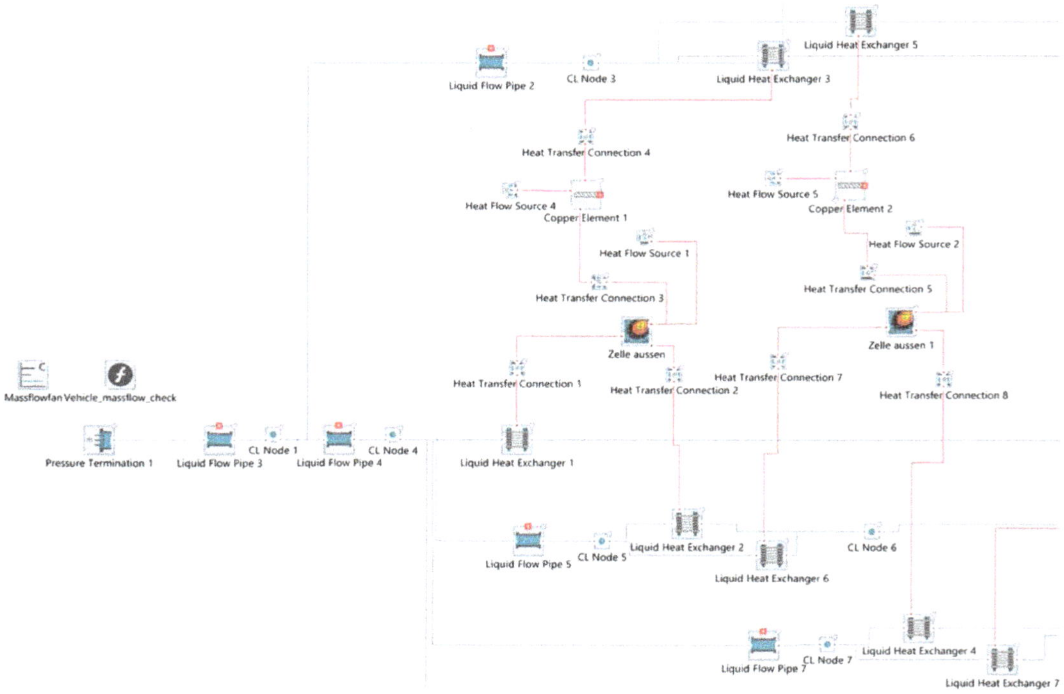

Figure 7. Section of the model of the battery with integrated air cooling.

To avoid calculation errors due to the compressibility of air and to enable a more stable solver run, a "liquid system" with properties of the medium "air" was used for the calculation due to the small pressure differences in the air cooling system. Figure 8 shows the temperature response of the most forward cell, the most rearward cell, and the connecting terminals during an endurance run.

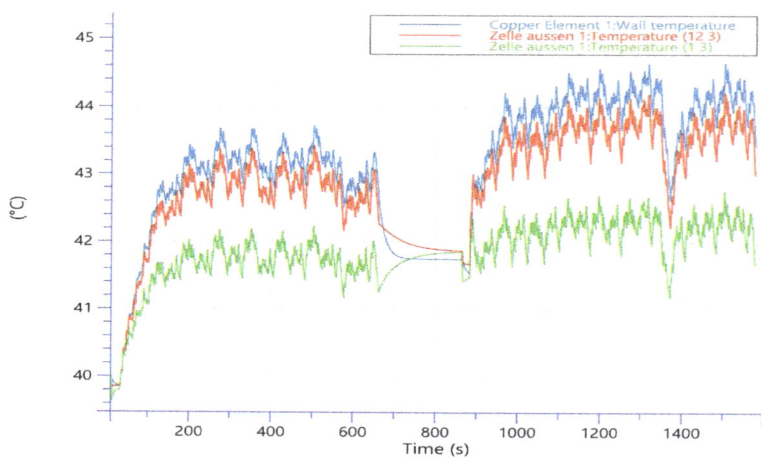

Figure 8. Thermal behavior of the battery cells and connecting terminals.

The verification in the simulation environment also allowed to investigate the critical load case with a maximum ambient temperature of 40 °C which could be classified as non-problematic.

6. Validation

For the first development, the simulations were able to provide important information about the vehicle, but the simulations must be verified to be representative. Because no test bench data or real driving data from previous developments are available, theoretical validation options had to be identified. It was possible to develop two methods for this validation. On the one hand, the energy consumption for an endurance run and on the other hand, the cooling power to be dissipated in relation to the competitors are used for validation. For both validation methods, competitor analyses were used to identify other vehicles that use the same drive-train package with a similar overall vehicle concept. To validate the simulation model, a speed profile of a Formula Student vehicle from a previous competition was simulated and the energy consumption and the cooling power to be dissipated were compared with a similarly performing vehicle from a competitor. Both validation methods showed that the simulation supplies results in the expected magnitude and can therefore be considered theoretically validated.

7. Optimization of the Vehicle Concept

This section describes the optimizations that can be performed on the drive-train concept using the AVL CRUISE^{TM} M simulation. Especially the coupling of the drive-train simulation with a self-developed lap time simulation, which is executed in Octave (GNU Octave, GNU General Public License), enables possibilities to optimize the concept of the electric race car. In the following, a study to determine the optimal gear ratio for the electric powertrain is explained. By coupling the two simulation tools, specific speed profiles are generated for the varying vehicle parameters by the lap time simulation, which are simulated and analyzed in AVL CRUISE^{TM} M (Figure 9).

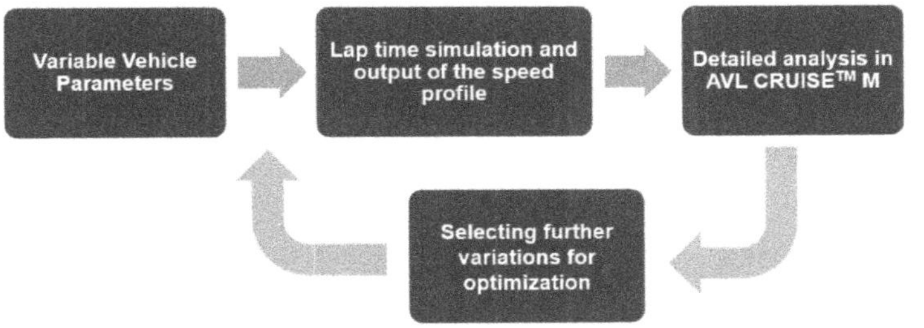

Figure 9. Optimization process using the lap time simulation and AVL CRUISE^{TM} M.

7.1. Optimization Process of the Gear Ratio

In this way, the influence of the transmission ratio on the vehicle can be evaluated in detail in a full-factorial study. The optimization process runs as described in the following:

1. Requirements-driven initial conception with a selection of a starting point, e.g., based on the top speed to be achieved or the maximally available package;
2. Modeling and simulation of the initial state to determine the influence on lap time, energy consumption or acceleration capability, etc.;
3. Variation of the gear ratio within the specified limits through a full factorial simulation study. The boundary conditions were determined based on certain design limits, such as the maximum package space or the desired maximum speed. Based on this, the gear ratios had to be varied between I = 10 and i = 15. Subsequently, the variation of the parameters and the analysis were performed by using a parametric model setup. Due to the changed gear ratio, a changed load profile with a different lap time for a

fast race lap was determined. This also led to a different drive train efficiency and a different energy consumption. These characteristics could be analyzed and optimized with the help of the determination of the minima as shown in Section 7.2;

4. The selected gear ratio concept was transferred to the detailed feasibility study in the FVA-Workbench (FVA-Workbench, FVA GmbH, Frankfurt, Germany) and the detailed 3D data were generated. These were finally checked with a load spectrum of the related endurance to investigate the durability of the chosen design.

7.2. Pareto Optimization

An appropriate method for the evaluation of computer experiments and related optimization processes is the so-called "Pareto optimization". For the determination of the Pareto frontier, all results of a full factorial study are recorded with the values of both factors to be minimized [26]. This method is useful for the appropriate selection of motor and gearbox combinations [18]. The results of the full factorial simulation of the gear ratios lead to the plotted distribution (Figure 10).

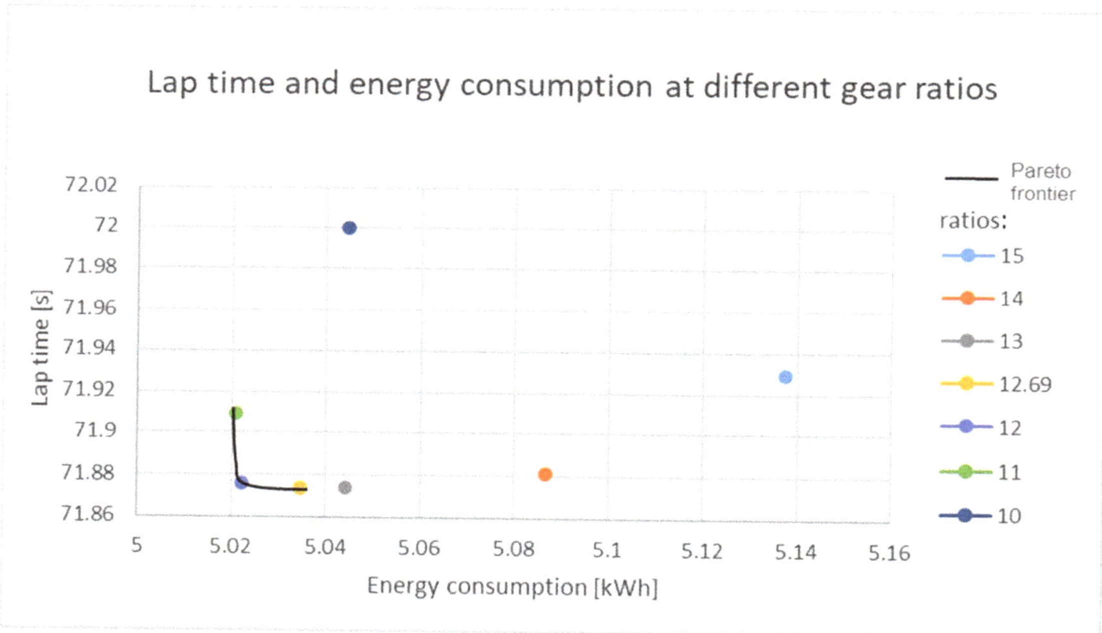

Figure 10. Full factorial study and resulting Pareto frontier to minimize the energy consumption and the reached lap time of the race car.

The generated Pareto frontier consists of the points i = 12.69, i = 12 and i = 11. These three points represent the "Pareto optimals" for the target parameters. The ratio i = 11 has the least energy consumption, compared to the ratio i = 12.69, which achieves the fastest lap time. The third possible result i = 12 offers a lower energy consumption compared to the fastest gear ratio but consumes more energy than i =11. The parameter of the lap time is rated higher than the energy consumption for a racing car. Therefore, the gear ratio with i = 12.69 is selected. After the gear ratio had been conceptually defined, a load profile was generated using the simulation tool for further optimization of the transmission part geometries. For this purpose, the torques were calculated according to the probability of occurrence for a lifetime of 40 operating hours (Figure 11).

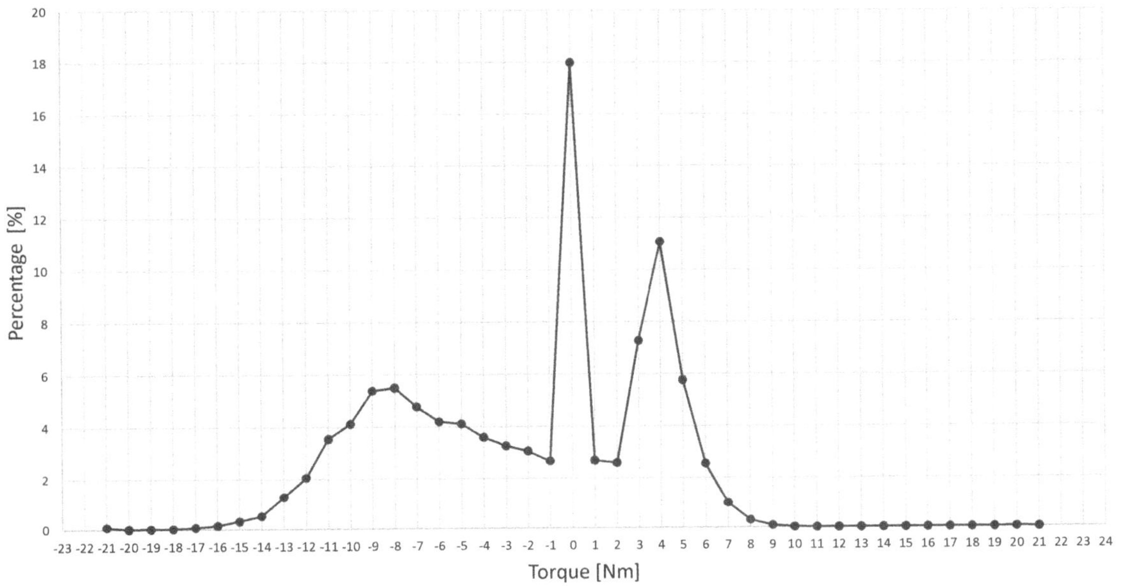

Figure 11. Occurrence of the torques over an Endurance applied to the gearbox input shaft.

To verify the load profile and to generate the geometry of the transmission components, the software FVA-Workbench was used.

8. Integration

For a holistic approach regarding the V-model, the AVL eSUITE™ provides the basis for a topology of different types of virtual, real-time prototype testing if AVL VSM™ (AVL VSM™, AVL List GmbH, Graz, Austria) is linked up with AVL CRUISE™ M [27].

The topology of the virtual prototype can represent a virtual vehicle by using both programs and the "AVL Model.CONNECT™" interface. In the first implementation level, both simulations are used to test the control of the ECU. This type of testing is called "software in the loop", as virtual systems are used to test dynamic vehicle control. In this case, the ECU is replaced by a model generated in "MATLAB-Simulink" (Simulink, MathWorks, Inc., Natick, MA, USA). This implementation level is used to develop the torque vectoring for the new race car.

In the second implementation stage, the virtual models of the motor and the ECU are replaced by real components. This offers the possibility to test the application of the components on the test bench before the vehicle is completed and to adjust it to the components in the best possible way (Figure 12, based on [27]). This testing method is known as "hardware in the loop". It would also be possible to test the planetary gearbox with virtual driving profiles resulting from the model coupling.

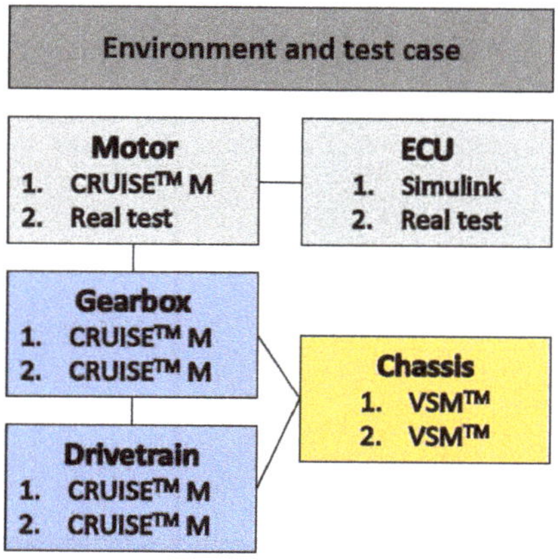

Figure 12. Topologies of virtual prototypes and different implementation levels.

9. Conclusions and Outlook

For combustion vehicles two dominant design variants can be identified:

- single-engine in front, camshaft in a longitudinal direction, rear-wheel drive and
- single-engine in front, camshaft in transverse direction front-wheel drive.

For electrical drive-trains, no clearly dominant design variants are obvious yet; maybe they will not appear because of lower physical restrictions. A holistic analysis-synthesis cycle of an electrical drive-train consequently has to enable a topological multitude in order to enhance the possibility to approach an optimum for a given set of requirements. This paper presents such a kind of analysis-synthesis cycle taking into account the current developments in the field of MBSE as well as simulation approaches. The main goal was the development and optimization of an electric drive-train. This task can be considered as accomplished; the Formula Student Team Weingarten will compete in the Formula Student Electric (FSE) category in the 2021 season with an electric powertrain based on these results and findings.

By employing extensive simulation, valuable insights into various components of the powertrain have been generated. For example, the design of the electrical components, such as the battery, could be verified, the mechanical components, for example, the planetary gearbox, could be optimized in terms of efficiency and performance, and finally, the fundamentals of the drive controller could be determined. The critical constituents identified during an FMEA were analyzed in detail. This was accompanied by a lap time simulation and a validation based on competitor analysis.

In the future, the lap time simulation can be used to evaluate numerous vehicle parameters regarding the entire drive-train performance and its influence on the lap time. Further research will aim to extend the linking of the different levels of the V-model possible at one software manufacturer using the design languages and DC 43 to other simulation tools from other manufacturers through FMI interfaces. This could lead to an automated and intelligent multidimensional product design.

Author Contributions: Conceptualization, S.S. (Sven Schumacher), S.S. (Stefan Schmid), P.W., R.S., and M.T.; methodology, S.S. (Sven Schumacher) and R.S.; design and dimensioning, S.S. (Sven Schumacher); simulation, S.S. (Sven Schumacher), S.S. (Stefan Schmid) and P.W.; analysis and evaluation, S.S. (Sven Schumacher); writing—original draft preparation, S.S. (Sven Schumacher) and R.S.; writing—review and editing, S.S. (Stefan Schmid), P.W. and M.T. All authors have read and agreed to the published version of the manuscript.

Funding: As a part of this research project, a master thesis was realized, which was supported by the Stifterverband für die Deutsche Wissenschaft e.V. (Donors' association for the promotion of humanities and sciences in Germany) fellowship "Master LAB #THENEWNORMAL".

Institutional Review Board Statement: Not applicable.

Informed Consent Statement: Not applicable.

Data Availability Statement: The data presented in this study are available on request from the corresponding author.

Acknowledgments: The authors wish to thank the administrative and technical staff of Ravensburg-Weingarten University (RWU), for the continuous support of the formula student team and all involved students for their immense contribution, see https://www.fsteamweingarten.de/ (accessed on 19 July 2021).

Conflicts of Interest: The authors declare no conflict of interest.

References

1. PwC. Digital Auto Report 2020: Volume 1. Available online: https://www.strategyand.pwc.com/de/de/studie/2020/digital-auto-report-2020.html (accessed on 20 May 2021).
2. VDI/VDE 2206—Entwurf. *Entwicklung Cyber-Physischer Mechatronischer Systeme (CPMS)*; Beuth: Berlin, Germany, 2020.
3. Gräßler, I.; Hentze, J. The new V-Model of VDI 2206 and its validation. *at-Automatisierungstechnik* **2020**, *68*, 312–324. [CrossRef]
4. Lu, J.; Chen, D.; Wang, G.; Kiritsis, D.; Törngren, M. Model-Based Systems Engineering Tool-Chain for Automated Parameter Value Selection. *IEEE Trans. Syst. Man Cybern. Syst.* **2021**, accepted for inclusion. Available online: https://ieeexplore.ieee.org/stamp/stamp.jsp?arnumber=9328223 (accessed on 20 July 2021).
5. Shaked, A.; Reich, Y. Using Domain-Specific Models to Facilitate Model-Based Systems-Engineering: Development Process Design Modeling with OPM and PROVE. *Appl. Sci.* **2021**, *11*, 1532. [CrossRef]
6. Gräßler, I.; Ole, C.; Scholle, P. Method for Systematic Assessment of Requirement Change Risk in Industrial Practice. *Appl. Sci.* **2020**, *10*, 8697. [CrossRef]
7. Walden, D.D.; Roedler, G.J.; Forsberg, K.; Hamelin, R.D.; Shortell, T.M. *Systems Engineering Handbook: A Guide for System Life Cycle Processes and Activities*, 4th ed.; Wiley: Hoboken, NJ, USA, 2015.
8. Pottebaum, J.; Gräßler, I. Informationsqualität in der Produktentwicklung: Modellbasiertes Systems Engineering mit expliziter Berücksichtigung von Unsicherheit. *Konstruktion* **2020**, *72*, 76–83. [CrossRef]
9. Laing, C.; David, P.; Blanco, E.; Dorel, X. Questioning integration of verification in model-based systems engineering: An industrial perspective. *Comput. Ind.* **2020**, *114*, 103163. [CrossRef]
10. Darpel, S.; Beckman, S.; Ferlin, T.; Havenhill, M.; Parrot, E.; Harcula, K. Method for tracking and communicating aggregate risk through the use of model-based systems engineering (MBSE) tools. *J. Space Saf. Eng.* **2020**, *7*, 11–17. [CrossRef]
11. Gräßler, I.; Wiechel, D.; Pottebaum, J. Role model of model-based systems engineering application. *Mater. Sci. Eng.* **2021**, *1097*, 012003.
12. Pfeifer, S.; Seidenberg, T.; Jürgenhake, C.; Anacker, H.; Dumitrescua, R. Towards a modular product architecture for electric ferries using Model-Based Systems Engineering. *Procedia Manuf.* **2020**, *52*, 228–233. [CrossRef]
13. Rudolph, S. Übertragung von Ähnlichkeitsbegriffen. Habilitationsschrift, Fakultät Luft- und Raumfahrttechnik und Geodäsie. Habilitation Thesis, Universität Stuttgart, Stuttgart, Germany, 2002.
14. IILS mbH. Design Compiler 43. Available online: https://www.iils.de (accessed on 4 May 2021).
15. Holder, K.; Zech, A.; Ramsaier, M.; Stetter, R.; Niedermeier, H.P.; Rudolph, S.; Till, M. Model-Based Requirements Management in Gear Systems Design Based on Graph-Based Design Languages. *Appl. Sci.* **2017**, *7*, 1112. [CrossRef]
16. Elwert, M.; Ramsaier, M.; Eisenbart, B.; Stetter, R. Holistic Digital Function Modelling with Graph-Based Design Languages. In Proceedings of the Design Society: International Conference on Engineering Design; Cambridge University Press: Cambridge, UK, 2019; Volume 1, pp. 1523–1532. [CrossRef]
17. Stetter, R. Approaches for Modelling the Physical Behavior of Technical Systems on the Example of Wind Turbines. *Energies* **2020**, *13*, 2087. [CrossRef]
18. Holder, K.; Schumacher, S.; Friedrich, M.; Till, M.; Stetter, R.; Fichter, W.; Rudolph, S. Digital Development Process for the Drive System of a Balanced Two-Wheel Scooter. *Vehicles* **2021**, *3*, 3. [CrossRef]
19. Formula Student Germany GmbH. Formula Student Rules 2020 Version 1.0. Available online: https://www.formulastudent.de/fsg/rules/ (accessed on 13 February 2021).
20. Trzesniowski, M. *Rennwagentechnik:Grundlagen, Konstruktion, Komponenten, Systeme*; ATZ/MTZ-Fachbuch; Springer Vieweg: Wiesbaden, Germany, 2014.
21. Pischinger, F.; Adomeit, P.; Dorenkamp, R.; Schindler, K.-P.; Baar, R.; Gumpoltsberger, G.; Greiner, J.; Sasse, C.; Steinel, K.; Lanzer, H.; et al. Antriebe. In *Vieweg Handbuch Kraftfahrzeugtechnik*; Braess, H.H., Seiffert, U., Eds.; ATZ/MTZ-Fachbuch; Springer Vieweg: Wiesbaden, Germany, 2013; p. 213. [CrossRef]

22. Morche, D.; Schmitt, F.; Genuit, K.; Elsen, O.; Kampker, A.; Deutskens, C.; Heimes, H.H.; Swist, M.; Maue, A.; vom Hemdt, A.; et al. Fahrzeugkonzeption für die Elektromobilität. In *Elektromobilität:Grundlagen einer Zukunftstechnologie*; Kampker, A., Vallée, D., Schnettler, A., Eds.; Springer Vieweg: Wiesbaden, Germany, 2018; pp. 181–277. [CrossRef]
23. Naunheimer, H.; Bertsche, B.; Ryborz, J.; Novak, W.; Fietkau, P. Systematik der Fahrzeuggetriebe: Konstruktive Grundkonzepte. In *Fahrzeuggetriebe:Grundlagen, Auswahl, Auslegung und Konstruktion*, 3rd ed.; Naunheimer, H., Bertsche, B., Ryborz, J., Novak, W., Fietkau, P., Eds.; Springer: Berlin/Heidelberg, Germany, 2019; pp. 139–273. [CrossRef]
24. Colla, A.; Katrašnik, T.; Walter, K.; Diwoky, F. Battery-electric vehicle efficiency potential assessment and thermal aging investigation. In *Der Antrieb von Morgen 2017*; Liebl, J., Ed.; Springer Vieweg: Wiesbaden, Germany, 2017. [CrossRef]
25. Erhard, S.V. Mehrdimensionale Elektrochemisch-Thermische Modellierung von Lithium-Ionen-Batterien. Dissertation, Technischen Universität Münche, München, 2017. Available online: https://mediatum.ub.tum.de/doc/1338266/1338266.pdf (accessed on 20 April 2017).
26. Siebertz, K.; van Bebber, D.; Hochkirchen, T. Optimierung. In *Statistische Versuchsplanung:Design of Experiments*; Siebertz, K., van Bebber, D., Hochkirchen, T., Eds.; VDI-/Buch; Springer: Berlin/Heidelberg, Germany, 2010; pp. 225–245. [CrossRef]
27. Puntigam, W.; Zehetner, J.; Lappano, E.; Krems, D. Integrated and Open Development Platform for the Automotive Industry. In *Systems Engineering for Automotive Powertrain Development*; Hick, H., Küpper, K., Sorger, H., Eds.; Powertrain; Springer: Berlin/Heidelberg, Germany, 2020; pp. 1–27. [CrossRef]

vehicles

MDPI

Article

Digital Development Process for the Drive System of a Balanced Two-Wheel Scooter

Kevin Holder [1,*], Sven Schumacher [2], Matthias Friedrich [3], Markus Till [2], Ralf Stetter [2], Walter Fichter [3] and Stephan Rudolph [4]

1 ZF Friedrichshafen AG, 88046 Friedrichshafen, Germany
2 Department of Mechanical Engineering, Ravensburg-Weingarten University (RWU), 88250 Weingarten, Germany; ss-191205@hs-weingarten.de (S.S.); markus.till@rwu.de (M.T.); ralf.stetter@rwu.de (R.S.)
3 Institute for Flight Mechanics and Controls, University of Stuttgart, 70569 Stuttgart, Germany; matthias.friedrich@ifr.uni-stuttgart.de (M.F.); fichter@ifr.uni-stuttgart.de (W.F.)
4 Design Theory and Similarity Mechanics Group, Institute of Aircraft Design, University of Stuttgart, 70569 Stuttgart, Germany; rudolph@ifb.uni-stuttgart.de
* Correspondence: kevin.holder@zf.com

Abstract: Graph-based design languages have received increasing attention in the research community, because they offer a promising approach to address several major issues in engineering, e.g., the frequent manual data transfer between computer-aided design (CAD) and computer-aided engineering (CAE) systems. Currently, these issues prevent the realization of machine executable digital design processes of complex systems such as vehicles. Promising scenarios for urban transportation include an interconnection of mass transportation systems such as buses and subways with individual vehicles for the so-called "last mile" transport. For several reasons, these vehicles should be as small and light as possible. A considerable reduction in weight and size can be achieved, if such vehicles are tailored to the individual size, weight and proportion of the individual user. However, tailoring vehicles for the individual characteristics of each user go beyond a simple building set and require a continuous digital design process. Consequently, the topic of this paper is a digital design process of a self-balanced scooter, which can be used as an individual last-mile means of transport. This process is based on graph-based design languages, because in these languages, a digital system model is generated, which contains all relevant information about a design and can be fed into any simulation tool which is needed to evaluate the impact of a possible design variation on the resulting product performance. As this process can be automated by digital compilers, it is possible to perform systematic design variations for an almost infinite amount of parameters and topological variants. Consequently, these kinds of graph-based languages are a powerful means to generate viable design alternatives and thus permit fast evaluations. The paper demonstrates the design process, focusing on the drive system of the respective balanced two-wheel scooter and highlights the advantages (data integration and possibility for machine execution).

Keywords: digital design process; urban vehicles; balanced two-wheel scooter

Citation: Holder, K.; Schumacher, S.; Friedrich, M.; Till, M.; Stetter, R.; Fichtner, W.; Rudolph, S. Digital Development Process for the Drive System of a Balanced Two-Wheel Scooter. *Vehicles* **2021**, *3*, 33–60. https://doi.org/10.3390/vehicles3010003

Received: 31 December 2020
Accepted: 19 January 2021
Published: 20 January 2021

1. Introduction

Vehicles for the so-called last mile are an essential component for individual mobility in inner-city logistics concepts (compare [1]). In the case of individual mobility, users can ideally switch directly between different means of transport—one possibility for this is a lightweight, balanced two-wheeled scooter. This paper is based on the thesis that a considerable reduction of size and weight could be achieved, if certain products are tailored to the weight, size and proportion of the user. For instance, within a usual population, a difference in size of 25% and in weight of more than a 100% is present. Obviously, a linear relationship between those parameters and the size and weight of a product, which these persons use, cannot be postulated. However, for lightweight transportation products such as bikes and scooters, a strong connection is probable. It is important to note that tailoring these kind of products for the individual size, weight and proportion of the user requires a continuous digital design process (compare [2]). Consequently, the focus of this paper is a requirement-oriented digital design process of a certain vehicle for last-mile transport—a self-balanced scooter. Through cross-domain modeling, the user individual requirements can be mapped throughout the entire design process. To create a digital and executable process, the individual steps are defined and domain specific data, such as product design, geometry creation, drive design and production planning are stored in a central data model, using an innovative approach applying graph-based design languages in combination with a design compiler. The scope of the research presented this paper are the requirements, the geometry synthesis and simulations (Figure 1).

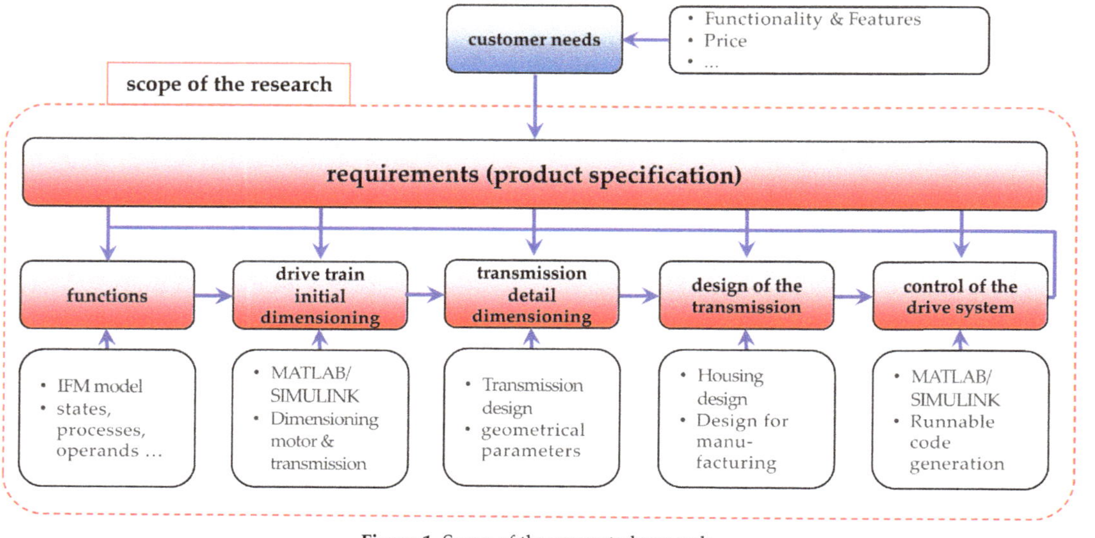

Figure 1. Scope of the presented research.

The following steps and domain models are integrated in this process: drive train rough estimation via SIMULINK (drive train simulation model—SIMULINK is a software from The MathWorks, Natick, MA, USA), gear-specific model (gear set oriented drive train model), geometry model, control design models, detail model for simulation of the vehicle and of its controller function and code generation from the simulation model. Determining a suitable solution to a given design problem essentially can be understood as an continuous interplay between reasoning about required properties and functions and the particular combination of solution elements which will be provided for this, as well as simulations of their behavior (compare [3,4]). This leads to an iterative process consisting of analyzing and gradually synthesizing the potential solution. It should be noted that some of the solution features or constraints might necessitate a partial or complete redefinition of the problem [5]. Such iterations are not limited to individual design stages, but carry through the entire design process. These iterations have to go hand-in-hand with a continuous update of the models, which represent the respective information [6]. Graph-based design languages offer a promising approach to address the issues of laborious manual transfer and incomplete, inconsistent linking. Such languages generate a digital meta- or system model, storing all relevant information about a design and feed this into any relevant computer-aided engineering (CAE) tool as needed to simulate and test the impact of any design variation on the resulting product performance. As this can be automated in digital compilers to perform systematic design variation for an almost infinite amount of parameters, such graph-based languages are a powerful means to generate viable design alternatives—and thus permit fast comparison and selection—much faster than any design team manually could. This paper investigates the underlying challenges and proposes a digital design process.

The main focus of this paper is on the digital design process. The object of this process is a drive system for a balanced two-wheel scooter. Due to the need to balance the vehicle and simultaneously to follow a given path, the use of two individual actuators on each wheel and an elaborate control system with a continuous tracking of the inclination angle is inevitable. Due to many aspects such as simplicity, the choice of electrical motors as actuators is logical. An initial design step was needed in order to determine whether a gear system is necessary and which transmission ratio would be beneficial for certain configurations. It became apparent that designs which use a transmission are more beneficial for the given set of requirements (including system costs), leading to the general drive system configuration shown in Figure 2.

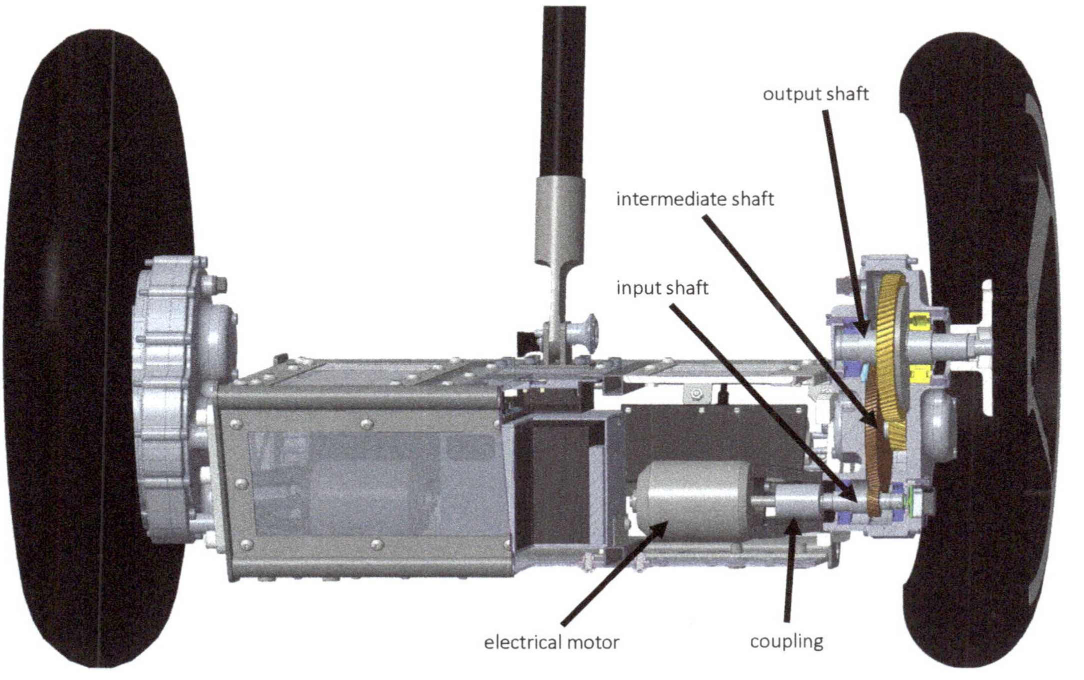

Figure 2. Scooter drive system—overview.

Visible in the cross-section shown in Figure 2 are one of the two electrical motors, a coupling and the shafts of the transmission system. Together with the inclination sensor and the control system, they represent the key technologies of the balanced two-wheel scooter. The chosen drive system is also similar to drive systems in mobile two-wheel applications (compare [7]). The integration of alternative technical solutions in the digital design process is discussed in the reflections section. The main research goal was the development of a digital design process for this and similar vehicles. This process is realized by applying graph-based languages. The underlying concept of graph-based design languages is described in the next section. Methods and tools for the continuous management of requirements are the contents of Section 3. In Section 4, an approach for functional representation is explained. The concrete dimensioning of the motor and transmission of the balanced two-wheel scooter is discussed in Section 5. Based on the general dimensions and detail, dimensioning of the transmission was possible; this is described in Section 6. One main challenge of balanced two-wheel scooters is the control system—the model-based development is explained in Section 8. In Section 9, the digital design process is reflected and advantages and challenges are discussed. The final section of this paper presents a conclusion and an outlook to further research topics and activities.

2. Graph-Based Design Languages

Graph-based design languages provide a powerful engineering framework through their automated, systematic variation and calculation features, and have the capability to feed into any relevant CAE tool. They map product information into vocabulary and rules typically based on the Unified Modeling Language (UML). These languages permit machine execution and the reuse of design and manufacturing knowledge, through automated model generation, calculation and simulation. They allow the generation of a large number of design alternatives based on parameter variation across multiple simulation tools, in a fast and automated manner [8–14]. Figure 3 shows some successful applications of graph-based design languages.

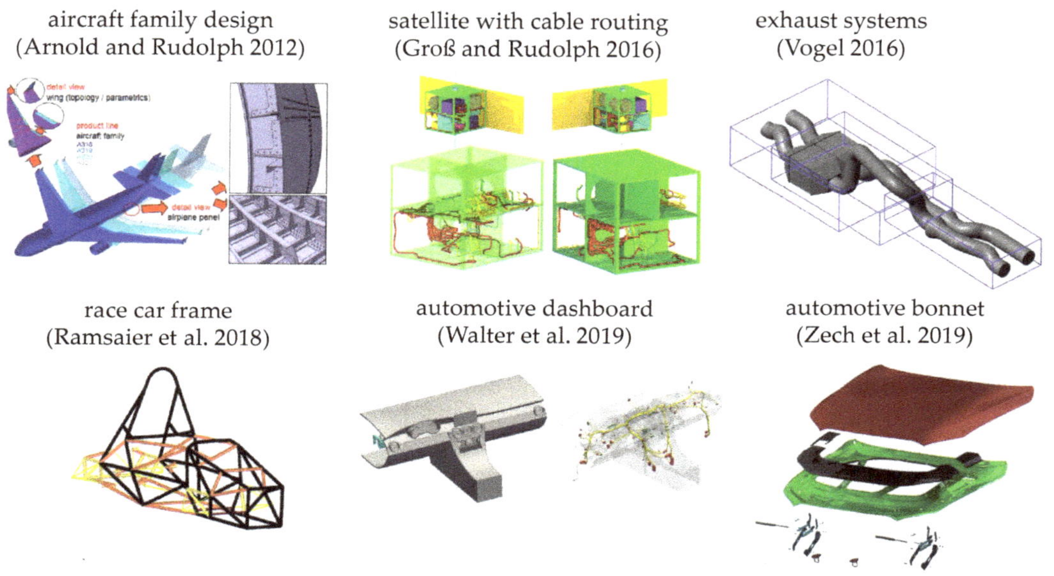

Figure 3. Successful applications of graph-based design languages.

In graph-based design languages, the geometry can be abstracted to its parts and their abstract geometry. The parts can be organized into the class diagram which represents the vocabulary of the design language. This vocabulary can be used together with rules in order to generate the geometrical model of the product. One important objective of the application of graph based languages is to accelerate innovation. A second central goal is to develop knowledge representations and, through this, to enable a processing of the product lifecycle that allows for more human-independent technological process integration, thereby improving knowledge transfer. The basic concept of a graph-based design language is illustrated in Figure 4.

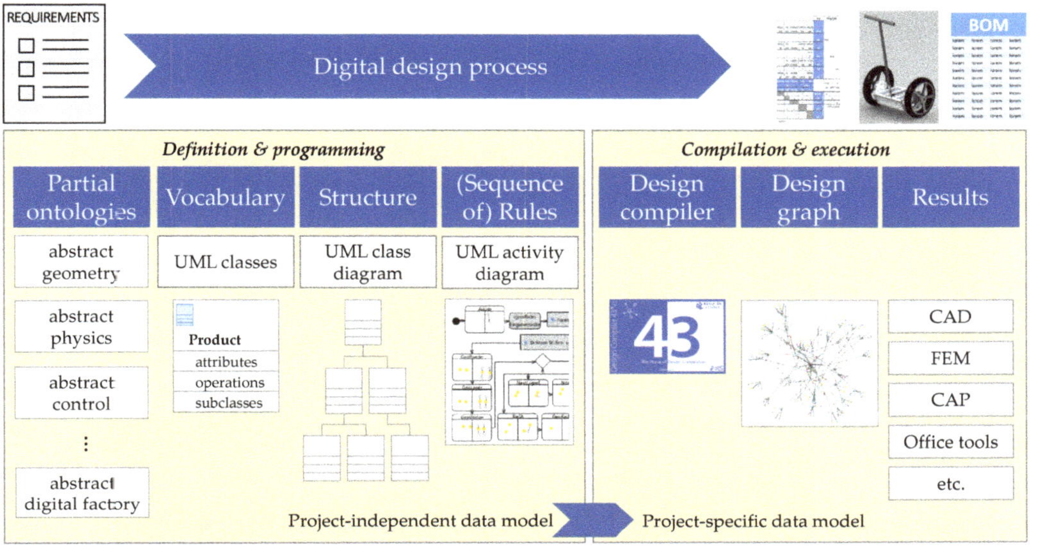

Figure 4. Concept of a graph-based design language.

Individual components of a graph-based design language are UML classes, which represent the objects of the product (vocabulary). These components can be arranged in ontologies (an ontology is an organized, annotated, related and/or hierarchical arrangement of engineering entities, which may represent aspects of engineering knowledge). These classes are connected in the UML class diagram by different associations and are instantiated via UML model transformations (rules). A sequence of such rules is modelled in an activity diagram. A central element of the digital design process is the Design Compiler 43™ (DC43 for short, see www.iils.de for details). DC43 uses the different sources of information to create the design graph. From this central meta- or system model, further models (for example computer-aided design (CAD) geometry models, simulation models like FEM (finite element method) or multi-body-simulation (MBS) models, descriptions of abstract physics, function representations, a bill of materials, etc.) can be generated in a fully automatic manner. In the balanced scooter project, several generic models could be generated in Figure 5.

The design compiler can process design languages on all abstraction levels to the full detail of the simulation models. This leads to a completely automated process within a unified framework. The design compiler executes the design rules and automatically derives the analysis models of the respective design domains (CAD, CFD (computational fluid dynamics), FEM, etc.). The information content is not limited to geometry, but can also include information concerning physical loads, materials, functions, etc. The work on "DesignCompiler43" is ongoing, but many important functionalities are already available in the current version. The starting points of the digital development process are requirements; their management is the topic of the subsequent section.

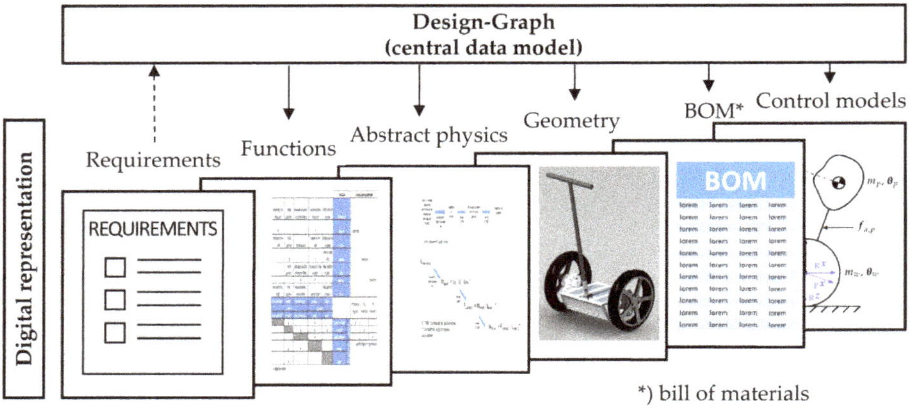

Figure 5. Design graph with digital models.

3. Requirements Management

Requirements are a decisive factor in system development projects in industry (compare e.g., [15]). Studies [16,17] show that four of ten top risks in projects are strongly connected with requirements and that only 52 percent of the originally intended requirements appear in the final released version of the product. An enormous amount of literature concerns requirements management and requirements engineering; a good overview can be found in [18–20]. Additionally, international standards exist: the ISO standard 29148 "Systems and software engineering—Life cycle processes—Requirements engineering" [21], the ISO standard 15288 [22] "Systems and software engineering—Systems life cycle processes" and the quality management norm ISO 9001 [23]. Today, several tools for requirements management are currently available; a comprehensive overview is given in [24]. Additionally, open source products are available, which can provide extensive functionality, such as the Eclipse (Eclipse is an open-source programming tool for developing software of various types which is managed by the Eclipse Foundation, Ottawa, Canada, a non-profit corporation with the mission to manage the Eclipse open source community and its projects)-based tool "ProR"; Figure 6 shows an example of the requirements of the balanced two-wheel scooter modelled in ProR.

In earlier research, a concept for model-based requirements management was developed [25]. The central objectives of the research were to demonstrate an integration of this model-based requirements management with a development process of gear systems by means of applying graph-based design languages. For the innovative approach, two models were developed—one model for generating various gear system design variants and a second model for requirements management (RM), which includes state of the art RM know-how. The RM-model, as well as the model for product design variants, has been implemented by means of graph-based design languages that are based on UML (Unified Modelling Language). For the explicit digital mapping, the definition of an interface between both models ("requirements" and "geometry synthesis and simulation") is needed. The model-based requirements management has, in very general terms, three main aspects, which could also be understood as dimensions (Figure 7).

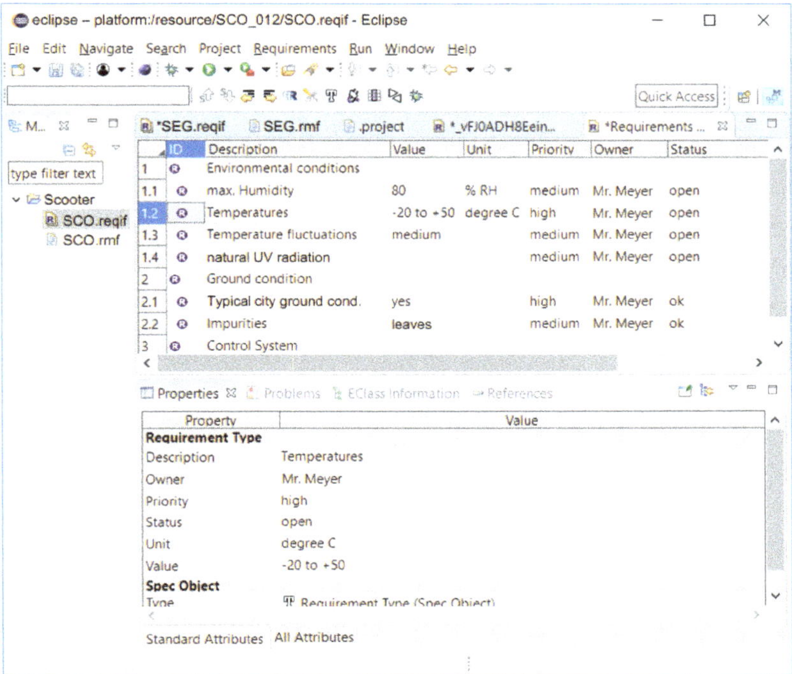

Figure 6. Requirements management with Eclipse "ProR"—example.

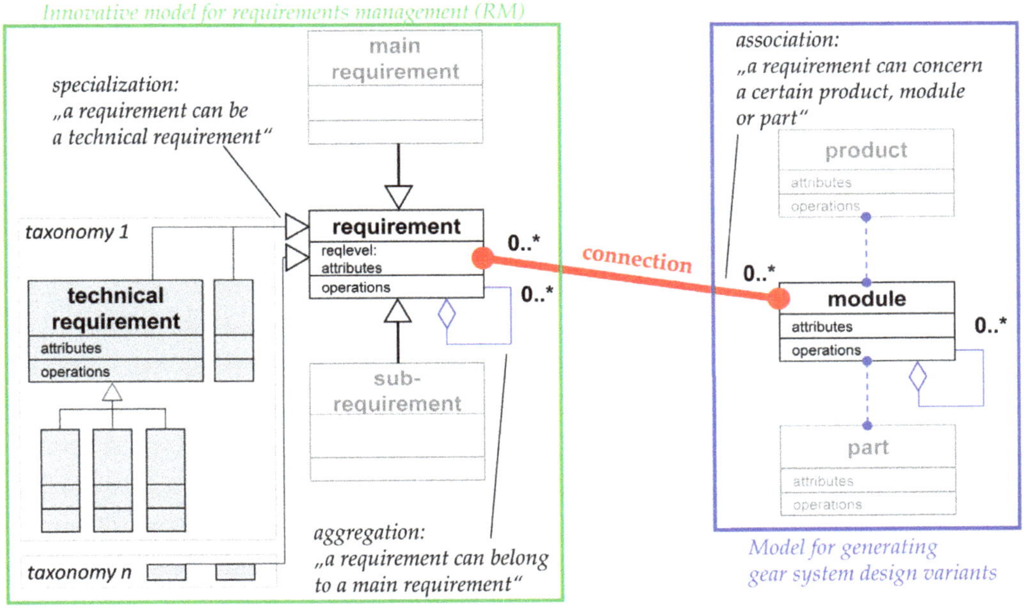

Figure 7. Product classes linked to requirement classes.

The first dimension consists of the model for generating gear system design variants (right blue frame). The second dimension consists of the model for requirements management (left green frame) and the third dimension consist of the linking of requirements to the product graph. This linking is represented by a red association link. For this, attribute "RelatedToPart" is introduced within the structuring of requirements. This expansion of the framework of graph-based languages allows a concrete assignment from a certain requirement to its related product, module or part. The next section discusses how functional representations can be included in this framework.

4. Functional Representation

The domain of product functions describes the solution concept of a product on an abstract level and supports the description and synthesis of complex, multi-domain products. This section describes possibilities to expand the applicability of graph-based design languages into the domain of product functions. These possibilities allow one to cohesively link integrative function modelling to product structures and intend to close the gap between the early, abstract stages and the systematic, concrete design generation and validation with relevant simulation tools. The approach is based on the integrated function modelling framework (IFM [26]), which is intended to provide development engineers with an integrated, cross-disciplinary approach for modelling system functionality. In this context, a function is defined as an intended or already perceivable behavior of a technical system to fulfil a specific task. The respective function information is presented in associated views; an example for the use case "drive balanced scooter" is given in Figure 8.

The state view (upper left) represents the states of operands, as well as their changes associated with related processes. The process flow view (upper right) qualitatively visualizes the flow of sequential or parallel (interaction or transformation) processes related to a specific use case. The interaction view (lower left) maps the bilateral impacts between actors and operands and represents the structure of the system. The actor view indicates the involvement of one or more actors in the realization of individual processes. For linking this approaches with the concept of graph based design languages, the class diagram of IFM has been re-modelled in UML in order to be able to use it in a design language [27]. The respective class diagram can be seen in Figure 9.

The entities, as they are represented in the IFM framework, are integrated with the classes and instantiations represented/modelled in DC43. Different from the IFM, for instance, DC43 can be used for requirements engineering as well. Consequently, the class "Requirement" has been added in Figure 5. The requirements are directly linked to use cases. This means, it is possible to identify, which use case fulfils which requirement and which requirements are not covered yet by any use case [25]. Furthermore, some links and interfaces had to be expanded and updated to match the DC43 UML environment. The existing UML model including requirements and functions could be used for the further steps—the dimensioning of motor and transmission as well as the design of the gear system and the control of the drive system.

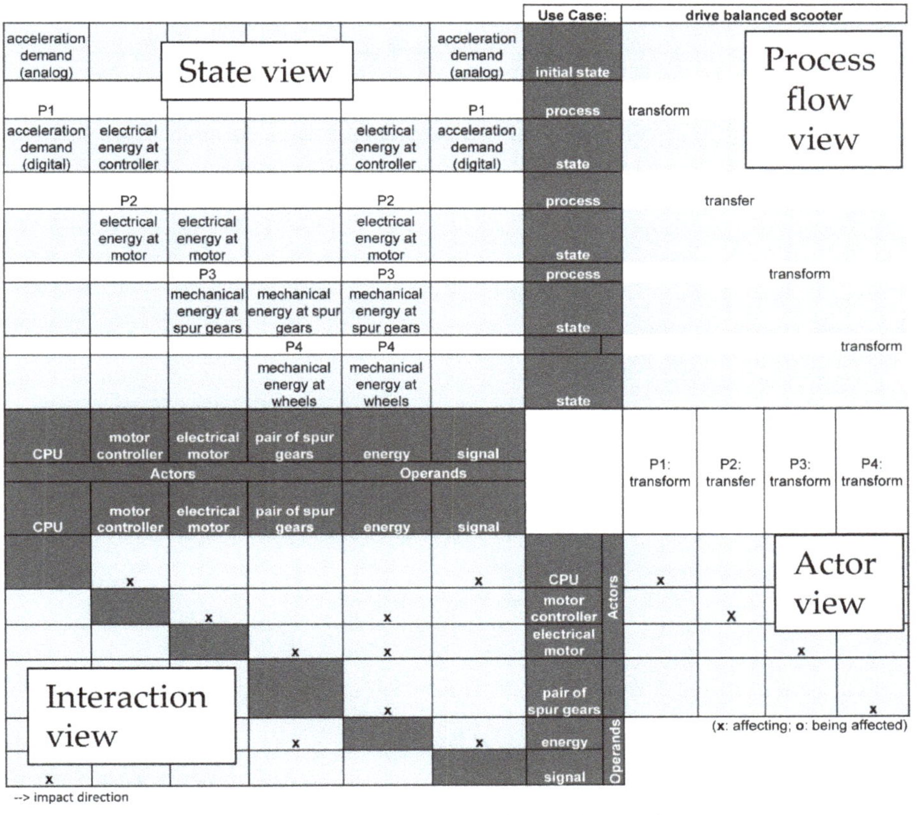

Figure 8. Associated views in the IFM framework.

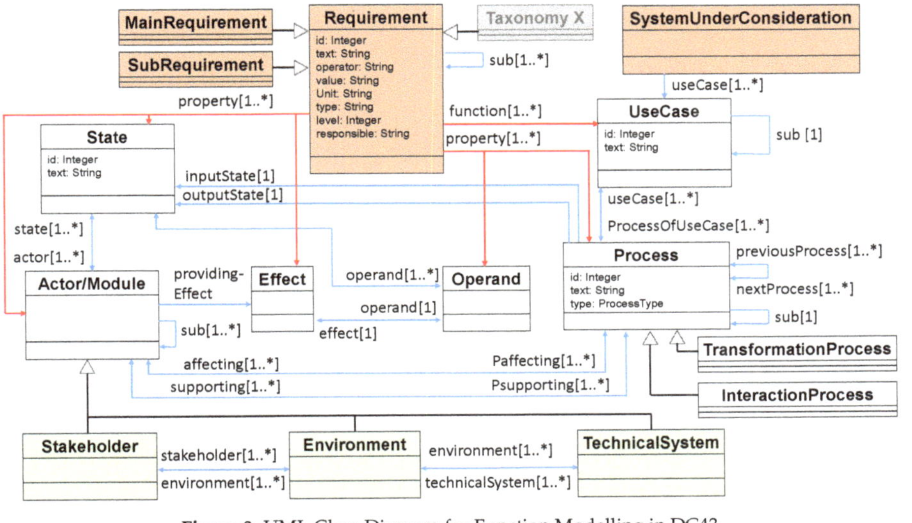

Figure 9. UML Class Diagram for Function Modelling in DC43.

5. Dimensioning of Motor and Transmission

For the synthesis of the drive system of the balanced two-wheel scooter, a process with steps shown in Figure 10 is implemented. The blue-colored steps are part of an automatic process chain which is implemented in a graph-based design language (GBDL). Each process step is linked to the main requirements of the vehicle or sub requirements derived from them and from the information, and therefore parameters from the previous process.

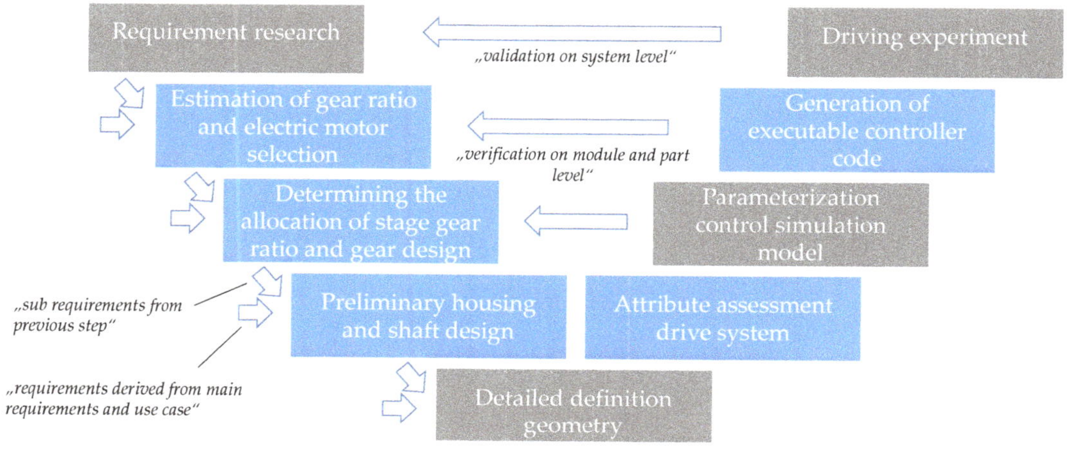

Figure 10. Process of drive train synthesis with validation.

For the design of the drive train of the self-balanced scooter, the following requirements could be identified out of the (main) vehicle requirements:

- high wheel torque in terms of the controllability of the vehicle (extinction of the pitch angle θ),
- accessibility of the desired rated vehicle speed of 20 km/h and
- acceleration capability at rated speed to ensure angle control at each operating point.

To achieve this, a drive system consisting of an electric motor and a gearbox is chosen. This chapter is about the characterization of the main performance features of the motor and the transmission. Because of the focus of the research, a suitable electric motor will be purchased and not designed. Therefore, a database with 49 motors from various manufacturers is available as the basis, containing all relevant motor data, such as armature resistance [Ohm], torque constant [Nm/A], armature inductance [H] etc. To evaluate a motor and transmission combination, a substitution model for the scooter is built. The ratio of the gear system is the parameter with the greatest influence on the dynamics of the vehicle and will be identified. Each motor is combined with different gear ratios via a single-track vehicle model, see Figure 11. The single-track simulation model contains a drive train, which consists of an electric machine (blue) with a gearbox (green) and a vehicle with traction wheel (red). This model includes basic driving resistances due to acceleration and rolling friction. When running the simulation, the vehicle mass is accelerated. The gear ratio of the drive is varied in the range from $i = 8$ to $i = 24$ in an adjustable step size. In the specific use case, this results in 7840 simulation runs with a step size of $\Delta i = 0.1$. The step response of the system with regard to a voltage jump is determined for each simulation.

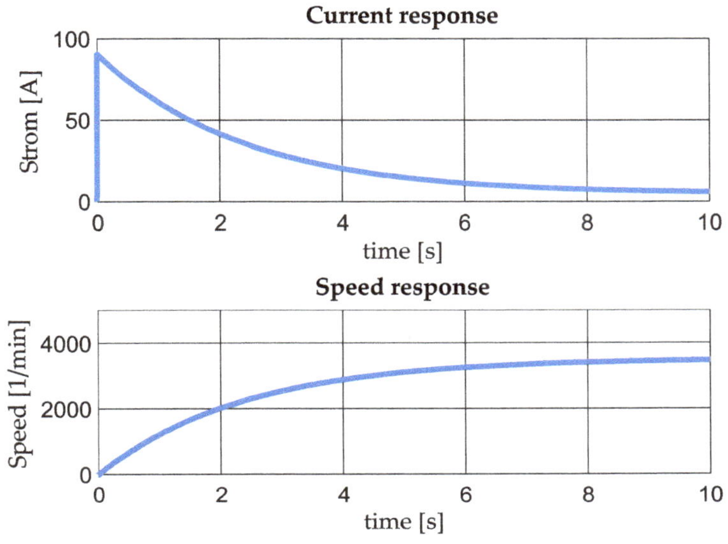

Figure 11. Step response as early dynamic evaluation approach.

The step response is evaluated with regard to the following criteria, which are assessed as relevant with regard to the requirements mentioned:

- time to rated speed,
- acceleration ability at start moment and
- acceleration at rated speed.

The step responses can be used to compare the various motors with their different gear ratios and to make a decision for a suitable combination. The result is the best performing motor with its appropriate gear ratio. Due to the substitution model, an overall ratio of $i = 17.5$ in combination with a low price BLDC motor "Turnigy Aerodrive SK3" [28] could be determined as an acceptable solution. After the precise design of the transmission geometry in Section 7, it is only then possible to assess the dynamics behavior in Section 8.

6. Detail Dimensioning of the Transmission

The process presented in this chapter generates a preliminary transmission design from the desired overall gear ratio and the drive train performance data. The process is divided into the following sub-steps:

- deriving the geometric requirements of the transmission,
- automatic synthesis of the gear set including parameter variation, as well as decision for one solution and
- generation of a geometry model, design of bearings and shafts, evaluation of the overall design.

For this purpose, the following compound requirements could be generated from previous process and the vehicle as a system:

- packaging restrictions—low center of gravity, therefore motor below the driver standing area; results in a center distance from traction wheel of the vehicle to motor shaft, $a_{ges} = 112.5\,\text{mm}$,
- limitation of the installation space due to the frame and the rim of traction wheel is converted into a maximum lateral deflection of the gear set, $h_{max} = 130\,\text{mm}$, see Figure 12,
- Performance data for motor and overall gear ratio (max input torque $T = 4\,\text{N m}$, max input speed $n = 7600\,\text{min}^{-1}$, overall ratio $i_{ges} = 17.5$).

There are further requirements for the design of the gear set, including parameters such as desired safety values, application factors, etc. For the synthesis of the gear geometry using GAP (Getriebeauslegungsprogramm—German for gear system dimensioning program) [29], a design criterion must also be defined, here, e.g., the requirement for a minimum mass of the gear set is used. Via GAP, the gear geometry is generated in an explicit way according to the requirements and stored in the graph. For this purpose, an interface for commanding GAP and feeding back the data from the gear set to the central data model is implemented. The design of the gear set is repeated for several parameters, a design space is fully factorial scanned. The geometric parameters are shown in Figure 12.

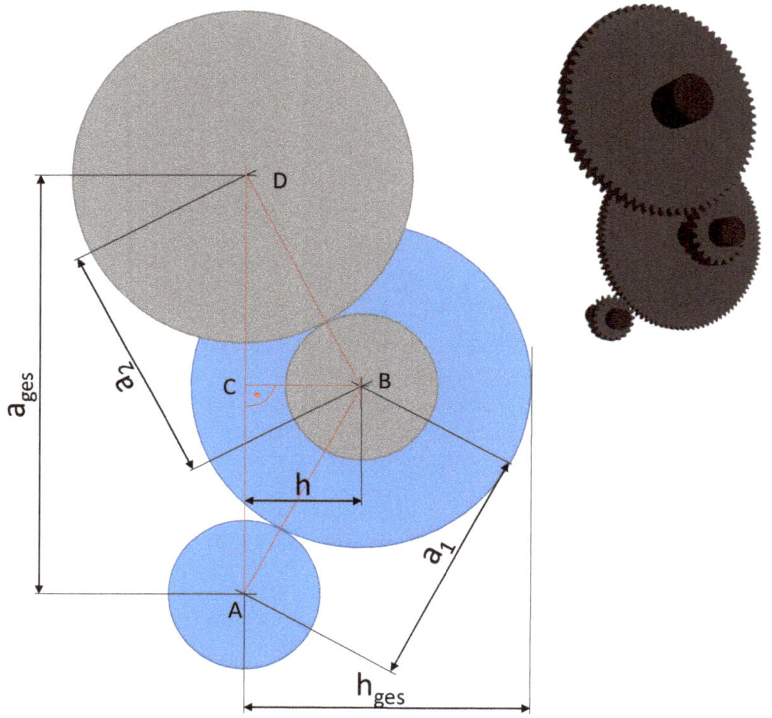

Figure 12. Geometrical parameters at a two-stage gear system.

The variable parameters for the exploration of the two-stage gear set scheme are the ratio distribution of the overall ratio to the two stages and the center distances of the two individual stages. When dividing the overall ratio, the ratio of the first stage i_1 is varied and the ratio of the second stage i_2 results from the interrelation:

$$i_2 = \frac{i_{ges}}{i_1} \tag{1}$$

The two center distances of the gear stages (a_1 and a_2) are varied in the range of 40–90 mm for the scooter gear system use case. The size of a bearing system for the shafts limits the minimal distance of the shafts at 40 mm. The upper bound of 90 mm is estimated in terms of the number of variants. In addition, a valid total center distance of 112.5 mm can only be achieved if the following geometrical relationship is given:

$$a_{ges} \leq a_1 + a_2 \tag{2}$$

The geometrical relation of the deflection h of the two-stage gear set can be formulated as follows with regard to the maximum height of the triangle ABD.

$$a_{ges} = \sqrt{a_1^2 + h^2} + \sqrt{a_2^2 + h^2} \tag{3}$$

Formula (3) solved for h:

$$h = \pm \frac{\sqrt{a_2^4 - 2a_1^2 a_{ges}^2 - 2a_1^2 a_2^2 + a_{ges}^4 - 2a_{ges}^2 a_2^2 - a_1^4}}{2a_{ges}} \tag{4}$$

with $a_{ges} \neq 0$.

If the radius of the larger wheel of the intermediate shaft is added to the h, the result is h_{ges}, whereby the requirements for every valid design must apply $h_{ges} \leq h_{max}$. These mentioned boundary conditions must be fulfilled for all generated variants in order to be eligible. By integrating the gear synthesis into the graph-based design language, the model can be used for subsequent design steps and further evaluations of the drive train. A simple geometry model is created for all generated gear sets, and the shaft-hub connections between the wheels and their shaft is calculated using a simplified approach referred to [30]. In addition, the bearing lifetime is calculated using an approach based on [31]. All the determined properties are saved in the central data model in the form of a graph. Figure 13 visualizes the central data model of a design in a simplified manner.

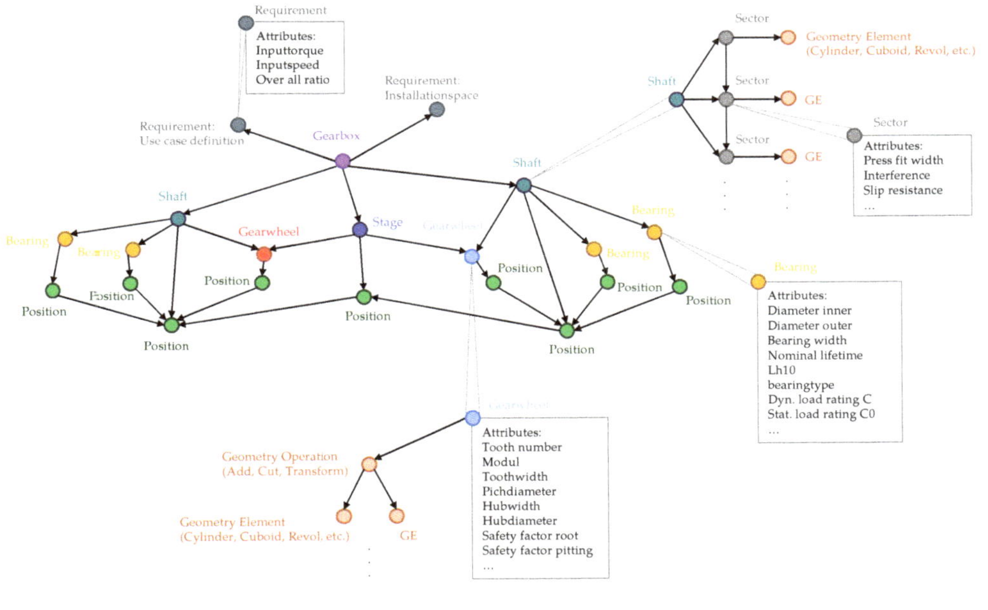

Figure 13. Design graph as a central data model for the drive train.

The properties of the gear set mass and the power loss are shown for the generated variants in Figure 14. The investigation for the scooter gearbox produces 13,892 valid geometrical variants of the gear set.

The left-lower limit of the variant point cloud represents the so-called pareto-front (orange) with regard to the two evaluation parameters (power loss and gear set mass). The selection of one progressed design variant happened from a procurement perspective. The gear geometry needed to be adapted to the on market available gears, as no gear production was possible within the scope of the research, see red point Figure 14. The parameter configuration of the final design is $i_1 = 5$; $i_2 = 3.5$ and $a_1 = 60$ mm; $a_2 = 67.5$ mm. This preliminary design with its central data model, including the evaluation it contains, is used to create a manufacturable design, this is presented in Section 7.

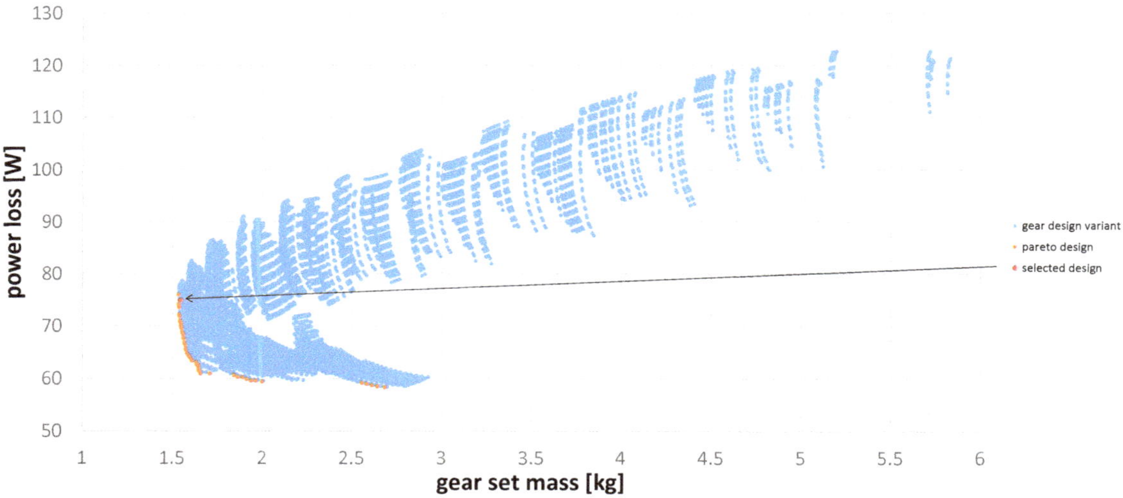

Figure 14. Design exploration and its evaluation according to power loss and gear set mass.

7. Design of the Gear System

This section describes the detail design of the gear system with the housing ready to be manufactured. Due to the results of the automatic modelled and dimensioned motor and transmission unit, a detailed CAD model of the electric drivetrain is designed. Of the very many requirements, including manufacturing requirements, only a few requirements are mentioned here:

- mechanical integrity—load cases for driving situations,
- mounting points for motor, rotor position sensor, shaft seals, etc.
- assemblability of the gear system—result in a splitting concept of the housing and
- geometrical requirements of the selected gear set.

Subsequently, the derivation of load cases and the associated evaluation is shown; also, the final design of the balanced two-wheel scooter drivetrain is presented. Due to the target weight of the scooter and a maximum carrying capacity of about 100 kg, and an overall weight of 145 kg as a main vehicle requirement it is possible to elaborate sub requirements for load cases. Besides the installation space model, the critical load cases are very important for the ongoing drivetrain design. One critical load to evaluate the mechanical integrity is the curbstone crossing. The structure of the housing and mountings should withstand such a curb crossing at a driving speed of 20 $\mathrm{km\,h^{-1}}$. It is assumed that the vehicle passes over a curbstone with a height of 80 mm. According to the following Equation (5) mentioned in [32], the peak vertical acceleration a_z during the curb crossing is calculated by the traction wheel radius r, the height of the curb h and the velocity of the vehicle v:

$$a_z = \frac{2}{2r - h} v^2 \tag{5}$$

The calculation leads to 13 g peak acceleration during the curb crossing. After considering the damping effects of the tire, the peak acceleration is reduced to approximately 10 g. Including the mass of the scooter and the estimated peak acceleration, a maximum vertical force has been calculated as the critical force caused by driving the scooter. The second critical load is referenced to the weight and torque of the electric motor. In order to be able to install and remove the drive system as a complete module, the motor is supposed to be connected to the transmission housing. According to the motor specification, the motor has an overall weight of 0.84 kg and a peak torque of 4 N m [28]. This means that the design of the drive unit has to support the torque and weight. The weight of the motor, similar to the total weight of the scooter, has to be held up with a permissible deformation for a curb crossing of 10 g. As shown in Figure 1, the previously mentioned requirements are taken into the design and verified and optimized by simulations.

According to the vehicle requirements, an installation space model for the drive unit is generated. The model shown in Figure 15 presents three different views, including the possible installation space for the drive unit (yellow marked geometry), which absorbs the previously specified interfaces to the gear set, such as the connection to the wheel hub and the attachment points of the frame. According to the requirements, it should be possible to install the drive unit on both sides of the balanced two-wheel scooter. For this reason, special care was taken to ensure that the attachment points and the motor, as well as the output shaft, are arranged symmetrically to each other. Based on the automatically generated gear set and the defined interfaces, a first model of the gearbox housing is created, limited by the installation space. The detailed model is created in the design software Creo Parametric 5.0. A finite element method (FEM) simulation is generated in Creo Simulate to verify the required load cases (Creo Parametric 5.0 as well as Creo Simulate are software tools of the PTC Inc. from Boston, MA, USA). After the boundary conditions are transferred into the model, the mesh size is investigated by a convergence and divergence study. For the studies shown in Figure 9, the elemental stresses are investigated according to von Mises. The divergence study is used to check the mesh size in the border areas and the convergence study of the mesh fineness in the unbounded area [33]. For each of the studies, one element is tested with different mesh sizes (visible in Figure 16).

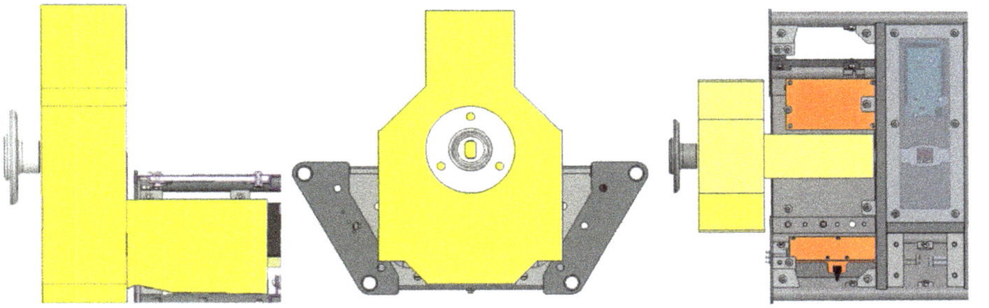

Figure 15. Space Model for the drivetrain view form front, side and above.

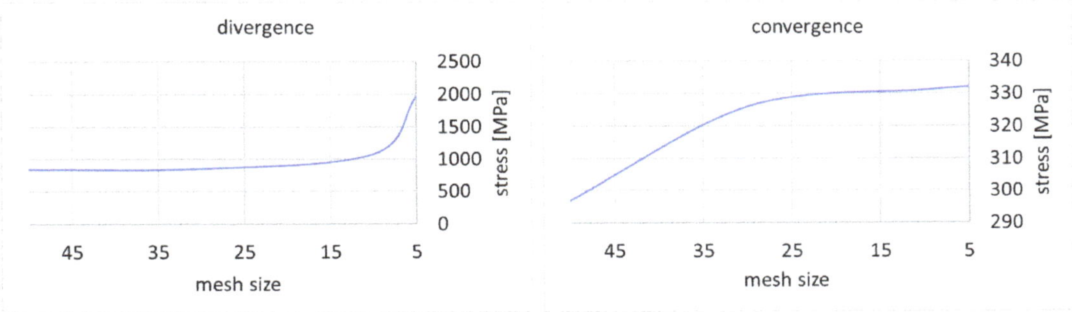

Figure 16. Convergence and divergence behavior for different mesh sizes.

The divergence study shows that no valid analysis of the stresses in the border areas can be achieved with a grid fineness of less than 10. The convergence study shows that no significant change in the stresses in the unbounded area can be detected from a mesh size of less than 25. Accordingly, a mesh size of 10 is used for the more detailed analysis of the load cases. A further advantage of this mesh size is the computing time, which is kept to a minimum. This is decisive for the optimization process, in order to simulate a high value of different geometries by achieving valid results. The above-mentioned optimization of the component geometries mainly concentrated on the component displacement, as the element stresses in all components are consistently lower than the critical material parameters allow. The comparison shown in Figure 17 represents the deflection of the drive unit in the Z-direction viewed from the side.

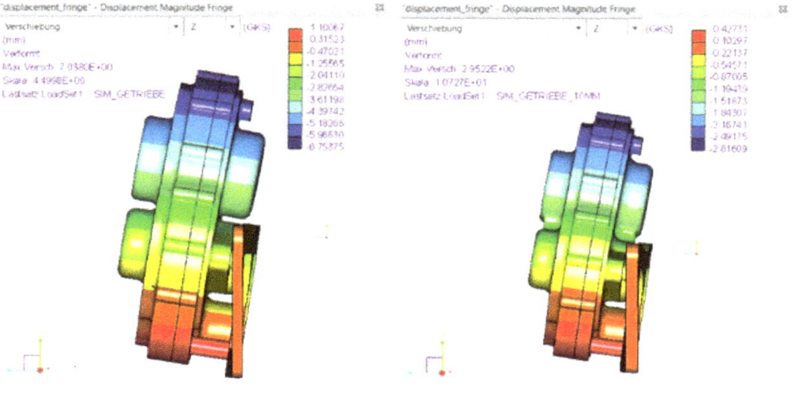

Figure 17. Displacement with the first design (**left**) and the optimized geometry (**right**).

On the left-hand side, the simulation result of the first created geometry is shown. On the right-hand side, the finally selected and optimized geometry is shown. During the optimization process, the component displacement could be reduced by about 60%. For stiffening the individual parts, for example, ribs are inserted between the individual bearing seats of the transmission housings. To increase the stiffness at equal weight, it is necessary to thicken the mounting plate to 10 mm and remove material according to the load paths. To further reduce the overall weight and costs, a geometry for additive manufacturing with polylactic acid (PLA) is designed. Therefore, a topology optimization is implemented for the motor connector. For the topology optimization and the following FEM simulation the structure is simulated considering the maximum motor torque and the impact of the motor weight during the curb crossing. Figure 18 shows the results of the FEM simulation.

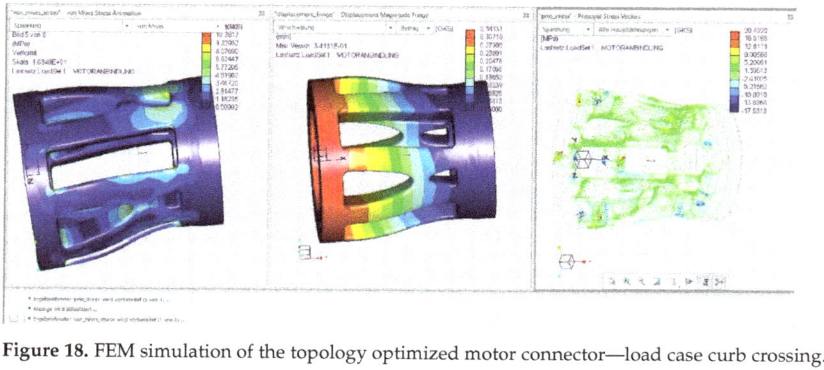

Figure 18. FEM simulation of the topology optimized motor connector—load case curb crossing.

The motor is coupled to the input shaft of the transmission with a flexible coupling. Therefore, two criteria can be defined to investigate the component. On the one hand, the critical material parameters of PLA are not allowed to be exceeded and on the other hand the displacement of the connection may not exceed the maximum possible flexible coupling deflection (max. angle of 5°) [34]. Otherwise, bending loads could damage the coupling and increase bearing loads. As shown by the simulation results, neither of the two criteria are reached or even exceeded by the critical loads.

Another crucial aspect was the supply of lubricant in the gear system. The requirements for the oil system included sealing, as well as ensuring a sufficient supply of lubricant to the gears and bearings. The first one is realized by using an O-ring seal cord and proper radial shaft seals. The second is checked by a fluid simulation. The purpose of the simulation is to determine the amount of oil required for reliable operation of the drive train and how well the individual bearings and gears are supplied with lubricant. Figure 19 shows a screenshot from the simulation set up in the software PreonLab. PreonLab is a fluid simulation tool developed by the FIFTY2 Technology GmbH from Freiburg, Germany. Based on the results of the simulation, a filling level of 50 mL is determined to be appropriate for the safe operation of the transmission.

Figure 19. Analysis of the splash lubrication system by using the software PreonLab.

The resulting drive system with its components can be seen in Figure 20.

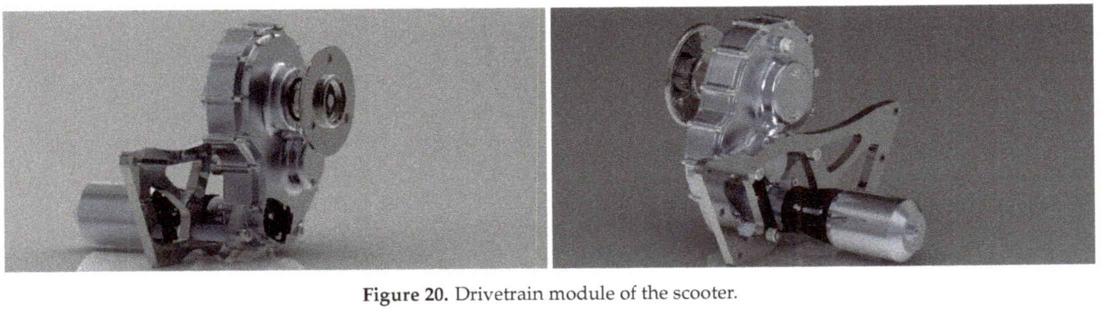

Figure 20. Drivetrain module of the scooter.

After the entire drivetrain is designed, verified and optimized according to the requirements, a detailed centre-of-gravity (COG) model is derived. This model includes the vehicle COG and the rotational moment of inertia of the drive train. These data are transferred back to the central model to be used as parameters for the automated controller configuration.

8. Control of the Drive System

In order to guarantee the functionality of the above derived scooter design, the dynamic properties stated in the product requirements must be checked. Typical requirements for the balanced two-wheeled scooter are, on the one hand, stability of the attitude dynamics and, on the other hand, a firmly defined transfer behavior of disturbances introduced by the human driver on the translational velocity of the scooter. These properties can be concretized in the design of the control system in the form of overshoot width Δ_o, overshoot time t_o and settling time t_s of the system for appropriately selected step responses, as show in Figure 21. The settling time refers to the case in which the setpoint remains within a specified value range Δ_s.

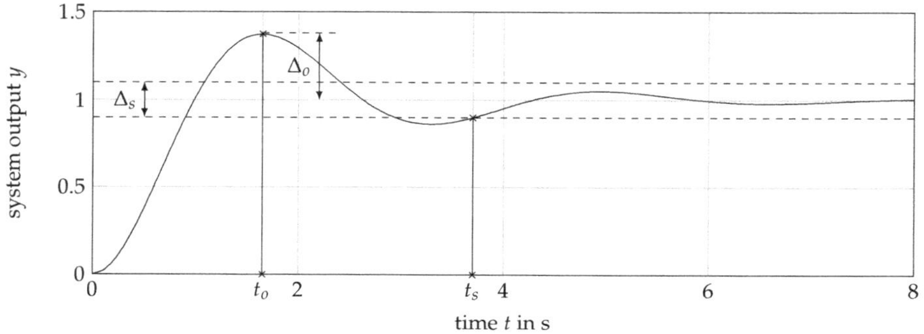

Figure 21. Schematic of a step response and its characteristics.

In case of the proposed scooter design, four different cases of simulation scenarios are proposed to define the dynamic behavior of the scooter. The first and most crucial specification is a demand on the stability of the attitude dynamic, thus stabilizing an unstable pole describing a pitch movement of the platform body around its rotation axis. The second specification specifies the suppression of input disturbances created by the human driver based on a shift of his center of gravity. The last two specifications make demands on the tracking behavior of the control system. They specify the dynamic properties of the tracking of a commanded velocity v_c and the tracking of the yaw rate $\dot{\psi}_c$.

A detailed listing of the specifications can be seen in Table 1.

Table 1. Specification of the dynamic properties of the scooter.

Type	Reference Value	Overshoot (t_o/Δ_o)	Transient (t_s/Δ_s)
1. Stability of the attitude dynamic	40°	2 s/15°	6 s/4°
2. Input disturbance rejection	125 N m	2 s/15°	6 s/4°
3. Tracking of v	3 m s^{-1}	2 s/0.5 m s^{-1}	6 s/0.2 m s^{-1}
4. Tracking of $\dot{\psi}$	10° s^{-1}	2 s/5°	6 s/2°

In order to check these requirements, the closed loop behavior of the scooter is simulated. This leads to a knowledge of the control variables, and thus the drive system defined above can be validated with respect to the required demands. Firstly, the dynamic model used to simulate the balanced two-wheeled scooter is introduced. A schematic of the plant model is shown in Figure 22.

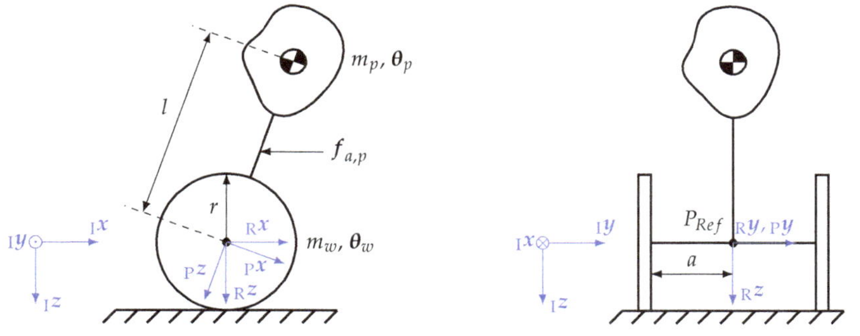

Figure 22. Schematic of the segway model with a heading $\psi = 0$.

As can be seen in Figure 22, a minimal set of independent variables q can be given by

$$q = \begin{pmatrix} x & y & \psi & \theta \end{pmatrix}^T \tag{6}$$

consisting of the position coordinates x and y and the yaw angle ψ and pitch angle θ. The dynamic system consists of three bodies, the main platform with the passenger and the two wheels. The two wheels are mounted on an axle with a length $2a$ and the center of mass of the wheel bodies lie in their respective rotation axis, thus the inertia tensor of each wheel is invariant to rotation around their respective rotation axis. The length of the lever of the center of gravity of the platform body with respect to its rotational degree of freedom is described by the parameter l. A simplification made in the modeling is that the wheels can only move in the body fixed x direction, thus narrowing the degrees of freedom in the derivatives of the independent variables $\dot{q}$, as depicted in Equation (7):

$$\dot{q} = \begin{pmatrix} v & \dot{\psi} & \dot{\theta} \end{pmatrix}^T. \tag{7}$$

Thus, for this problem, the kinematic description of the wheel movement of the reference point P_{Ref} in the center of the wheel axle based on [35] can be given in the form

$$_I\dot{r}_{Ref} = \begin{pmatrix} v \cos \psi \\ v \sin \psi \\ 0 \end{pmatrix} \tag{8}$$

with the help of the translational velocity v and the heading ψ of the scooter. Based on the translational velocity of the reference point P_{Ref}, depicted in Equation (8), the translational velocity $\dot{r}_{w,n}$ of the n-th wheel body can be expressed in the inertial frame:

$$_I\dot{r}_{w,n} = {_I\dot{r}_{Ref}} + \underbrace{{_I T_R}{_R\dot{l}_{w,n}}}_{=0} + {_I T_R}[{_R\omega_r}]_x {_R l_{w,n}}. \tag{9}$$

The vector $l_{w,n}$ used in Equation (9) describes the position of the wheel with respect to the reference point and only depends on the axle length a, if described in the R-frame. The transformation matrix $_R T_I$ describes the transformation of the inertial frame to the R-frame and can be fully defined by a elementary rotation with the angle ψ around the $_I z$ axis. In order to describe the change in the orientation of the R-frame with respect to the body fixed frame, the angular rate ω_r is introduced as show in Equation (10).

$$_R\omega_r = \begin{pmatrix} 0 & 0 & \dot{\psi} \end{pmatrix}^T. \tag{10}$$

Similar to the translational velocity of the wheels, the translational velocity $\dot{\boldsymbol{r}}_p$ of the platform can be formulated with respect to the reference point P_{ref}, as shown in Equation (11).

$$ {}_{\mathrm{I}}\dot{\boldsymbol{r}}_p = {}_{\mathrm{I}}\dot{\boldsymbol{r}}_{Ref} + \underbrace{{}_{\mathrm{I}}\boldsymbol{T}_{\mathrm{P}}{}_{\mathrm{P}}\dot{\boldsymbol{l}}_p}_{=0} + {}_{\mathrm{I}}\boldsymbol{T}_{\mathrm{P}}\left[{}_{\mathrm{P}}\boldsymbol{\omega}_p\right]_x {}_{\mathrm{P}}\boldsymbol{l}_p. \tag{11} $$

The vector $\boldsymbol{l}_p$ describes the position of the center of mass of the platform body, which only depends on the arm length l and is a time invariant quantity, if described in the P-frame. The transformation matrix ${}_{\mathrm{P}}\boldsymbol{T}_{\mathrm{I}}$ describes the orientation of the inertial frame with respect to the P-frame and is defined by two sequential elementary rotations. The first rotation describes a rotation with the angle ψ around the ${}_{\mathrm{I}}\boldsymbol{z}$ axis and the second rotation describes a rotation with the angle θ around the $\boldsymbol{y}$ axis of the new coordinate system. In order to fully define the translational velocity of the platform depicted in Equation (11), the rotational velocity $\boldsymbol{\omega}_p$ is needed

$$ {}_{\mathrm{P}}\boldsymbol{\omega}_p = \begin{pmatrix} 0 & \dot{\theta} & 0 \end{pmatrix}^T + {}_{\mathrm{P}}\boldsymbol{T}_{\mathrm{R}}\begin{pmatrix} 0 & 0 & \dot{\psi} \end{pmatrix}^T, \tag{12} $$

which describes the change of orientation of the platform system with respect to the inertial frame. To fully describe the motion of the individual bodies of the scooter, it is necessary to link the translational velocities v and the yaw rate $\dot{\psi}$ with the rotational velocity $\dot{\alpha}_n$ of the wheel bodies. Based on the fact that the scooter does not experience any slip, this relationship can be expressed in the following equations:

$$ v = -\frac{r}{2}(\dot{\alpha}_1 + \dot{\alpha}_2) \tag{13} $$

$$ \dot{\psi} = -\frac{r}{2a}(\dot{\alpha}_2 - \dot{\alpha}_1). \tag{14} $$

Based on the above introduced Equations, (13) and (14), the rotational velocity of the wheel can be expressed as illustrated in Equation (15).

$$ {}_{\mathrm{R}}\boldsymbol{\omega}_{w,n} = \begin{pmatrix} 0 & \dot{\alpha}_n & \dot{\psi} \end{pmatrix}^T = \begin{pmatrix} 0 & -\frac{1}{r}v + \frac{a}{r}\dot{\psi} & \dot{\psi} \end{pmatrix}^T \tag{15} $$

In order to describe the dynamic model based on d'Alemberts principle, the whole momentum of the system needs to be in a state of equilibrium, thus leading to the following relationship:

$$ \begin{aligned} &\boldsymbol{J}_{T,p}^T\left({}_{\mathrm{I}}\dot{\boldsymbol{p}}_p - {}_{\mathrm{I}}\boldsymbol{f}_{e,p}\right) + \sum_{n=1}^{2} \boldsymbol{J}_{T,w,n}^T\left({}_{\mathrm{I}}\dot{\boldsymbol{p}}_{w,n} - {}_{\mathrm{I}}\boldsymbol{f}_{e,w,n}\right) + \dots \\ &\dots + \boldsymbol{J}_{R,p}^T\left({}_{\mathrm{P}}\dot{\boldsymbol{l}}_p - {}_{\mathrm{P}}\boldsymbol{\tau}_{e,p}\right) + \sum_{n=1}^{2} \boldsymbol{J}_{R,w,n}^T\left({}_{\mathrm{R}}\dot{\boldsymbol{l}}_{w,n} - {}_{\mathrm{R}}\boldsymbol{\tau}_{e,w,n}\right) = 0 \end{aligned} \tag{16} $$

In order to describe the momentum $\dot{\boldsymbol{p}}_p$ of the platform, the mass m_p of the platform and the translational acceleration in the inertial frame are needed, which can be easily calculated based on Equation (11). Similarly, the momentum $\dot{\boldsymbol{p}}_{w,n}$ of the n-th wheel can be expressed based on its mass m_w and the time derivative of its translational velocity in the inertial frame depicted in Equation (9). To describe the rotational momentum of the platform $\dot{\boldsymbol{l}}_p$ or of the n-th wheel body $\dot{\boldsymbol{l}}_{w,n}$ in its respective body fixed frame, the inertial tensor $\boldsymbol{\Theta}_p$ of the platform and $\boldsymbol{\Theta}_{w,n}$ of the wheel and their angular acceleration in the body fixed frame are needed. These can be evaluated based on the description of their angular velocities described in the Equations (12) and (15). The Jacobian matrices, $\boldsymbol{J}_{T,p}$, $\boldsymbol{J}_{T,w,n}$, $\boldsymbol{J}_{R,p}$ and $\boldsymbol{J}_{R,w,n}$, describe, with respect to the choosen independen variables, the directions in which the systems body is allowed to move. They can be expressed by the partial derivatives of the translational and rotational velocities of each body with respect to the vector $\dot{\boldsymbol{q}}$. To describe the external influences on the system, the external forces and moments acting on the system needs to be defined as follows:

$$_{\mathrm{I}}\boldsymbol{f}_{e,p} = m_p {_{\mathrm{I}}\boldsymbol{g}} + {_{\mathrm{I}}\boldsymbol{T}_{\mathrm{PP}}}\boldsymbol{f}_{a,p} \tag{17}$$

$$_{\mathrm{I}}\boldsymbol{f}_{e,w,n} = m_w {_{\mathrm{I}}\boldsymbol{g}} \tag{18}$$

$$_{\mathrm{P}}\boldsymbol{\tau}_{e,p} = \begin{pmatrix} 0 & \tau_{M,1} + \tau_{M,2} & 0 \end{pmatrix}^T \tag{19}$$

$$_{\mathrm{R}}\boldsymbol{\tau}_{e,w,n} = \begin{pmatrix} 0 & -n_r \tau_{M,n} & 0 \end{pmatrix}^T. \tag{20}$$

The moments produced by the two motors are taken into account by the quantities $\tau_{M,n}$. The dynamic model depicted in Equation (16) can be reformulated in the mass matrix form:

$$\boldsymbol{M}(\boldsymbol{q})\ddot{\boldsymbol{q}} + \boldsymbol{g}(\boldsymbol{q}, \dot{\boldsymbol{q}}) = \boldsymbol{f}. \tag{21}$$

The external torques acting on the platform system and on the wheels mainly consist of the torques $\tau_{M,n}$ produced by the n-th motor of the scooter. Taking the dynamic properties of the motor into account [36], the full dynamic model reads

$$\begin{pmatrix} \dot{\boldsymbol{q}} \\ \ddot{\boldsymbol{q}} \\ \dot{\boldsymbol{i}} \end{pmatrix} = \begin{pmatrix} \begin{pmatrix} v\cos\psi & v\sin\psi & \dot{\psi} & \dot{\theta} \end{pmatrix}^T \\ \boldsymbol{M}(\boldsymbol{q})^{-1}(\boldsymbol{f} - \boldsymbol{g}(\boldsymbol{q}, \dot{\boldsymbol{q}})) \\ \frac{1}{L}\left(\boldsymbol{u} - \psi^\star \left(\dot{\theta}\begin{pmatrix} 1 & 1 \end{pmatrix}^T - \dot{\boldsymbol{\alpha}}\right) - R\boldsymbol{i}\right) \end{pmatrix} \tag{22}$$

with the state vector

$$\boldsymbol{x} = \begin{pmatrix} \boldsymbol{q} & \dot{\boldsymbol{q}} & \boldsymbol{i} \end{pmatrix}^T. \tag{23}$$

Parameters of the motor model are the electrical resistance R, the inductivity L and the magnetic flux $\psi^\star$ of the motor. In order to automatically generate the simulation model, the inertia properties and the positions of the center of masses of the different bodies with respect to the reference point need to be extracted. This can be easily done with the help of a random CAD-Kernel. A possibility to model a multibody system in a design language is shown in Figure 23.

The central class of this library is the "Body" class, which inherits a set of inertia properties needed for a multibody simulation and each body can be disassembled into an infinite set of elements to be able to represent different assemblies, which can be combined to one body in the sense of a multibody simulation. To take care of the coordinate sytems in which the different quantities are expressed, the "Frame" class is introduced. It describes the rotation and translation to the inertial coordinate frame, thus all quantities can be expressed in a frame which is chosen with regard to the topology of the multibody model. The interface class to the geometric representation is the "Component" class, a class originating from a DC43 library which represents instances with a geometric shape. In this development process, the automated design process is inturrepted in the detailed design of the gear system, which is depicted in Section 7. If this design process could be automated too, the multibody simulation could be triggered automatically based on the model proposed in the class diagram shown in Figure 23.

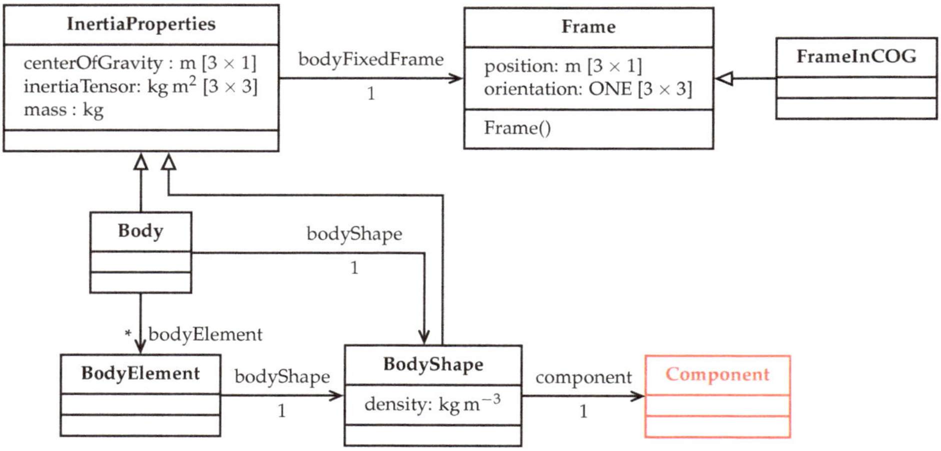

Figure 23. Class diagram of a multibody library.

The last step of the design process would be the closed loop simulation of the plant model. For this purpose, Matlab/Simulink was used. The controller design chosen for the scooter design is depicted in Figure 24.

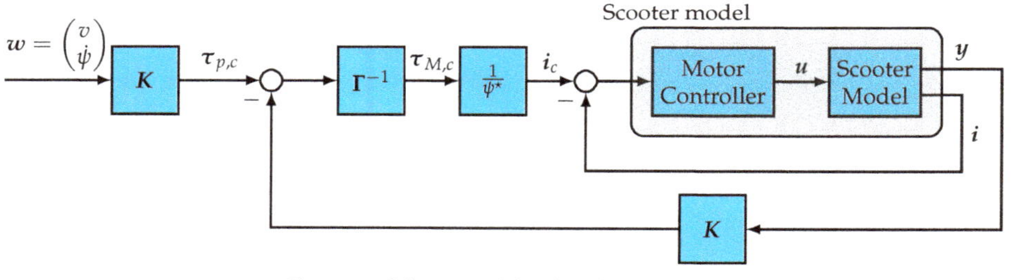

Figure 24. Schematic of the closed loop model.

The controller is based on a simple output feedback controller for the pitch and velocity dynamic and a separated state feedback controller for the yaw dynamics. This controller design and its decoupled nature between the pitch and the yaw dynamic is extensively discussed in [37–41]. The output vector $\boldsymbol{y}$ consists of the output $\boldsymbol{y}_v$ for the pitch and velocity controller and the output $\boldsymbol{y}_{\dot{\psi}}$ for the yaw rate controller. Similar to this, the gain matrix $\boldsymbol{K}$ is a block diagonal matrix consisting of the vector $\boldsymbol{k}_v^T$ of the pitch and velocity controller and the vector $\boldsymbol{k}_{\dot{\psi}}^T$ of the yaw rate controller. In order to map the torques calculated in the two decoupled controllers on the actual actuators, a simple allocation scheme

$$\begin{aligned}\boldsymbol{\tau}_p &= \boldsymbol{\Gamma}\boldsymbol{\tau}_M \\ \begin{pmatrix}\tau_v \\ \tau_{\dot{\psi}}\end{pmatrix} &= \begin{pmatrix}1 & 1 \\ -a & a\end{pmatrix}\begin{pmatrix}\tau_{M,1} \\ \tau_{M,2}\end{pmatrix}\end{aligned} \tag{24}$$

is used, which is uniquely defined for all combinations. To define the gains of the control system in order to fit the specified requirements, a linear design model based on a subset of the nonlinear model depicted in Equation (22) is used. It reads

$$\dot{\boldsymbol{x}}_v = \boldsymbol{A}_v\boldsymbol{x}_v + \boldsymbol{b}_v u_v \tag{25}$$

$$y_v = \boldsymbol{c}^T\boldsymbol{x}_v \tag{26}$$

with the state vector $\boldsymbol{x}_v$ and the output vector $\boldsymbol{y}_v$

$$\boldsymbol{x}_v = \begin{pmatrix} e_v & \theta & v & \dot{\theta} & e_i & i_v \end{pmatrix}^T \tag{27}$$

$$\boldsymbol{y}_v = \begin{pmatrix} e_v & \theta & v & \dot{\theta} \end{pmatrix}. \tag{28}$$

In this design model, the dynamic of the current controller of the motor is taken into account too, because in order too stabilize the unstable pole of the pitch dynamic, a very fast response is needed and thus unwanted coupling effects with the motor dynamic should be avoided. The control law for the current controller reads

$$i_v = k_i(i_{v,c} - i_v) + k_{e_i} e_i \tag{29}$$

which creates a current output i_v. It can be seen that the model is extended by the state e_i, which describes the integrated current error of the controller. The output feedback control law for the pitch and velocity dynamic reads

$$i_{v,c} = \frac{1}{\psi^\star}\tau_v = \frac{1}{\psi^\star}\boldsymbol{k}_v^T \boldsymbol{y}_v, \tag{30}$$

which generates a commanded value $i_{v,c}$ based on the state of the pitch and velocity dynamic. In order to achieve steady state accuracy, the integrated error of the velocity command is taken into account by the state e_v. The control laws shown in Equations (29) and (30) combined with the simplified linear design model of Equation (25) can be used for pole placement [42] to achieve the specified dynamic properties. The poles are placed in order to satisfy the condition

$$\det\left(s\boldsymbol{I}_{6\times6} - \left(\boldsymbol{A}_v - \boldsymbol{b}_v \boldsymbol{k}_v^T \boldsymbol{c}_v\right)\right) = \left(s^2 + 2\zeta_v \omega_v s + \omega_v^2\right)^3, \tag{31}$$

thus specifying the eigenfrequency ω_v and damping ζ_v of the closed loop model. If the chosen controller does not satisfy the required dynamic properties in the nonlinear simulation, the eigenfrequency and damping of the system can be adapted automatically. Similar to velocity and pitch dynamic, the yaw rate controller can be designed. In this case, the motor dynamic is neglected. Thus, a simple linear model for the design of the controller can be expressed by

$$\dot{\boldsymbol{x}}_\psi = \boldsymbol{A}_\psi \boldsymbol{x}_\psi + \boldsymbol{b}_\psi \tau_\psi \tag{32}$$

$$\boldsymbol{y}_\psi = \boldsymbol{x}_\psi \tag{33}$$

with the state vector $\boldsymbol{x}_\psi$

$$\boldsymbol{x}_\psi = \begin{pmatrix} e_\psi & \dot{\psi} \end{pmatrix}^T. \tag{34}$$

The state feedback control law reads

$$\tau_\psi = \boldsymbol{k}_\psi^T \boldsymbol{x}_\psi \tag{35}$$

and generates a torque command τ_ψ based on the state $\dot{\psi}$ and the integrated error e_ψ of the yaw rate. The pole placement [42] can be achieved by satisfying the following condition

$$\det\left(s\boldsymbol{I}_{2\times2} - \left(\boldsymbol{A}_\psi - \boldsymbol{b}_\psi \boldsymbol{k}_\psi^T\right)\right) = s^2 + 2\zeta_\psi \omega_\psi s + \omega_\psi^2, \tag{36}$$

with the help of the eigenfrequency ω_ψ and the damping ζ_ψ of the yaw rate dynamic. Similar to the pitch and velocity dynamic, the controller can be tuned to satisfy the specified dynamic properties, by changing the damping and the eigenfrequency in the design model if the nonlinear validation model does not show the desired effects. In Figure 25, the critical case for the motor design is shown, which is depicted by an error in the pitch angle for the here chosen dynamic properties shown in Table 1.

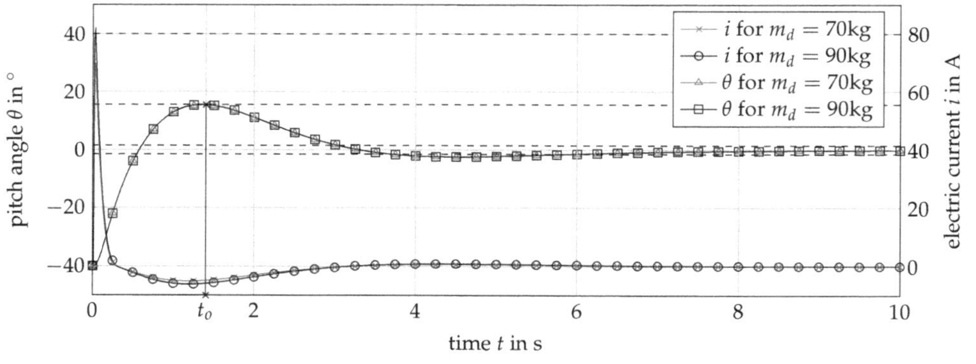

Figure 25. Simulation data of the stability specification.

An initial estimate for the maximum motor moment or current input was done in Section 5 in order to design the transmission. It can be seen that the trajectories of the pitch angle satisfy the conditions of the max overshoot, overshoot time and settling time for the case of a passenger of 70 kg and 90 kg, but in the second case, the maximum motor current $i_{\max} = 80$ A is violated. For the second case, a redesign of the scooter is necessary; for the first case, a valid scooter design is found, which satisfies all requirements. With the second case, a redesign could be done by using the maximal needed current of the closed loop model as an input in the transmission design.

9. Reflection of the Digital Design Process

The described control system design process is an integral part of the digital design process of the balanced two-wheel scooter, which was realized by means of an integrated engineering framework based on graph-based design languages. All stages are fed with updated input information from a model based requirements management. The processes, operands and states are identified and connected in an integrated function model and actors are connected to the modular product structure under development. Crucial for the developed scooter is the dimensioning of the motor and transmission of the drive system. For the determination of the main performance features of the motor and transmission, an automated synthesis process with integrated validation was presented. The synthesis led to a choice of motor and transmission ratio—this was subsequently the input information for the detail dimensioning of the transmission. This detail dimensioning consists of an automatic synthesis of the gear set using parameter variation and pareto-front-based selection. The concrete design of the gear set included detailed finite element and lubrication analyses. For all kinds of balanced vehicles, the control system will determine the general usability and the driving comfort and safety. A detailed close loop behavior simulation was carried out, based on a controller design with a simple output feedback controller for the pitch and velocity dynamics and a separate state feedback controller for the yaw dynamics.

The central advantages of the presented digital design process in comparison to conventional design process is the possibility for machine execution, i.e., all steps can be carried out in an automatic manner. This allows the synthesis and analysis of a large number of possible solutions and increases the probability of generating and choosing an optimum solution. Additionally, compliance with all product design requirements can only be verified after the design is complete. This applies in particular to the evaluation of the dynamic properties of the closed-loop and the realizability of the control commands. Based on the fully automated design process presented in this paper, a suitable product design can be found by manipulating the input data without additional development work.

Another important advantage is that through a large part of the design process a central model can be used in the engineering framework. This reduces the risk of inconsistencies. In the opinion of the involved design engineers, the resulting design process is more robust in case of design changes.

The main challenge in the application of this engineering framework is the necessity of a paradigm shift; the design engineers are, during a large share of the process, not designing a single product, but programming a product variety. In the early stages, a larger investment in terms of working hours and critical thinking is necessary, but the involved engineers tend to believe that the advantages merit this investment.

One common issue in the application of automated design processes is the question of whether innovative solution concepts are possible within this process. One may ask the critical question, if this kind of process limits creativity and will exclude solution possibilities, which are different from an initial concept. This risk is reduced by the possibility to integrate alternative configurations in this engineering framework, which goes beyond digital building blocks, as they are present in conventional design automation systems. The engineering framework allows mutual interdependent components, as well as a large solution variety on functional and physical levels. For the balanced two-wheel scooter example, solutions with completely different transmission systems could still be part of a consistent product family.

10. Conclusions and Recommendations for Further Research

This paper demonstrated several stages of the digital design process of a balanced two-wheel scooter. An integrated engineering framework based on graph-based design languages serves as back-bone for this process. The result of the digital development process is the scooter shown in Figure 26.

The main advantages of the presented process are the linking of synthesis and analysis steps and the high degree of interconnected, model-based processes on different levels of abstraction and concerning different domains. The engineering framework based on graph-based design languages avoids double inputs, independent generic models and inconsistencies. Conventional tools are mostly only used to evaluate manually defined designs; very few can also be used to generate the models (synthesis). The process shown in this paper still has some gaps in automation, but represents a closed requirements-driven development chain. The gaps in the automation capability are particularly due to the sheer number of requirements—especially in the area of the geometry of the housing. Still, a primarily digital requirement driven design process and an enrichment of requirements along the design process could be demonstrated. Future work is aiming at a further increase of the automation level and the application for even more complex systems. Further possibilities for presenting abstract physics will be included in the engineering framework, and simulation possibilities, as well as synthesis possibilities such as topology, will be expanded. Additionally, the scientific foundation for a stronger integration with model-based system engineering systems will be addressed.

Figure 26. Build balanced two-wheel scooter

Author Contributions: Conceptualization, K.H., M.T., R.S., W.F. and S.R.; methodology, R.S.; design and dimensioning, S.S.; simulation, K.H. and M.F.; control design, M.F.; analysis and evaluation, K.H., S.S. and M.F.; writing—original draft preparation, K.H., S.S., M.F. and R.S.; writing—review and editing, M.T., W.F. and S.R. All authors have read and agreed to the published version of the manuscript.

Funding: This research was funded in the scope of the Digital Product Life Cycle (ZaFH) project (information under: https://dip.rwu.de/), which is supported by a grant from the European Regional Development Fund and the Ministry of Science, Research, and the Arts of Baden-Württemberg, Germany (information under: https://efre-bw.de/).

Institutional Review Board Statement: Not applicable.

Informed Consent Statement: Not applicable.

Data Availability Statement: The data presented in this study are available on request from the corresponding author.

Acknowledgments: The authors wish to thank the administrative and technical staff of the Ravensburg-Weingarten University (RWU), for the continuous support and all involved students for their immense contribution.

Conflicts of Interest: The authors declare no conflict of interest.

References

1. Gumpert, K.; Wiese, J. Analyse des Potenzials der Elektromobilität im Hinblick auf das logistische Problem der letzten Meile. *Mobil. Glob. World* **2017**, *16*, 55.
2. Burkhart, J.; Breckle, T.; Ramsaier, M.; Till, M. A digital modelling approach for design configuration and manufacturing exemplified by a self-balancing scooter. *Procedia CIRP* **2020**, *91*, 214–219. [CrossRef]
3. Gero, J.S.; Kannengiesser, U. The function-behaviour-structure ontology of design. In *An Anthology of Theories and Models of Design*; Springer: Berlin/Heidelberg, Germany, 2014; pp. 263–283.
4. Chakrabarti, A.; Bligh, T.P. A scheme for functional reasoning in conceptual design. *Des. Stud.* **2001**, *22*, 493–517. [CrossRef]
5. Braha, D.; Reich, Y. Topological structures for modeling engineering design processes. *Res. Eng. Des.* **2003**, *14*, 185–199. [CrossRef]

6. Eckert, C.; Albers, A.; Bursac, N.; Chen, H.X.; Clarkson, J.; Gericke, K.; Gladysz, B.; Maier, J.; Rachenkova, G.; Shapiro, D.; et al. Integrated product and process models: Towards an integrated framework and review. In Proceedings of the ICED 2015, Seoul, Korea, 13–16 April 2015.
7. Rubio, F.; Valero, F.; Llopis-Albert, C. A review of mobile robots: Concepts, methods, theoretical framework, and applications. *Int. J. Adv. Robot. Syst.* **2019**, *16*, 1729881419839596, doi:10.1177/1729881419839596. [CrossRef]
8. Rudolph, S. *Übertragung von Ähnlichkeitsbegriffen*; Universität Stuttgart: Habilitationsschrift, Germany, 2002.
9. Arnold, P.; Rudolph, S. Bridging the gap between product design and product manufacturing by means of graph-based design languages. In Proceedings of the TMCE 2012 Symposium, Karlsruhe, Germany, 7–11 May 2012.
10. Gross, J.; Rudolph, S. Geometry and simulation modeling in design languages. *Aerosp. Sci. Technol.* **2016**, *54*, 183–191. [CrossRef]
11. Vogel, S. Über Ordnungsmechanismen im Wissensbasierten Entwurf von SCR-Systemen. Doctoral Thesis, Universität Stuttgart, Stuttgart, Germany, 2016.
12. Ramsaier, M.; Stetter, R.; Till, M.; Rudolph, S. Modelling and Simulation of a Race-Car Frame Using Graph-Based Design Languages. In Proceedings of the International Conference on Engineering Optimization, Lisboa, Portugal, 17–19 September 2018; pp. 789–800.
13. Walter, B.; Kaiser, D.; Rudolph, S. From Manual to Machine-executable Model-based Systems Engineering via Graph-based Design Languages. In Proceedings of the MODELSWARD, Prague, Czech Republic, 20–22 February 2019; pp. 201–208.
14. Zech, A.; Stetter, R.; Holder, K.; Rudolph, S.; Till, M. Novel approach for a holistic and completely digital represented product development process by using graph-based design languages. *Procedia Cirp.* **2019**, *79*, 568–573. [CrossRef]
15. Bernard, R.; Irlinger, R. About watches and cars: Winning R&D strategies in two branches. In Proceedings of the International Symposium "Engineering Design The Art of Building Networks", Munich, Germany, 4–5 April 2016.
16. Hruschka, P. *Business Analysis und Requirements Engineering: Produkte und Prozesse Nachhaltig Verbessern*; Carl Hanser Verlag GmbH Co KG: München, Germany, 2019.
17. Ebert, C.; Jastram, M. ReqIF: Seamless requirements interchange format between business partners. *IEEE Softw.* **2012**, *29*, 82–87. [CrossRef]
18. Darlington, M.; Culley, S. A model of factors influencing the design requirement. *Des. Stud.* **2004**, *25*, 329–350. [CrossRef]
19. Jiao, J.; Chen, C.H. Customer requirement management in product development: A review of research issues. *Concurr. Eng.* **2006**, *14*, 173–185. [CrossRef]
20. Zhang, Z.; Li, X.; Liu, Z.L. A Closed-loop Based Framework for Design Requirement Management. In Proceedings of the ISPE CE, Beijing, China, 8–11 September 2014; pp. 444–453.
21. ISO. *Systems and Software Engineering—Life Cycle Processes—Requirements Engineering*; International Organization for Standardization: Geneva, Switzerland, 2011.
22. ISO. *Systems and Software Engineering—System Life Cycle Processes*; Standard ISO/IEC 15288; International Organization for Standardization: Geneva, Switzerland, 2008.
23. ISO. *Quality Management Systems—Requirements*; Standard ISO 9001; International Organization for Standardization: Geneva, Switzerland, 2015.
24. De Gea, J.M.C.; Nicolás, J.; Alemán, J.L.F.; Toval, A.; Ebert, C.; Vizcaíno, A. Requirements engineering tools: Capabilities, survey and assessment. *Inf. Softw. Technol.* **2012**, *54*, 1142–1157. [CrossRef]
25. Holder, K.; Zech, A.; Ramsaier, M.; Stetter, R.; Niedermeier, H.P.; Rudolph, S.; Till, M. Model-based requirements management in gear systems design based on graph-based design languages. *Appl. Sci.* **2017**, *7*, 1112. [CrossRef]
26. Eisenbart, B.; Gericke, K.; Blessing, L.T.; McAloone, T.C. A DSM-based framework for integrated function modelling: Concept, application and evaluation. *Res. Eng. Des.* **2017**, *28*, 25–51. [CrossRef]
27. Ramsaier, M.; Holder, K.; Zech, A.; Stetter, R.; Rudolph, S.; Till, M. Digital representation of product functions in multicopter design. In Proceedings of the DS 87-1 21st International Conference on Engineering Design (ICED 17) Volume 1: Resource Sensitive Design, Design Research Applications and Case Studies, Vancouver, BC, Canada, 21–25 August 2017; pp. 369–378.
28. Hobbyking. Turnigy Aerodrive SK3-6374-168kv Brushless Outrunner Motor. Available online: https://hobbyking.com/de_de/turnigy-aerodrive-sk3-6374-192kv-brushless-outrunner-motor.html (accessed on 7 December 2020).
29. Parlow, J.C. *FVA-Nr. 421 IV Erweiterung Getriebeauslegungsprogramm: Benutzeranleitung Getriebeauslegungsprogramm*; Forschungsvereinigung Antriebstechnik e.V.: Frankfurt, Germany, 2016.
30. Deutsches Institut für Normung. *Pressverbände—Teil 1: Berechnungsgrundlagen und Gestaltungsregeln für zylindrische Pressverbände*; Standard DIN 7190; Deutsches Institut für Normung: Berlin, Germany, 2017.
31. ISO. *Wälzlager—Dynamische Tragzahlen und nominelle Lebensdauer*; Standard DIN ISO 281; International Organization for Standardization: Geneva, Switzerland, 2010.
32. Schimmelpfennig, K.H.; Uphoff, T. Unfallrekonstruktion: Näherungsgleichung für die Berechnung von Fahrwerksbelastung durch Schlaglöcher. *VRR Verkehrsrechtsreport* **2011**, *5*, 174–178.
33. Gebhardt, C. *Praxisbuch FEM mit ANSYS Workbench: Einführung in die Lineare und Nichtlineare Mechanik*; 2. Überarbeitete Auflage Editon; Carl Hanser: München, Germany, 2014.
34. MÄDLER GmbH. Artikel 60272600—Ausgleichskupplung KA aus Aluminium max. Drehmoment 4.0 Nm Gesamtlänge 31.75 mm Aussendurchmesser 25.40 mm Beidseitig Bohrung 8 mm. Available online: https://www.maedler.de/product/1643/1622/1945/ausgleichskupplungen-ka-kurz-aluminium?result=60272600 (accessed on 18 December 2020).

35. Muir, P.F.; Neuman, C.P. Kinematic modeling of wheeled mobile robots. *J. Robot. Syst.* **1987**, *4*, 281–340. [CrossRef]
36. Isermann, R. *Mechatronische Systeme: Grundlagen*; Springer: Berlin/Heidelberg, Germany, 2007.
37. Grasser, F.; D'arrigo, A.; Colombi, S.; Ruffer, A.; Mobile, J.A. Inverted pendulum. *IEEE Trans. Ind. Electron.* **2002**, *49*, 107–114. [CrossRef]
38. Jian-hai, H.; Shu-shang, Z.; Ji-shun, L.; Hang, L. Research on developed parallel two-wheeled robot and its control system. In Proceedings of the 2008 IEEE International Conference on Automation and Logistics, Qingdao, China, 1–3 September 2008; pp. 2471–2475.
39. Nawawi, S.W.; Ahmad, M.N.; Osman, J.H.S. Development of a two-wheeled inverted pendulum mobile robot. In Proceedings of the 2007 5th Student Conference on Research and Development, Selangor, Malaysia, 11–12 December 2007; pp. 1–5.
40. Feng, T.; Liu, T.; Wang, X.; Xu, Z.; Zhang, M.; Han, S.C. Modeling and implementation of two-wheel self-balancing robot equipped with supporting arms. In Proceedings of the 2011 6th IEEE Conference on Industrial Electronics and Applications, Beijing, China, 21–23 June 2011; pp. 713–718.
41. Li, J.; Gao, X.; Huang, Q.; Matsumoto, O. Controller design of a two-wheeled inverted pendulum mobile robot. In Proceedings of the 2008 IEEE International Conference on Mechatronics and Automation, Takamatsu, Japan, 5–8 August 2008; pp. 7–12.
42. Friedland, B. *Control System Design: An Introduction to State-Space Methods*; Courier Corporation: North Chelmsford, MA, USA, 2012.

vehicles

MDPI

Article

Transfer of Statistical Customer Data into Relevant Parameters for the Design of Vehicle Drive Systems

Raphael Mieth [1,2], Frank Gauterin [1,*], Felix Pauli [2] and Harald Kraus [2]

1 Institute of Vehicle System Technology, Karlsruhe Institute of Technology (KIT), 76131 Karlsruhe, Germany; raphael.mieth@daimler.com

2 Mercedes-Benz AG, 70327 Stuttgart, Germany; felix.pauli@daimler.com (F.P.); harald.kraus@daimler.com (H.K.)

* Correspondence: frank.gauterin@kit.edu

Abstract: Vehicle drive systems are often oversized for common customer operation in order to cover the high demands of rare driving events such as towing a trailer, high acceleration or steep inclines. This high torque and power requirement affects the efficiency map and the highest efficiency is around the area of increased torque and speed. However, in everyday use, drive systems are mostly driven by customers at low speed and load, and therefore are not operating in the most efficient area. Designing a drive system that only covers the area of highest customer operation can increase efficiency by moving the sweet spot of efficiency to the relevant area, and thus reduce energy consumption. Therefore, customer data need to be analyzed in order to identify customer requirements and to localize the area of greatest operation. The method presented in this paper analyzes customer data in order to identify design-relevant parameters for a customer-specific drive system design. The available customer data results from event-based counts and are submitted as a statistical frequency distribution. These statistics are compared with discrete time series recorded during test drives in order to derive representative time series that correspond to customer behavior. By applying the time frame-based load analysis to these relevant time series, the desired design-relevant parameters are pointed out.

Keywords: statistical customer data; design-relevant parameters; design of vehicle drive systems

Citation: Mieth, R.; Gauterin, F.; Pauli, F.; Kraus, H. Transfer of Statistical Customer Data into Relevant Parameters for the Design of Vehicle Drive Systems. *Vehicles* **2022**, *4*, 137–144. https://doi.org/10.3390/vehicles4010009

Academic Editors: Ralf Stetter, Udo Pulm and Markus Till

Received: 25 January 2022

Accepted: 8 February 2022

Published: 10 February 2022

1. Introduction

Increasing the efficiency of vehicle drive systems is one of the highest goals in the automotive industry. By reducing energy consumption, further benefits such as an increase in electric range or reduced vehicle mass can be realized. Since drive systems are occasionally oversized in terms of performance for the actual requirements, efficiency can be increased by adjusting the performance [1–4]. This increase in efficiency can be achieved by reducing the performance of a drive system to the requirements that customers mostly need in daily operation. Furthermore, a second drive system is required in order to enable high demands, which occur rather rarely.

As part of a dissertation, this publication gives an insight into the method for analyzing customer-relevant driving requirements and converting them into relevant target values for the design of overall drive systems. In the context of the dissertation, an overall drive system is designed which combines increased efficiency by focusing on the core area of customer needs with high functionality by covering high requirements.

2. Methods

This paper is divided into two methodical approaches as shown in Figure 1: first, time series data (TS) from endurance testing are compared with statistical customer data (CLC) with regard to identifying customer-relevant time series. In the second step, the root mean square (rms) is calculated for different widths of time frames and the time frame-based analysis (TFBA) is used to derive design-relevant parameters for the overall system design.

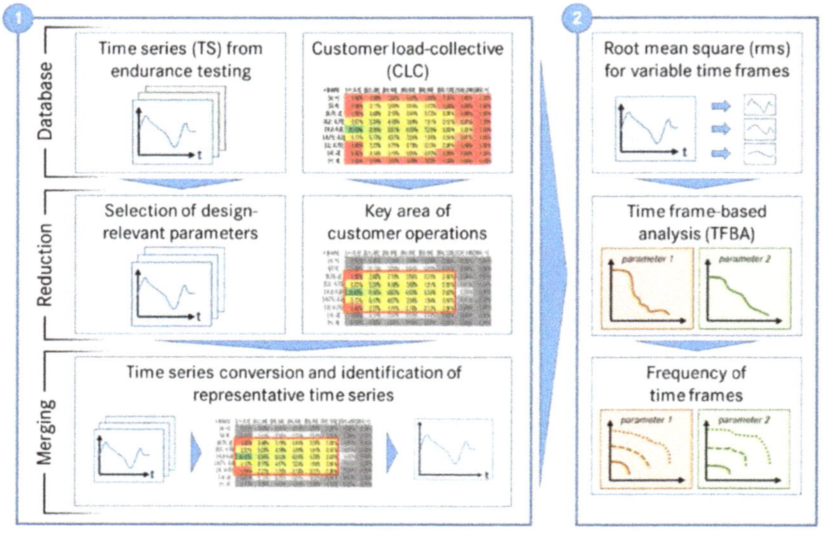

Figure 1. Scheme of the methodological structure.

The following Sections 2.1–2.3 explain the first methodical approach, shown on the left side in Figure 1. Sections 2.4 and 2.5 are then dedicated to the second approach.

2.1. Database

The first approach is based on two types of data, namely time series and customer load collective. The available time series data are records of various signals from endurance testing, from which the relevant load variables that are required for the drive system design must be selected. The main design-relevant parameters are the mechanical power and the torque on the wheel. In addition, the vehicle speed, respectively the wheel speed as well as the acceleration, is also used to identify representative time series. Dependent on the available customer data, other parameters may be considered.

The available customer data are the results of event-based counts in the control unit of the customer vehicles, so-called *customer load collective*. These counts provide information about the frequency distribution, i.e., how often certain events occur. They are shown in Figure 2 using the example of the vehicle speed and acceleration distribution while the drive system is active. The color scheme represents the accumulation of the customer's driving behavior. According to the color bar, the vehicles in customer operation actually do not drive at all about 25% of the time despite the ignition being switched on.

The largest accumulation in these statistical distribution matrices of around 95% covers the range between approximately 0 and 130 kph as well as −2 and 2 m/s^2. This conspicuity is also reflected in [5], where the driver behavior of the 34 participants is analyzed based on recorded real drives over a distance of around 35,000 km. The evaluation shows that the highest accelerations actually driven in the speed range up to approximately 100 kph are below 1.5 m/s^2 and at speeds of up to 140 kph are mostly not higher than approximately 1.0 m/s^2. The limitation of the speed range to a maximum of 130 kph is derived from [6].

According to this report, the average top speed on German highways without speed limit is between 110 and 130 kph. Abroad, this value is lower due to speed limits. Based on these findings, accelerations between −2 and 2 m/s^2 and speeds of up to 130 kph can initially be assumed to be customer relevant. Consequently, this area can be identified as the focus of customer driving operations and used as the basis for a customer-specific drive system design.

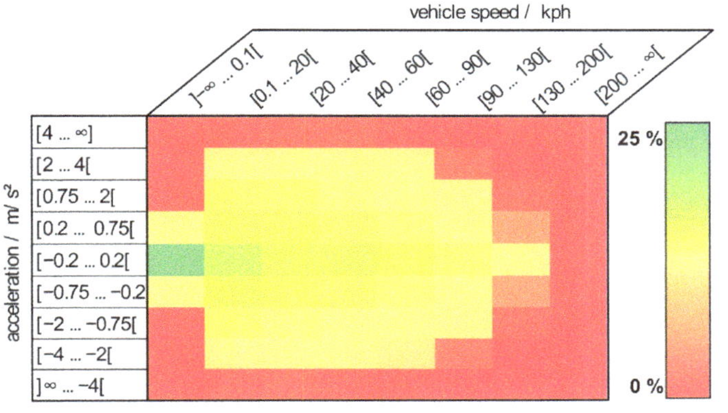

Figure 2. Exemplary heat map for customer operation.

2.2. *Comparison of the Data Bases and Derivation of Representative Time Series*

After selecting appropriate, design-relevant signals from the time series data and determining the focus of customer driving operations, the next step is to merge these two databases. For this purpose, distribution matrices for all time series are created analogous to the customer data. The aim is to describe the time series statistically by transferring them into heat maps such as acceleration versus vehicle speed, as shown above in Figure 2, as well as motor torque versus motor speed with the same clustering as the CLC data.

The deviation of these matrices from the CLC data is the measure of the customer relevance of the time series. First, the cell-specific deviation $|X_{i,j}|$ as absolute value of the cell-by-cell difference between time series $cell_{TS}$ and CLC data $cell_{CLC}$ is calculated according to Equation (1).

$$|X_{i,j}| = cell_{TS} - cell_{CLC} \tag{1}$$

The relevance of each time series is evaluated based on the mean total deviation. In this respect, the time series, whose deviation is the smallest, represents the best approximation to the customer distribution matrices. As shown in Equation (2), the arithmetic mean of the total deviation $\overline{X}$ is the sum of all cell-specific deviation values $|X_{i,j}|$ divided by the number of all cells n in the above-mentioned area of greatest customer operation.

$$\overline{X} = \frac{1}{n}\sum_{i=1}^{n}|X_{i,j}| \tag{2}$$

This approach allows the time series to be sorted in ascending order by their total mean deviation. According to the method of [7], time series are combined with one another to minimize the total deviation, starting with those two, whose deviation is the lowest. Increasing the number of combined time series leads to a decrease in the mean deviation, as shown by the blue curve as well as the yellow trend line in Figure 3. The red markers highlight exactly that time series whose combination leads to the smallest deviation of the key area pointed out in Section 2.1.

In this case, 1000 time series are evaluated, of which the combination of all 45 red-marked time series shows minimal total deviation. In other words, combining exactly these 45 out of the 1000 time series, which the markers indicate, reduces the deviation the most and leads to the minimum of 3.13 in mean deviation.

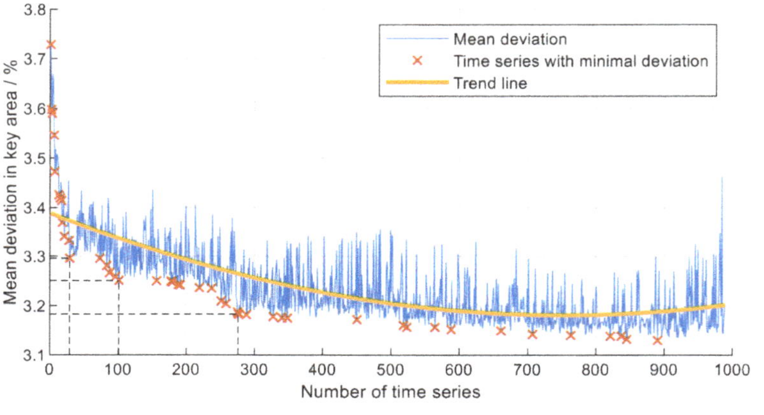

Figure 3. Exemplary regression curve: mean deviation versus number of time series.

Following the steep drop, the slope decreases with an increasing number of time series and the mean deviation decreases only slightly. Due to the trade-off of reducing the mean deviation as much as possible while at the same time minimizing the computational effort, the red regression curve shows several optima displayed by the dashed black lines. According to these, considering 12 specific time series, indicated by the first 12 markers from the left, would lead to the first big decrease in the mean deviation from about 3.7 to 3.3. The second and third dashed lines show the consideration of 17 and 28 time series, respectively, with a decrease in the mean deviation to 3.25 as well as 3.18, respectively. In this respect, considering these numbers of time series represents optimums regarding computational effort and deviation in order to describe customer behavior in the best possible way.

2.3. Validation of the Purposed Method

This methodological approach must take into account that the derivation of certain time series from statistics such as CLC data is ambiguous due to the lack of the time reference. In this case, the statistical frequency distribution can only be approximated by selecting suitable time series. In contrast, the reverse method of statistical representation of a time series represents a clear correlation.

Despite the lack of unambiguity, the procedure is admissible, as this method aims to approximate, rather than exactly represent, customer requirements by combining several time series.

In order to ensure the plausibility of the method, a given time series is first transferred into a distribution statistic for unambiguous assignment. In the next step, the method from [7] is used to analyze imported time series, one of which is the given one, with the aim of deriving the initial time series. The result of this analysis was a calculated deviation of zero. This shows that the procedure is able to re-identify the originally read-in time series, thus proving its functionality. To further check the plausibility, this process is repeated twice on top of that. Both different as well as the same time series are combined several times with one another and the method is applied, leading to the same result both times.

In addition, the available customer data represent a solid database due to the extensive amount of data. Using the example of an investigated model series, the data contains more than 15,000 vehicles with a total mileage of more than 100 million km. These selected vehicles all have a certain minimum mileage to evaluate only representative customer driving profiles. Reverse driving is also taken into account. This signifies a meaningful statistic, both in terms of different customer types and markets.

Furthermore, as already pointed out in Section 2.1, the entire heat map of customer data is not of equal interest, but rather the focus of the research is on the area of greatest accumulation.

2.4. Time Frame-Based Analysis (TFBA)

The second methodological aspect introduces the time frame-based analysis according to [8]. This method is used to identify design-relevant loads based on time-dependent criteria and to further evaluate their intensities, durations and frequencies, which is necessary for the design of drive system components.

For illustration, Figure 4 schematically shows two loads, z_1 and z_2, which are identical with respect to their classic statistical properties such as the root mean square (rms) z_{rms}, the minimum value z_{min} and maximum value z_{max} as well as their distribution functions.

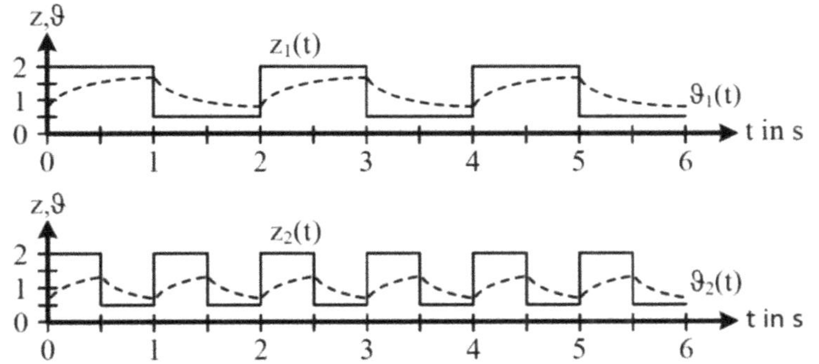

Figure 4. Comparison of two statistically equal loads. Modified from [9], Technische Universität Dresden, 2016.

Since these parameters do not provide any time-related information such as duration and frequency or the time interval between specific loads, the characteristics of these two loads cannot be compared. This circumstance is illustrated by the different temperature curves ϑ_1 and ϑ_2, which leads to different temperature gradients and thus, different thermal stresses on the components.

Therefore, the time frame-based load analysis is needed, in which time frames τ_{tf} with continuously increasing width are defined and shifted over the time series [9]. According to Equation (3), the maximum rms value is calculated for each time frame width.

$$Z(\tau_{tf}) = \max\left(\sqrt{\frac{1}{\tau_{tf}} \cdot \sum_{t}^{t+\tau_{tf}} z(t)^2 \cdot \Delta t}\right) \quad (3)$$

For this purpose, the conditions pointed out in Equation (4) need to be considered: The time frame width τ_{tf} must be at least equal to the time step width Δt. Additionally, the overall time series length T must be extended to length t_{total} to ensure considering sufficient values for the largest time frame $\tau_{tf,max}$ at the last time step $t = T$. Finally, the time frame is moved over the whole time series from $t = 0$ to $t = T$.

$$\begin{gathered} \tau_{tf} \geq \Delta t \\ T + \tau_{tf,max} \leq t_{total} \\ 0 \leq t \leq T \end{gathered} \quad (4)$$

Figure 5 shows the resulting time-related continuous load curves for the two loads mentioned above. The diagram displays the maximum values for both loads z_1 and z_2 and each time frame width, which indicates that the respective rms value is demanded at least once for that frame.

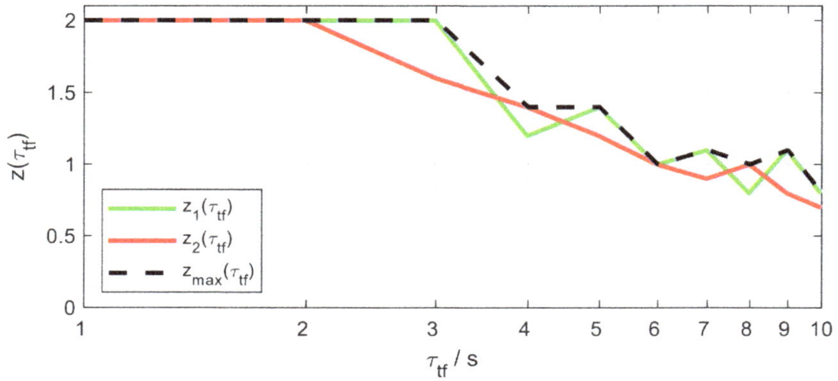

Figure 5. Exemplary time-weighted continuous load curves. Modified from [9], Technische Universität Dresden, 2016.

As expected from the temperature curves ϑ_1 and ϑ_2 above, the comparison between the two signals at $\tau_{tf} = 3$ s indicates that the rms value z_1 is greater than z_2. This leads to the envelope curve z_{max}, which represents the overall highest value of all loads. These highest values indicate maximum thermal load and are therefore crucial for the system design.

This method allows the analysis of different time series, regardless of their length. However, only those physical quantities that are directly proportional to the losses of a component can be analyzed [9]. With respect to the design of drive systems, power and torque are suitable quantities.

2.5. Frequency of Time Frames

Finally, the analysis of the frequency of time frames evaluates the correlation of duration and frequency of occurrence. By displaying the time frame frequency over the time frame width, as shown in Figure 6, different loads can be compared by evaluating how often and for how long certain loads occur.

The curves visualize the duration and frequency of the rms value of both loads z_1 and z_2. First of all, the black graph shows that all rms values are positive for both loads, so, as expected, the frequency of the rms value is 100%. Additionally, the figure displays the curves for an exemplary threshold value of 60, so the curves show the rms values $z_{rms} \geq 60$. In this case, $z_{rms,1}$ is greater than $z_{rms,2}$, both in terms of duration and frequency. This not only means that the time frame $\tau_{tf,1}$ for this specific rms value is longer, but also the frequency of occurrence in this load has a higher share in the overall process. This leads to z_1 being critical for the system design. With the help of these analyses, loads of different duration, intensity and characteristics can be compared and design-critical loads can be identified.

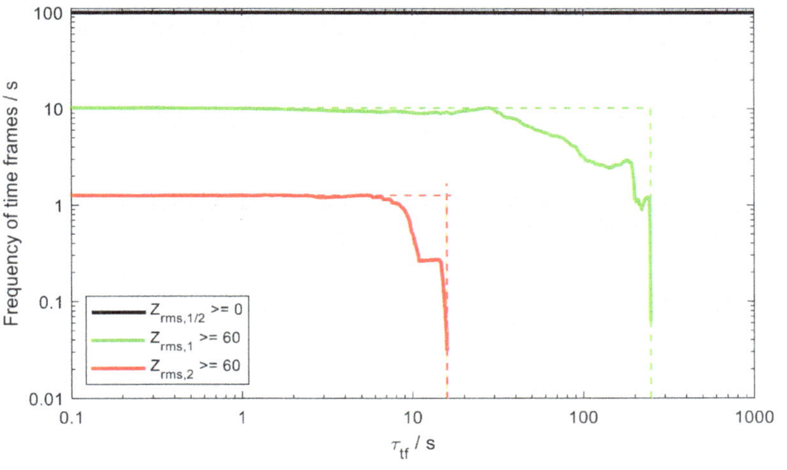

Figure 6. Exemplary frequency of time frames.

3. Conclusions and Future Work

The method introduced in this work allows the derivation of relevant parameters for the design of vehicle drive systems based on statistical customer data. For this purpose, time series are analyzed by comparing them with customer data. The aim is to identify customer-representative time series. Furthermore, these representative time series are investigated using the time frame-based load analysis. This procedure of describing the time-dependent characteristics of a time series allows the detection of design-critical loads and generates relevant parameters for the system design.

These relevant parameters such as peak values and continuance values for torque and power as well as working points and sweet spots represents the input for the design of overall drive systems. By limiting this input to the most frequently required customer needs, the efficiency of the overall system can be increased.

Author Contributions: Conceptualization, methodology, software, validation, formal analysis, investigation, resources, data curation, writing—original draft preparation, writing—review and editing, visualization, project administration, funding acquisition, R.M.; supervision, F.G., F.P. and H.K. All authors have read and agreed to the published version of the manuscript.

Funding: This research is funded by the Publication Fund of the Karlsruhe Institute of Technology.

Institutional Review Board Statement: Not applicable.

Informed Consent Statement: Not applicable.

Conflicts of Interest: The authors declare no conflict of interest.

References

1. Binder, A. *Technische Optimierungspotentiale bei Elektrischen Antriebssystemen, Vortrag*; Institut für Elektrische Energiewandlung: Darmstadt, Germany, 2009. (In German)
2. Mayer, A. Analyse Gesamtsystemischer Zusammenhänge von Hybriden Antriebsstrangkonzepten Bezüglich Kundenrelevanter Anforderungen. Doctoral Dissertation, Universität Siegen, Siegen, Germany, 2021. (In German)
3. SHS plus GmbH. How You Can Halve the Energy Consumption of Your Extruder Drive. 2018. Available online: https://www.extrusion-training.de/en/wie-sie-den-energieverbrauch-ihres-extruderantriebes-halbieren-koennen/ (accessed on 4 January 2022).
4. Motor Challenge. Replacing an Oversized and Underloaded Electric Motor. 2014. Available online: https://www.energy.gov/sites/prod/files/2014/04/f15/mc-2463.pdf (accessed on 4 January 2022).
5. Heinrich, D. Modellierung des Fahrverhaltens zur Ermittlung von Bauteilbelastungen im Fahrzeugantriebstrang. Doctoral Dissertation, Karlsruher Institut für Technologie, Karlsruhe, Germany, 2016. (In German with English Abstract).

6. Löhe, U. *Geschwindigkeiten auf BAB in den Jahren 2010 bis 2014*; Bundesanstalt für Straßenwesen (BaSt): Bergisch Gladbach, Germany, 2016; Available online: https://www.bast.de/DE/Publikationen/Fachveroeffentlichungen/Verkehrstechnik/Downloads/Geschwindigkeiten-BAB-2010-2014.pdf?__blob=publicationFile&v=4 (accessed on 8 December 2021). (In German)
7. Plum, N. Methode zur Analyse kundenrelevanter Fahranforderungen und Überführung in relevante Zielgrößen für die Auslegung von Gesamtantriebssystemen. Bachelor's Thesis, Fachhochschule Aachen, Aachen, Germany, 2021. (In German)
8. Von Lingen, J.; Schmidt, P. Zeitgewichtete Belastung und Belastbarkeit und deren Nutzung für die Dimensionierung von Bahnenergieversorgungsanlagen. *Elektrie* **1994**, *48*, 448–457. (In German)
9. Pauli, F. Allgemeingültige Beschreibung der Lebensdauerbelastung elektrischer Antriebskomponenten in Pkw auf Basis von Mess- und Simulationsdaten. Dissertation, Technische Universität Dresden, Dresden, Germany, 2016. (In German)

vehicles

MDPI

Article

Enhanced Derivation of Customer-Specific Drive System Design Parameters with Time Frame-Based Maximum Load Analysis

Raphael Mieth [1,2,†], Fabian Markschies [2,3,†], Ruixin Zhou [2,3,†], Frank Gauterin [1,*] and Arnd Stephan [3,*]

1 Institute of Vehicle System Technology, Karlsruhe Institute of Technology, 76131 Karlsruhe, Germany
2 Mercedes-Benz AG, 70327 Stuttgart, Germany
3 Institute of Railway Vehicles and Railway Technology, TU Dresden, 01069 Dresden, Germany
* Correspondence: frank.gauterin@kit.edu (F.G.); arnd.stephan@tu-dresden.de (A.S.)
† These authors contributed equally to this work.

Abstract: Only a small part of the high performance of electric drive systems in vehicles is used in everyday operation by customers. As a result, most drives are not operated in the optimum efficiency range. Designing a suitable drive system, whose performance is aligned with actual customer requirements, presents the potential to increase efficiency. Based on the findings of previous research, this paper serves to complement an existing method, which already introduced the basic method of transferring statistical customer data into relevant parameters for the design of a customer-specific drive system. In order to improve the method, further criteria for the selection of relevant time series come into place. Furthermore, the impact on maximum loads resulting from various sequences of the selected time series is identified and evaluated with time frame-based analysis. A new approach for the effective computation of maximum design-relevant loads in the admissible time frame range is introduced and validated. By taking this approach, the sensitivity of the derived design parameters regarding various time series sequence is evaluated in the context of selected datasets. In addition, concatenations of time series are identified which may have a relevant influence on the maximum loads. Consequently, the design process is safeguarded thoroughly against potential maximum loads as well as the associated thermal stresses.

Keywords: design of vehicle drive systems; analysis of customer data; sequence of time series

Citation: Mieth, R.; Markschies, F.; Zhou, R.; Gauterin, F.; Stephan, A. Enhanced Derivation of Customer-Specific Drive System Design Parameters with Time Frame-Based Maximum Load Analysis. *Vehicles* **2023**, *5*, 306–320. https://doi.org/10.3390/vehicles5010017

Academic Editors: Ralf Stetter, Udo Pulm and Markus Till

Received: 5 December 2022
Revised: 15 February 2023
Accepted: 15 February 2023
Published: 24 February 2023

1. Introduction

Electrical drive systems in vehicles are usually designed according to the requirements, e.g., extracted from endurance testing. The maximum loads occurring in the recorded time series data define the target performance of the developed drive system. In contrast to these requirements, the loads actually occurring in customer operation and thus the real customer requirements are quite different. In this respect, the analysis of vehicle- and market-specific customer data provides valuable information about the use of driving systems. In system design, this discrepancy in requirements affects the efficiency of the powertrain, among other things. Maximizing the efficiency of the overall system can lead to an increase in range and thus customer benefit.

By applying a previously published method [1], statistical customer data are transformed into a customer-oriented driving cycle from which customer-relevant requirements can be derived. In contrast to already existing work, whose methods also deal with the synthesis of representative driving cycles, the methodical approach of this work pursues different goals and therefore differs in detail. The customer data used in [2] are based only on statistical averages, while the database of this work contains statistics recorded from real customer usage. Compared to [3], in this work, there is no categorization of subcycles in different driving styles or driving environments and no further subdivision of time series, but the approximation of distribution statistics. Due to the limited database of only 175 measurement series from real drives, the method according to [4] does not fully capture the breadth of customer driving behavior and, in contrast to the customer data analyzed here, does not represent different markets. In summary, these methods do not represent a suitable or adequate approach for the ambition of this work.

The methodological approaches for derived design-relevant parameters presented in this paper are based on the methods from [1] and serve to extend the existing method. Due to the ambiguity in the derivation of time series from statistical customer data, multiple criteria are required for the selection of relevant time series. This enables a customer-specific drive system design with the potential to increase efficiency compared to existing drives.

The basic cycle generation methods could be categorized into the following variants according to [5]: (1) micro-trip based cycle construction, (2) segment-based cycle construction, (3) cycle construction with pattern classification, and (4) modal cycle construction. Despite of different methodical approaches, the sequencing of selected time series of all above-mentioned variants are based on stochastic or random principles. The influence of various subcycle sequences has been neglected. Due to the fact that varying the chronological order of loads and breaks affects the thermal behavior of powertrain components differently, it is necessary to investigate the sequence of the selected relevant time series regarding to their maximum loads and to identify the concatenations that may affect these loads most. Based on the importance of powertrain thermal safeguarding, it is necessary to investigate and evaluate the impact of different time series sequences regarding to the occurring loads. In this presented work, an extensive method developed on time frame-based analysis is introduced to assess varying load sequences. By doing this, a consequent safeguarding process against thermal stresses is achieved and the corresponding relevant design parameters are derived.

2. Improvement of Customer-Specific Drive Cycle Derivation

According to the method in [1], statistical customer data are transformed into a combined driving cycle whose distribution statistics are as close as possible to those of the customer data. This cycle further serves for the derivation of design-relevant parameters for electric powertrains.

The following sections deal with the extension of the existing method. Additionally, it is important to mention that the actual method must be considered separately from the available database. While the quality of the data may vary, the method as such must be self-sufficiently applicable. In the context of the exemplary application in this work, it is assumed that the data used are plausible and contain the required information.

2.1. Evaluation of the Intersection of Multiple CLC-Data

The following investigations are based on customer load collective (CLC) data, which are event-based counts in the control unit of real customer vehicles. These data are stored in a cloud and can be retrieved and analyzed on a vehicle- or market-specific basis. Based on the available CLC data, only the distribution of acceleration versus velocity has been considered by the method so far. In order to increase the accuracy of the derived conclusions from the analysis of the customer data, additional CLC data will be evaluated. Depending on the availability of the CLC data for the specific vehicle, customer statistics, such as torque versus speed, as well as the mechanical power of the machine, can also be taken into account. Considering these data reduces the deviation to the customer data and thus increasing the quality of the derived conclusions by including additional information, e.g., acceleration and torque are not necessarily proportional; instead, torque includes the influence of slope.

Thereby, the structure of the torque versus speed map is analogous to the acceleration versus velocity, whereas the mechanical power is already time-weighted in the CLC data. As the example in Figure 1 shows, the classes in y-direction reflect ascending time frames τ, and the power classes in the x-direction represent certain power ranges. The number of power classes and time frames are chosen here as an example and influence the accuracy of the CLC data. Finally, the color scheme symbolizes the frequency of occurrence and thus represents the accumulation of the customers driving behavior.

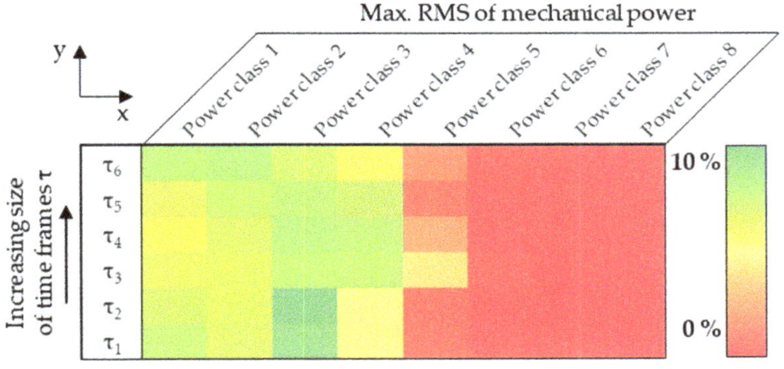

Figure 1. Exemplary distribution map of the mechanical power of an electric SUV.

The highest accumulation, represented by the green color, can be localized in this particular case around power class 3 for small time frames and becomes significantly lower for larger time frames. This tendency is caused by the maximum occurring root mean square value (RMS), which becomes smaller for larger time frames, e.g., due to breaks between the loads. In addition, the red marked area of the map shows that there are basically no operating points in customer use cases above power class 5.

As mentioned above, the method considers the distribution of acceleration vs. velocity as well as torque vs. speed and can finally be extended to take into account the frequency of the time-averaged mechanical power. As pointed out in [1], the TFBA (time frame-based load analysis) provides additional information, such as duration and frequency of specific loads, and therefore increases the accuracy of the method. The deviation is calculated for those maps according to the method mentioned in [1] and finally results in the total deviation, which serves as a measure for the relevance of each time series. This dimensionless variable becomes even smaller the more representative the respective time series is.

The determination of the time series is now based on the intersection of the criteria mentioned. An investigation shows that the algorithm considers fewer time series for the approximation to the customer data after implementation of these new criteria. At the same time, the length of the derived cycle as well as the deviation from the CLC data decreases. This increases the accuracy considerably, as more appropriate time series are selected. This extension of the method allows an improved approximation of customer behavior.

2.2. Concatenation of the Time Series

In order to combine time series with each other, it must be ensured that the signal courses are still continuous. Implausible signal characteristics, such as jumps or kinks, must be avoided. For example, this would be the case if both time series have different velocities at the junction. However, as Figure 2 shows, it is not sufficient that the signal values of both time series match at the end and beginning, respectively. The gradient must also match to avoid kinks.

To ensure this, the method combines only time series that start and end at standstill, so the velocity is zero. This means that signals, such as speed, torque, or power, can be strung together without any problems and result in plausible signal progressions. Both speed and torque become zero at standstill, whereas auxiliary units or, e.g., the electrical DC-power of the drive system, contains the standby losses of the inverter and thus does not become zero.

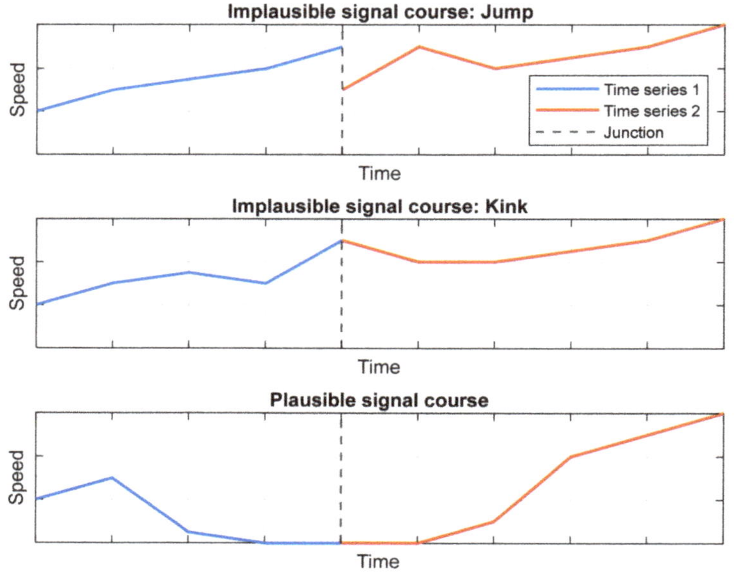

Figure 2. Implausible vs. plausible signal characteristics, e.g., for speed.

The only exceptions regarding the concatenation of time series are state variables such as temperature or state of charge (SOC), since these quantities are not independent of temporal repeatability. For example, a certain load case, which runs several times in succession, causes the same signal trace for torque or power. In contrast, signals, such as temperature or SOC, can change significantly over the time series and consequently do not fit together. With signal types, a distinction must therefore be made between load and state variables. The methodical procedure is nevertheless permissible, since only mechanical load variables are considered in the context of this method in order to approximate the customers driving behavior.

2.3. Limitation of the Total Length

By limiting the total length of the combined time series, another extension of the method comes into play. The problem is caused by the algorithm combining multiple time series, creating a new cycle that extends over several hours in duration and hundreds of kilometers in distance. Due to the limited range, especially for electric vehicles, such a profile would not be drivable continuously. Therefore, building a cycle that exceeds the limit in range of vehicles needs to be avoided.

This derived cycle is used for the design of the electric drive components, such as power electronics, electrical machine, and transmission, which does not include the battery. Nevertheless, the battery capacity of the specific vehicle is requested by the method in order to limit the cycle and thus must first be stored. The algorithm then integrates the battery power of each time series, as exemplified in the top diagram in Figure 3, and records the total energy in kWh. Finally, the concatenation of the relevant time series runs until the energy consumption of the combined cycles reaches, but does not surpass, the total capacity of the battery. The graph of the total energy consumption of the composite cycle can be seen in the exemplary bottom diagram in Figure 3.

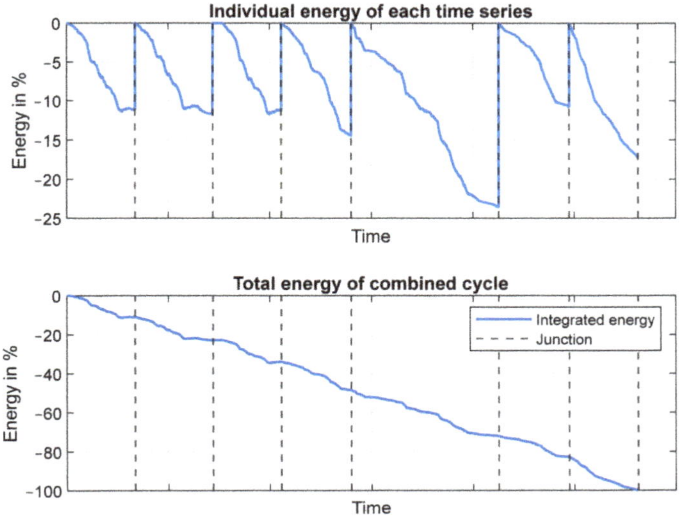

Figure 3. Limitation of the cycle length by the battery capacity.

3. Analysis of Load Characteristics under Identical Conditions of Statistics

The authors in [1] established a method including the criteria introduced in Section 2, which delivers a customer-specific drive cycle without a specified chronological order. The benchmark of this method is a statistical parameter, which indicates the overall difference between customer CLC data and the combined time series in the region of interest. This benchmark parameter and all further statistics, e.g., the velocity and acceleration distributions, remain unchanged under the condition of varied cycle sequences.

However, time series chronological order could have a significant impact due to variable load peaks and breaks arising in different sequences, which results in varied thermal stresses and therefore influences various design factors of target drive system, e.g., cost, packaging, weight and manufacture. Therefore, the transition between the subcycles must be examined more closely. It is assumed that the selected time series can be combined in any order and all loads can occur at any point in time in the derived cycle.

The time frame-based analysis provides a key framework to understand the varying load characteristics caused by subcycle permutations for infinite, periodic recurring time series. Under the scope of the presented work, the constraint of time series recurrence is applied.

Section 3.1 introduces the phenomenon, that the varying cycle sequences give rise to different load characteristics. Section 3.2 provides a novel method to identify the maximum design-relevant load within the valid time domain. Section 3.3 defines a generalized metric to quantify the possible impact of subcycle permutations within the scope of the underlying work. This analysis provides essential information for the design and thermal protection of the components.

3.1. Impact of Time Series Sequences on Load Characteristics

Three exemplary synthetic profiles A, B, and C are used, which represent drive events with loads occurring in different intensity, length, and position, such as alternating current in electric motor. Figure 4 demonstrates the cause and effect between arising loads and the temperature profiles of a drive system component.

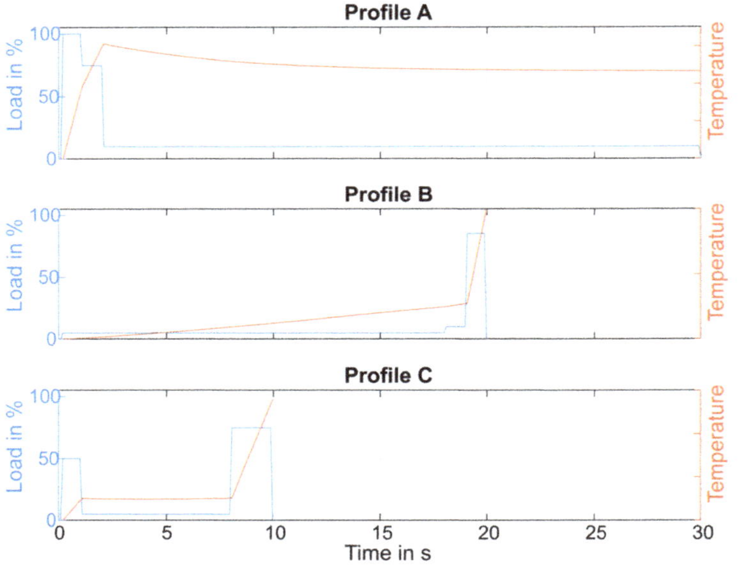

Figure 4. Load and temperature profiles of synthetic profiles A, B and C.

Under the premise that all requirements from Section 2.2 are met, six composite cycles ABC, ACB, BAC, BCA, CAB, and CBA are generated by varying the sequences of given time series. The number of composite cycles corresponds to the factorial of the profiles count. In order to examine the load characteristics of all concatenations generated by n time series, the time frame-based analysis needs to be performed $n!$ times.

The authors in [6,7] introduced time-rated load periods curve, also known as time-dependent continuous load curves in [1], depicted for the composite cycles in Figure 5. Different loads resulting from subcycle sequences could be identified for specific time frames.

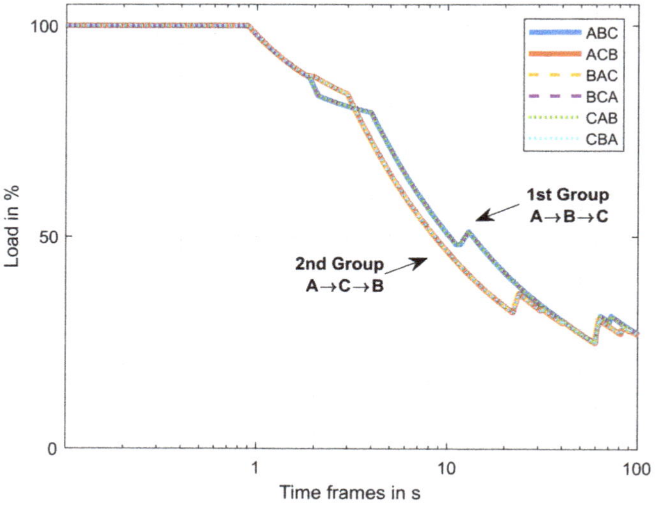

Figure 5. Characteristic of time frame-based loads for all time series permutations.

As Figure 5 shows, two different groups of time-dependent continuous load curves could be identified out of six composite cycles. The first group of curves contains the combination of ABC, BCA, and CAB, while the second one contains ACB, BAC, and CBA. This observed two-way partition of composite cycles gives insight about the time frame-based load characteristics using the state-of-the-art TFBA algorithm from [8]. The concatenations in each array of curves possess a unambiguous sequence, while the first subcycle could vary, e.g., ABC, BCA, and CAB correspond to sequence A $\rightarrow$ B $\rightarrow$ C and their starting time series are interchangeable. The observed results could be generalized, as concluded in Theorem 1.

Theorem 1. *Once a unambiguous sequence of n given time series is identified, all composite cycles within this sequence exhibit the same load characteristics in the meaning of TFBA, while the selection of the starting subcycle is arbitrary.*

Proof of Theorem 1. Given the periodic nature of time frame-based analysis, as well as the incorporated sliding window mechanism in TFBA algorithm, which are both thoroughly described and evaluated in [8], consider the two following cases: (1) A time series is periodically expanded and fixed on the time axis. A sliding time window is moved toward the positive direction on the time axis until each section of the given time series is examined. (2) A time window is fixed on the time axis. A periodically expanded time series slides pass the given time window until the end of time series is reached. While case 1 characterizes the TFBA algorithm, case 2 corresponds to the situation described in Theorem 1. As long as each section of the given time series is examined within the given time window, these two cases are equivalent. □

Using Theorem 1, the number of concatenations of n input time series could be reduced by a factor of n, i.e., the number of composite cycles is effectively limited to factorial of $n-1$. In order to identify the relevant parameters for a customer-specific drive system in the given example with three synthetic profiles, the TFBA only needs to be applied for two unambiguous cycle sequences in total.

In addition, a further reduction can be achieved based on an analysis of transitions between time series, which is particularly relevant for the case with large number of total target cycles. The corresponding approach will be presented in the following section.

3.2. Identification of Maximum Design-Relevant Loads

A high overall load occurs especially when high loads occur in a quick succession with very short breaks or even without breaks at all. For example, the components experience a higher thermal stress when one time series, where a high load occurs at the end, is combined with another time series, where a high load occurs right at the beginning. For this reason, the transition between two cycles is closely examined in this section with time frame-based analysis.

In order to obtain maximum time frame-based loads, the analysis procedure introduced in Section 3.1 could be deployed. For n time series, the required number of TFBA to be performed on the entire composite cycle would be $(n-1)!$, which is computationally inadmissible once n becomes larger. The optimization of the introduced approach should be considered. On the one hand, the number of TFBA to be executed could be restricted. On the other hand, the target time frame-based analysis domain could also be narrowed down. The maximum design-relevant loads derived from the entire composite cycle originate from two type of events: (1) Subcycle events, i.e., the entire events and the related peak loads occur only within each original time series. In fact, this type of event could be detected in advance before the composite cycle is derived. The required number of TFBA is in linear relation to the number of input cycles. (2) (n-way) Transition events, which span across n time series. This type of event contains at least one transition between two time series. The TFBA results of subcycle events are also implicitly contained in the results of transition events, since each original time series will be analyzed at least once during the examination of all types of transition events.

Assuming high loads only occur in the transition between two input cycles, i.e., only two-way transition events are considered, a new approach to identify the maximum design-relevant loads with respect to cycle sequences is purposed based on Theorem 2.

Theorem 2. *For a composite cycle with n original time series, assume the design-relevant maximum loads are only related to subcycle events and two-way transition events. The peak loads of the composite cycle for each time frame up to the minimal time series duration could be represented by the envelope of time-dependent continuous load curves of all two-way combination based on original time series.*

Proof of Theorem 2. Each time frame-based analysis within each two-way combination generates one time-dependent continuous load curve, which represents the maximum loads inside each two-way combination and therefore covers all subcycle events and two-way transition events based on the initial assumption in Theorem 2. The constraint of the minimal target time series duration is resulted by this assumption. Consider a time window longer than the minimal target cycle duration. Thus, it would be possible for that time window to contain the minimal input cycle as well as the ending part of previous cycle and the beginning of next cycle simultaneously. This type of load corresponds to the case of three-way transition events. The initial assumption is therefore violated. □

Using Theorem 2, the necessary number of TFBA executions is reduced from factorial, i.e., $(n-1)!$, down to quadratic order of the binomial coefficient $\binom{n}{2}$, i.e., $\frac{n}{2}\cdot(n-1)$. By iteration of the proposed procedure, the limitation of minimal subcycle duration could eventually be lifted, while the required number of TFBA to be executed remains under the factorial order. The time frame-based analysis results of all two-way combinations in accordance with Theorem 2 are shown in Figure 6.

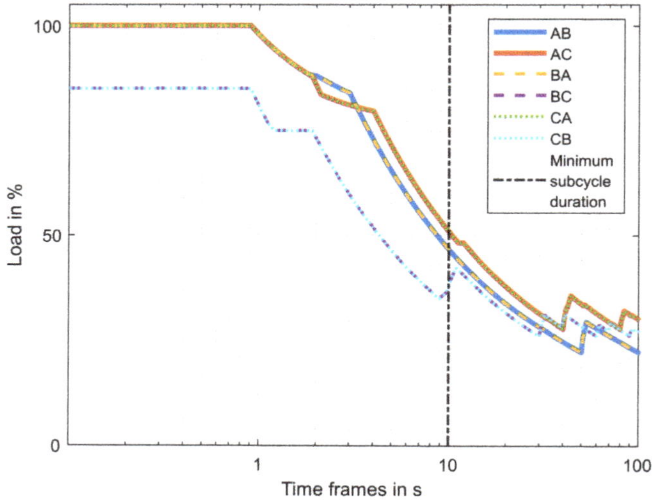

Figure 6. Characteristic of time frame-based loads for two-way combinations.

Comparing the results in Figure 5 within the constraint of minimum input cycle duration, all peak loads could be identified based on the envelope of time-dependent continuous load curves. The approach based on Theorem 2 substantially reduces the computational complexity and is therefore practically applicable for the use case of maximum design-relevant loads identification for up to n time series.

Assuming a certain two-way combination possesses an overall maximum continuous load curve, the corresponding time series could be classified as most critical part of the derived cycle regarding to thermal stresses. Therefore, there is also an option of selecting a two-way concatenation and adjusting it properly so that it generates a maximum continues load curve over all time frames, while the representative customer drive behaviors should be retained. With this approach, relevant parameters for thermal safeguarding can be derived.

3.3. Estimation of Influence Based on Subcycle Concatenations

According to the definition in [8,9], four TFBA metrics W_p, $\tau_{p,90}$, W_c, τ_c (peak load, 90% peak time window, continuous load, and continuous time window) are derived from each input time series.

Under the scope of the presented work, a simplified and extensible metric is defined to estimate the maximum influence for each two-way combination. Therefore, continuous value and time window from the predecessor time series are combined with peak value and 90% time window of the successor time series. Equation (1) shows the calculation of maximum value estimation ΔW for the impact due to the varying sequences. The worst-case assumption, that the peak of the second time series directly appears after the first time series end, matches to the upper bound of the estimated values.

$$\Delta W = \sqrt{\frac{\tau_{p,90} \cdot {W_p}^2 + \tau_c \cdot {W_c}^2}{\tau_{p,90} + \tau_c}} - W_c \tag{1}$$

By using this parameter as well as selecting an appropriate threshold, all potentially critical two-way combinations could be identified and pre-selected for further examination.

4. Results

4.1. Derivation of the Composite Cycle

To illustrate the results of the presented method, Figure 7 shows a comparison between the accumulation of the CLC-Data and the derived cycle. As already known, the color scheme of these diagrams represents the frequency distribution. These charts of acceleration vs. velocity serve as an example. Additional comparisons are included in Figure A1 in the Appendix A.

To generate the distribution table of the composite cycle, each of the time series used is first converted into a statistic and then merged depending on the proportion of time in the total cycle.

In order to quantify the distribution of the combined cycle from the initial customer data, the root mean square error (RMSE) of the deviation is calculated. This value indicates the accuracy with which the derived cycle approximates the customer statistics. Thus, a small value points out a good approximation. In this particular evaluation, the RMSE values of the deviation between both maps are approximately 0.4 for acceleration vs. speed, about 0.6 for torque vs. speed and about 2.6 overall. In contrast, the RMSE value of the deviation of only one criterion (see Figure 3 in [1]) was greater than 3. Compared to that, the RMSE value has decreased significantly. This smaller value results from an even smaller deviation from the CLC-data and thus indicates an even better approximation of the customer behavior.

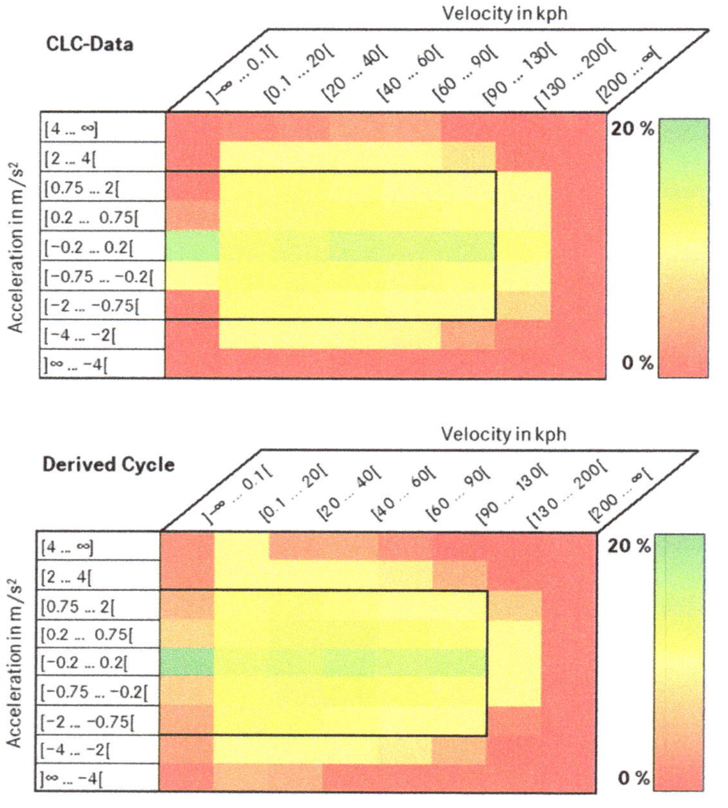

Figure 7. Comparison of the frequency distribution of acceleration vs. velocity.

4.2. Time Frame-Based Analysis of the Composite Cycle

The following diagram in Figure 8 shows the time-dependent continuous load curves of all considered time series (>20,000) using the example of mechanical power. These figures result from the time frame-based load analysis according to [6] and display the maximum occurring RMS value for each time frame width as described in [1].

As mentioned above, the derived cycle results from the concatenation of time series in order to reduce the deviation as much as possible. This cycle, represented by the red dashed line, is shown in Figure 8 in comparison, both to the time series used for the derived cycle in the foreground and to all time series considered transparently in the background.

This graph reflects the load of the derived cycle compared to all time series and it becomes evident that the composite cycle does not include all of the high, infrequent loads. The reason for this can be seen in the frequency distribution maps, e.g., in Figure 7, as high loads, such as high accelerations and velocities, occur too rarely to be considered in this cycle.

As the diagram shows, the derived cycle contains all combined time series. Therefore, as expected, the TFBA of the composite cycle represents an envelope curve of the combined time series.

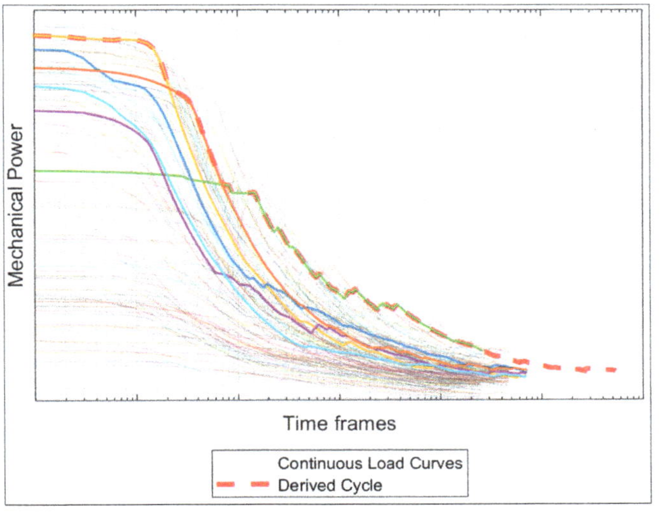

Figure 8. Derived cycle vs. considered time series: Comparison of the mechanical power.

4.3. Investigation of Subcycle Sequences Variation

Based on the previous analysis, the sequence of the selected subcycles has an impact on the course of the continuous load curves. With n selected cycles, there are $n!$ possibilities for a concatenation of these cycles. Therefore, there are 720 (*6!*) possible concatenations based on the six selected cycles shown in Figure 8, but only 120 (*5!*) compounds are unambiguous (cf. Section 3.1). Figure 9 shows the corresponding continuous load curves. The time-dependent continuous load curves of the unambiguous concatenation are shown in turquoise and represent the possible range between maximum and minimum loads.

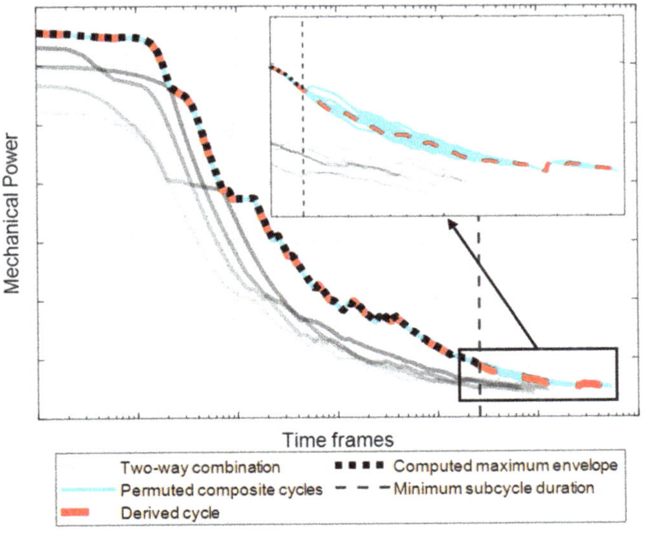

Figure 9. Comparison of derived cycle, maximum envelope and time series permutations.

In addition, the continuous load curves of the two-way combinations and the time frame of the minimal subcycle duration are shown in Figure 9. As explained in Section 3.2, the relevant loads for the time frame up to the duration of the minimum subcycle can be identified by analyzing the two-way combinations. The computed enveloping continuous load curve is also visible. This curve corresponds to the maximum of all permuted composite cycles in the admissible range of the minimum subcycle duration.

It can be seen that in this example the sequence of the cycles has low influence on the course of the continuous load curves. The maximum relative deviation between the derived cycle and the enveloping continuous load curves of the two-way combination is less than 0.2% in the admissible range.

Moreover, using the method from Section 3.3, the potentially relevant combinations are identified, which could critically influence the maximum time frame-based loads. Figure 10 illustrates the result.

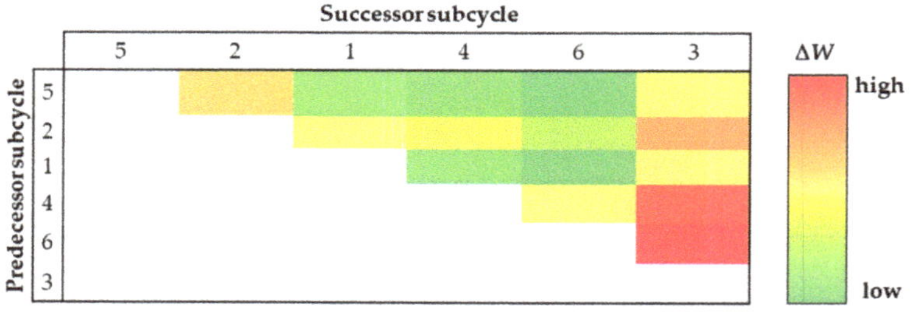

Figure 10. Estimation of influence based on various two-way combinations.

For this purpose and to place the focus on the influence of the peak of the following cycle, the input cycles are sorted in descending order with regard to their RMS value and only combinations where the following subcycle has a lower RMS are considered. Potentially relevant concatenations can be identified, and their impacts are estimated via the level of ΔW, assuming that a cycle can only occur once in the combination. In further analysis, these pre-selected combinations can be examined more precisely.

5. Conclusions and Future Work

The methods presented in this paper transfer statistical customer data into discrete time series in order to approximate the original distribution statistics as closely as possible. A composite cycle is derived and evaluated against related maximum time frame-based loads to identify the relevant design parameters.

The existing derivation method is based on a first criterion for the selection of suitable time series, the distribution statistics of acceleration vs. velocity. Consideration of additional CLC data will increase the accuracy in selecting appropriate time series. Therefore, the method is extended by the distribution statistics of torque vs. speed as well as the mechanical power of the machine. This reduces the number of relevant time series and increases the approximation of actual customer behavior. Moreover, when concatenating the time series, it must be ensured that the signal is still continuous; kinks or even jumps need to be avoided. Finally, limiting the total length of the composite cycle to the range of the vehicle is another innovation of this method.

These enhancements to the method ultimately result in a very small deviation of the derived cycle from the original customer data, as the results show. The RMSE of the deviation serves as measure for the approximation of the derived cycle to the customer statistics. This value ranges approximately between 0.4 and 0.6 for all customer data considered which indicates a very small deviation and thus a good approximation of customer behavior. However, as mentioned above, the method requires a valid and plausible database, which was assumed here.

Furthermore, the correlation between subcycle sequences and time frame-based load characteristics are investigated. Both subcycle and transition events within the composite cycle could generate loads on certain time window. The latter could result in varying time frame-based maximum loads, which do not occur in this form within a subcycle. Assuming that these loads can actually occur connected in reality, this leads to differences in design-relevant parameters, which could significantly affect the overall cost factors of a customer-specific drive system. The introduced approach of subcycle sequence analysis provides an evaluation framework regarding this issue and quantifies the impact of sequences as a deviation between the derived cycle and the computed maximum envelope of time frame-based loads.

In order to reduce complexity of the evaluation process, two theorems are introduced. Theorem 1 points out the unambiguous one-to-one relation between time series sequences and time frame-based load characteristics. Theorem 2 provides a foundation for the efficient computation of maximum envelope of relevant time frame-based loads in their admissible range using two-way combination. As a result, the required number of TFBA executions decreases from factorial down to quadratic order. Hence, it is possible to integrate the evaluation of a subcycle sequence's influence on the mentioned derivation process of design-relevant parameters based on a different dataset.

The application of the evaluation method shows no significant deviation in the admissible range based on the selected dataset, so that the derivation of relevant parameters based on the original composite cycle is justified. However, if the evaluation process demonstrates a relatively large deviation between the composite cycle and the overall maximum time frame-based loads, certain thermal reserve must be considered within the derivation process of design-relevant parameters.

By applying the simplified metric from Section 3.3, the impacts of potentially critical combinations in terms of maximum load can be estimated and selected for further analysis.

In summary, the aforementioned derivation and evaluation methods generate a contribution to the improved customer-specific drive system design, which is thoroughly safeguarded against potential thermal stresses according to the region of interest of customer behaviors.

Future work for these methods could include the quantitative analysis between time frame-based loads and the related component-specific thermal behavior, e.g., using thermal simulations. Furthermore, the impact of frequency of time frames should be further investigated. Based on these aspects, an extended algorithm for the identification of maximum thermal stresses over the total length of the composite cycles could be implemented.

Author Contributions: Conceptualization, methodology, software, validation, formal analysis, investigation, resources, data curation, writing—original draft preparation, writing—review and editing, visualization, project administration, funding acquisition, R.M., F.M. and R.Z.; supervision, F.G. and A.S. All authors have read and agreed to the published version of the manuscript.

Funding: This research is funded by the Publication Fund of the Karlsruhe Institute of Technology, funding number 02034100761.

Data Availability Statement: No publicly archived datasets analyzed.

Conflicts of Interest: The authors declare no conflict of interest.

Appendix A

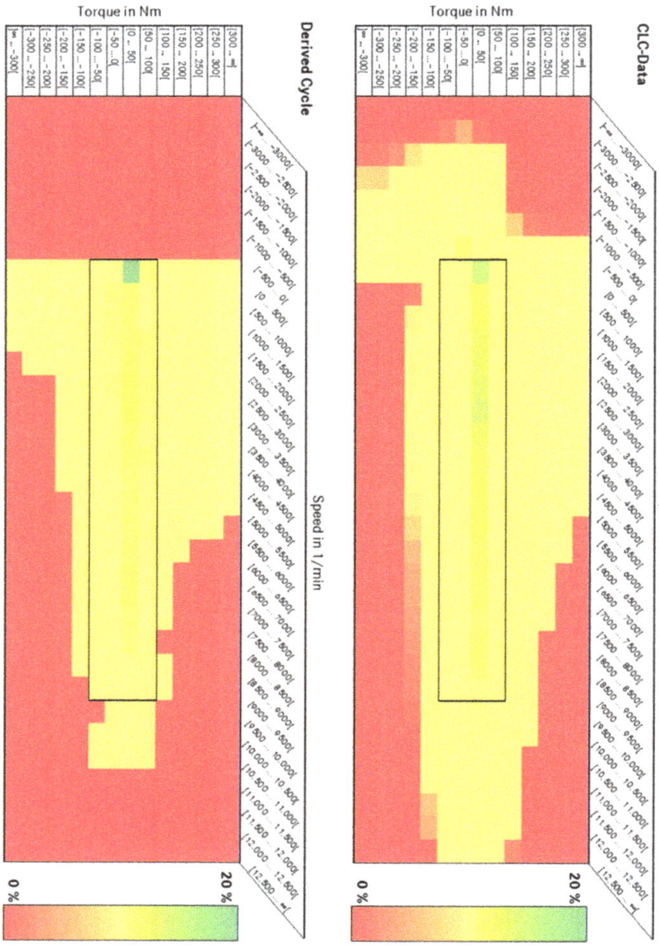

Figure A1. Comparison of the frequency distribution of Torque vs. Speed.

References

1. Mieth, R.; Gauterin, F.; Pauli, F.; Kraus, H. Transfer of Statistical Customer Data into Relevant Parameters for the Design of Vehicle Drive Systems. *Vehicles* **2022**, *4*, 137–144. [CrossRef]
2. Pfriem, M. Analyse der Realnutzung von Elektrofahrzeugen in kommerziellen Flotten zur Definition einer bedarfsgerechten Fahrzeugauslegung. Doctoral Dissertation, Karlsruher Institut für Technologie, Karlsruhe, Germany, 2015. (In German).
3. Förster, D.; Inderka, F.; Gauterin, F. Data-driven identification of characteristic real-driving cycles based on k-means clustering and mixed-integer optimization. *IEEE Trans. Veh. Technol.* **2019**, *69*, 2398–2410. [CrossRef]
4. Tewiele, S. Generierung von repräsentativen Fahr- und Lastzyklen aus realen Fahrdaten batterieelektrischer Fahrzeuge. Doctoral Dissertation, Universität Duisburg-Essen, Duisburg, Germany, 2020. (In German).
5. Dai, Z.; Niemeier, D.; Eisinger, D. Driving cycles: A new cycle-building method that better represents real-world emissions. *U.C. Davis-Caltrans Air Qual. Proj.* **2008**, *66*, 37.
6. Von Lingen, J.; Schmidt, P. Zeitgewichtete Belastung und Belastbarkeit und deren Nutzung für die Dimensionierung von Bahnenergieversorgungsanlagen. *Elektrie* **1994**, *48*, 448–457. (In German)
7. Hofmann, G.; Röhlig, S. Zeitgewichtete Belastungsdauerkurve bei elektrische Bahnen. *Elektr. Bahnen-EB* **1997**, *95*, 272–279. (In German)

8. Pauli, F. Allgemeingültige Beschreibung der Lebensdauerbelastung elektrischer Antriebskomponenten in Pkw auf Basis von Mess- und Simulationsdaten. Dissertation, Technische Universität Dresden, Dresden, Germany, 2016. (In German).
9. Pauli, F.; Mueller, K. Statistische Untersuchung von zeitfensterbasierten Analyseergebnissen. *Elektr. Bahnen-EB* **2015**, *113*. (In German)

vehicles

MDPI

Article

Design of a Hybrid Electric Vehicle Powertrain for Performance Optimization Considering Various Powertrain Components and Configurations

Manh-Kien Tran [1], Mobaderin Akinsanya [2], Satyam Panchal [2], Roydon Fraser [2] and Michael Fowler [1,*]

[1] Department of Chemical Engineering, University of Waterloo, Waterloo, ON N2L 3G1, Canada; kmtran@uwaterloo.ca

[2] Department of Mechanical and Mechatronics Engineering, University of Waterloo, Waterloo, ON N2L 3G1, Canada; makinsan@uwaterloo.ca (M.A.); satyam.panchal@uwaterloo.ca (S.P.); rafraser@uwaterloo.ca (R.F.)

* Correspondence: mfowler@uwaterloo.ca; Tel.: +1-519-888-4567 (ext. 33415)

Citation: Tran, M.-K.; Akinsanya, M.; Panchal, S.; Fraser, R.; Fowler, M. Design of a Hybrid Electric Vehicle Powertrain for Performance Optimization Considering Various Powertrain Components and Configurations. *Vehicles* **2021**, *3*, 20–32. https://doi.org/10.3390/vehicles3010002

Received: 9 December 2020
Accepted: 28 December 2020
Published: 31 December 2020

Publisher's Note: MDPI stays neutral with regard to jurisdictional claims in published maps and institutional affiliations.

Abstract: Emissions from the transportation sector due to the consumption of fossil fuels by conventional vehicles have been a major cause of climate change. Hybrid electric vehicles (HEVs) are a cleaner solution to reduce the emissions caused by transportation, and well-designed HEVs can also outperform conventional vehicles. This study examines various powertrain configurations and components to design a hybrid powertrain that can satisfy the performance criteria given by the EcoCAR Mobility Challenge competition. These criteria include acceleration, braking, driving range, fuel economy, and emissions. A total of five different designs were investigated using MATLAB/Simulink simulations to obtain the necessary performance metrics. Only one powertrain design was found to satisfy all the performance targets. This design is a P4 hybrid powertrain consisting of a 2.5 L engine from General Motors, a 150 kW electric motor with an electronic drive unit (EDU) from American Axle Manufacturing, and a 133 kW battery pack from Hybrid Design Services.

Keywords: hybrid electric vehicles; powertrain design; powertrain configurations; powertrain components; fuel economy optimization; vehicle performance optimization

1. Introduction

One of the most significant issues that the world is currently facing is climate change, which is largely caused by the ever-increasing amount of greenhouse gas (GHG) emissions [1]. The transportation sector accounts for a large portion of GHG emissions. In Canada, the transportation sector was responsible for 24% of Canada's GHG emissions in 2016 [2]. Because of this, there have been some efforts from various governments and the automotive industry to reduce the GHG emissions from vehicles, such as increased regulations on vehicle emissions standards [3]. The transportation sector has always been characterized by trends such as new technological developments, government mandates, varying regulations, environmental concerns, or changes in the global economic status [4]. Recent trends show that there has been a gradual decline in the popularity of fossil-fuel-powered internal combustion engine vehicles (ICEVs), while electric vehicles (EVs) and other zero-emission vehicles (ZEVs) are becoming increasingly more popular as alternatives [5,6]. A growing number of federal governments have announced aggressive timelines for the elimination of ICEVs, leading to a shift of focus toward EVs and battery technology by global automakers [7]. It was recently reported that the automotive industry will spend a minimum of $300 billion in the development of EVs over the next 10 years [8].

However, the transition from ICEVs to battery EVs (BEVs) has not been smooth because the battery technology development is still in its early stages [9]. Hybrid electric vehicles (HEVs) have proven to be a necessary bridge into the eventual complete BEV transition [10]. Advancements in electrified powertrain technology have also helped decrease the costs of HEVs, leading to their increased prevalence on the road. It was estimated that HEVs will achieve price parity with ICEVs by 2024 and become cheaper by 2025 [11]. The presence of an engine in HEVs also alleviates the range anxiety concern that still plagues the BEV segment [12]. Furthermore, with more resources being invested in EV technology, some of the recent HEVs are showing better performance and lower costs compared to ICEVs.

The term electrified powertrain is often used to describe several powertrain configurations that utilize electrical energy to produce propulsive torque [13]. Electrification within vehicles can take place in many different forms, including mild hybrid, strong hybrid, plug-in hybrid, and full battery EVs [14,15]. (1) Mild hybrid vehicles have the engine as the primary power source and use an electric motor with a small battery pack to produce electrical energy, which is used to assist with the engine output [16]. These vehicles usually do not have a dedicated driving mode that allows for propulsion via electrical power only, but the addition of electrification still helps reduce their fuel consumption in comparison to ICEVs. (2) Strong hybrid vehicles, also known as HEVs, use a combination of an engine and a battery-powered electric motor to drive the vehicle [17]. They have a more complex vehicle architecture and physical packaging requirements than mild hybrid and conventional vehicles. HEVs offer significant improvements in fuel consumption, as well as superior overall performance compared to similar conventional vehicles. (3) Plug-in hybrid electric vehicles (PHEVs) are similar to HEVs, with the main distinction being the PHEVs' larger battery packs that can be recharged directly from grid electricity via a plug-in charger [18]. This allows PHEVs to have a larger EV-only range when compared to HEVs. (4) BEVs do not have an engine or any of the related internal combustion components. Instead, they solely utilize the battery-powered electric motor, which often comes with a very large battery pack, to provide propulsive torque to drive the vehicles. Like the PHEVs, the BEVs can also be recharged via a plug-in charger [19].

Several research works have been conducted to help develop and improve the design of electric powertrain in EVs. Dagci et al. [20] utilized planetary gear sets (PGs) to develop an automated design process for PG-based HEV systems focusing on both fuel economy and performance. The design process consisted of five major stages, and their case study results showed that a light-duty truck's performance requirements could be fulfilled by various two-PG HEV designs without sacrificing fuel economy if the appropriate synthesis techniques for exploring the entire design space are developed. Kabalan et al. [21] investigated the potential of efficiency improvement of the simple series-parallel HEV powertrain using topology modification, which was the addition of gears for the components or a gearbox with a few numbers of ratios. The findings showed an efficiency decrease in one variant and an efficiency improvement in another variant with a fuel consumption result that was comparable to the standard Toyota Hybrid System. Vora et al. [22] introduced a model-based framework that incorporated powertrain simulation and battery degradation models to predict fuel consumption, electrical energy consumption, and battery replacements. These results were combined with economic assumptions to enable the exploration of a larger design space to provide better insights to vehicle integrators, component manufacturers, and buyers of HEVs. Lei et al. [23] demonstrated a novel approach for designing an electric powertrain to optimize energy consumption while maintaining vehicle performance and ride comfort. The requirements for power performance, energy consumption, and ride comfort were generated on the vehicle level. Subsequently, the generated requirements were applied to the subsystem level, where torque outputs, motor efficiency, and vehicle weight were the corresponding requirements. A multi-objective global optimization was carried out on the subsystem level while a constrained energy approach was proposed for the vehicle level. The final solution had a lightweight ratio of 93.5% and motor efficiency of 92%. Zhou et al. [24] outlined an optimal selection methodology for PHEV powertrain configuration utilizing optimization and a comprehensive evaluation of powertrain design schemes. To determine the performance potential of each configuration, a multi-objective powertrain optimization design was proposed and applied to series, parallel pre-transmission, output power-split, and multi-mode power-split powertrain configurations. The results suggested that the parallel pre-transmission configuration could be selected for optimal acceleration capacity, the multi-mode power-split configuration could be selected for optimal electric energy efficiency, and the output power-split configuration could be selected for optimal fuel economy.

In this work, the design and optimization of an HEV powertrain are examined in the context of a vehicle development competition. The University of Waterloo Alternative Fuels Team (UWAFT) is participating in the EcoCAR Mobility Challenge, sponsored by the United States (US) Department of Energy, General Motors (GM), and MathWorks, and managed by Argonne National Laboratory [25]. The competition tasks 12 North American universities to apply advanced propulsion systems, electrification, and vehicle connectivity to improve the energy efficiency of a 2019 Chevrolet Blazer while balancing factors such as emissions, safety, and consumer acceptability. This program provides the opportunity to apply the model-based design methodology using software-in-the-loop and hardware-in-the-loop to evaluate and optimize the HEV powertrain. This study outlines the process of developing an HEV powertrain using MATLAB and Simulink that is optimized for performance, fuel economy, and emissions. Powertrain configuration selection, as well as powertrain component (engine, motor, and battery) selection and sizing, is examined to evaluate the potential benefits and drawbacks for each layout. This study focuses on the first two steps in the model-based design methodology, as shown in Figure 1. The contribution of this study is the in-depth description of a powertrain design process using the model-based design methodology and software modeling and simulation to electrify a conventional vehicle, as well as the optimization of the hybrid powertrain performance by considering different design parameters.

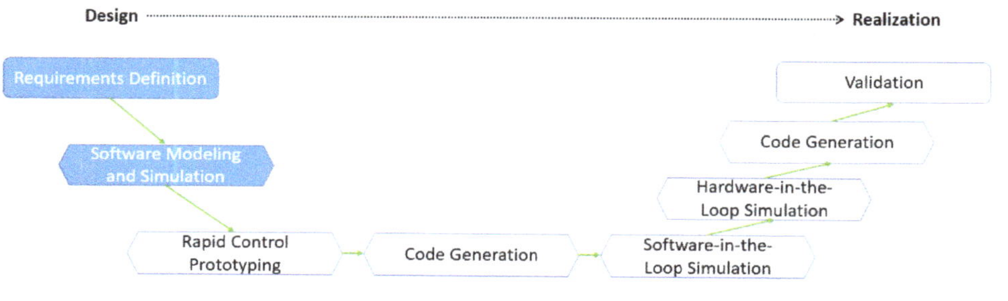

Figure 1. Schematic of the model-based design methodology.

The rest of this paper is organized as follows: Section 2 describes various available hybrid powertrain configurations; Section 3 provides the design criteria and constraints; Sections 4 and 5 outline the powertrain design process and analyze the simulation results of the built-in vehicle model in MATLAB and Simulink; Section 6 provides some concluding remarks.

2. Hybrid Powertrain Configurations

From a vehicle architecture standpoint, HEV powertrains can be classified into three main categories, which are series, parallel, and series-parallel split. These categories are defined by the vehicle's overall power flow and torque path.

In a series HEV powertrain, the engine does not provide propulsive torque to drive the vehicle. Its main function is to convert potential energy from fuel to mechanical energy which is then converted to electrical energy using a generator. The electrical energy is used to propel the motor via an inverter. This configuration allows for the engine speed to be controlled independently from the vehicle speed, which means that the engine can be controlled to run at the optimal speed to minimize losses incurred in the electricity generation process [26]. The electric motor used to drive the vehicle receives power from the engine or the battery pack as shown in Figure 2.

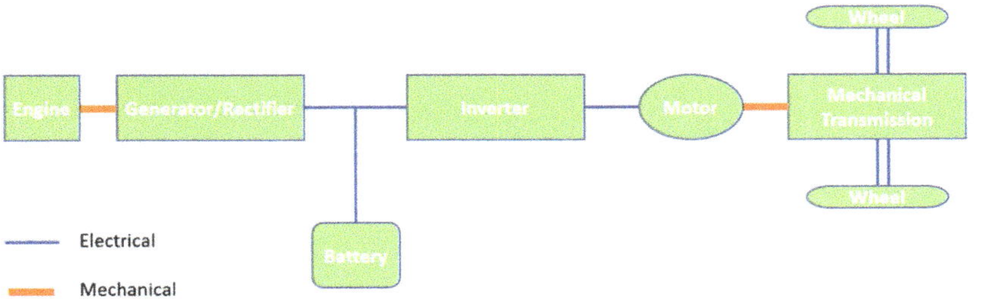

Figure 2. Schematic of a series hybrid powertrain.

In a parallel HEV powertrain, the engine, similar to conventional vehicles, provides propulsive torque directly to the wheels to drive the vehicle. An electric motor, powered by a battery pack, is also mechanically coupled to the driveline, allowing it to boost the power output of the engine. A mechanical coupler combines the torques generated from the engine and motor and delivers the resulting torque to the wheels. The engine torque and the motor torque can be controlled individually, but the speed of the engine and the motor each have a fixed proportion to the overall vehicle speed. An example of the parallel HEV powertrain configuration is shown in Figure 3.

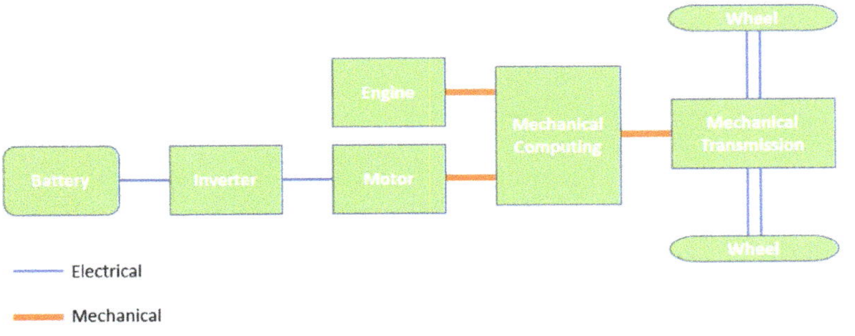

Figure 3. Schematic of a parallel hybrid powertrain.

The series-parallel HEV powertrain, shown in Figure 4, is a significantly more complex configuration as it allows for both series and parallel driveline functionality, optimizing the vehicle for various driving scenarios [26]. This is enabled by a mechanical coupling component that can either connect or disconnect the power output of the engine from the vehicle's main driveline.

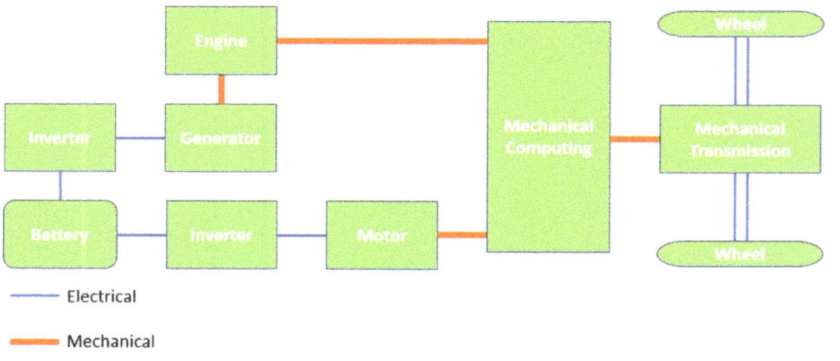

Figure 4. Schematic of a series-parallel split hybrid powertrain.

3. Powertrain Design Requirements and Constraints

3.1. Vehicle Performance Metrics

The first step in the HEV powertrain designing process involves establishing design objectives and evaluation criteria. The vehicle technical specifications (VTS) were established by the EcoCAR competition as the key performance indicators used to evaluate the potential designs in terms of overall performance. Table 1 lists all the VTS metrics targeted for the powertrain.

Table 1. Summary of the target vehicle technical specifications.

Criteria	Units	Competition Targets
Acceleration 0–60 mph	s	≤7
Acceleration 50–70 mph	s	≤6.5
Braking 60–0 mph	ft	≤138.4
Total range	mi	≥250
Combined fuel economy	mpg	≥33.5
Total emissions	g/mi	≤373

3.2. Available Options for Powertrain Configurations and Components

Aside from the VTS, other technical requirements and limitations were imposed to ensure the creation of a viable propulsion system design that aligns with the core vision of the competition. For instance, P0 and P4 powertrain architectures, shown in Figure 5, were pre-approved by the EcoCAR organizing committee and, hence, teams were encouraged to use these architectures. In the P0 architecture, an electric motor is connected with an internal combustion engine through a belt, on the front-end accessory drive. In the P4 architecture, the electric motor, located in the rear axle drive, is connected through a gear mesh on the rear axle of the vehicle and decoupled from the internal combustion engine.

The competition also limits the options for powertrain components such as the engine, motor, and battery to ensure a certain level of safety control. Due to confidential reasons, the exact details of these components, despite being used in the MATLAB/Simulink simulation, are not disclosed in this study; instead, the information given is the high-level specifications of the components. Table 2 shows two engine options, which are manufactured by GM.

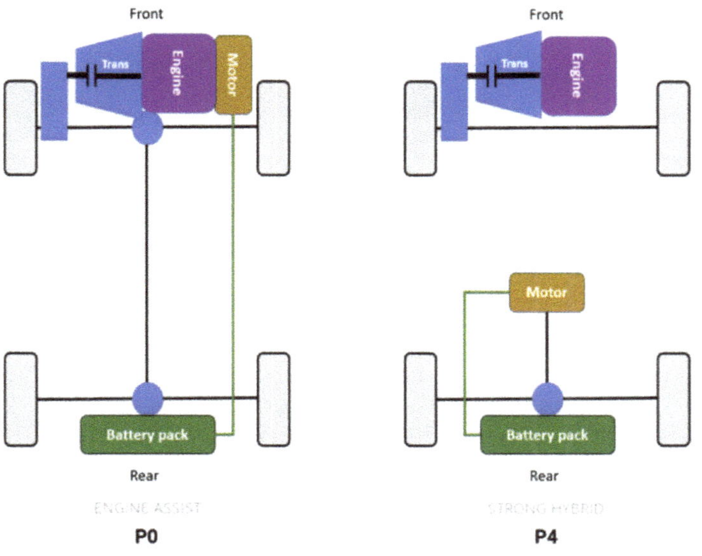

Figure 5. Viable powertrain architectures.

Table 2. Overview of engine options.

Code	Displacement	Intake System
LYX	1.5 L	Turbocharged
LCV	2.5 L	Naturally Aspirated

The electric system, consisting of an electric motor and a battery pack, must be designed to ensure that components are not only compatible with each other but also satisfy the objective to produce a vehicle optimized for fuel economy. The power levels of the two systems should be matched such that the battery pack's discharge power capabilities meet the requirements of the motor. Figure 6 outlines two available battery packs and four corresponding motors. The two options for battery include a smaller pack, the Malibu hybrid battery (HEV4), with a maximum discharge power of 53 kW made by GM, and a larger pack with a maximum discharge power of 133 kW custom-made by Hybrid Design Services (HDS). The HDS pack consists of 768 2-Ah Samsung cells (lithium nickel manganese cobalt oxide), with 96 cells in series and eight in parallel. The smaller GM HEV4 pack is compatible with two motors which are the American Axle Manufacturing (AAM) EDU2 and the Emrax 228, while the larger HDS pack is compatible with the Phi Power 217 s and the AAM EDU4.

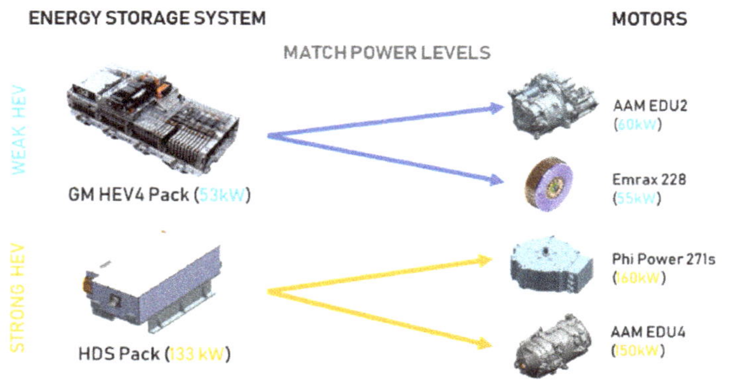

Figure 6. Electric system component options.

Energy recovery capability should also be considered to optimize fuel economy. The energy recovery is usually maximized by using regenerative braking as the primary means of slowing the vehicle. The charging capability of the battery is the main factor determining the amount of energy that can be recovered through regenerative braking. Table 3 shows the regenerative capabilities for the two battery options.

Table 3. Battery regenerative braking capability. (GM, General Motors; HDS, Hybrid Design Services.)

Battery Pack	Peak Charge Power	Regenerative Braking Capability (Deceleration Rate of 3 m/s^2)
GM HEV4	65 kW	45 km/h
HDS	133 kW	93 km/h

The HDS pack can bring the vehicle, with an assumed mass of 1730 kg, to a complete stop from 93 km/h using only regenerative braking at a deceleration rate of 3 m/s^2, whereas the HEV4 pack can only slow the vehicle from 45 km/h before the use of mechanical brakes is required.

4. Powertrain Modeling in MATLAB/Simulink

This section details the modeling and simulation environment setup and the baseline performance results of several powertrain architectures. MATLAB/Simulink, specifically the Powertrain Blockset™, was utilized as the primary tool in the development of the powertrain model to explore various architectures consisting of different components. The powertrain model is derived from a Simulink model for the stock 2019 Chevrolet Blazer provided by MathWorks. The model has four unique subsystems, including the powertrain, drivetrain, controllers, and driver input, as shown in Figure 7. These systems were developed in parallel with component selection and architecture refinement to achieve the best results. MATLAB scripts were developed to efficiently evaluate various architectures and perform component selection sweeps.

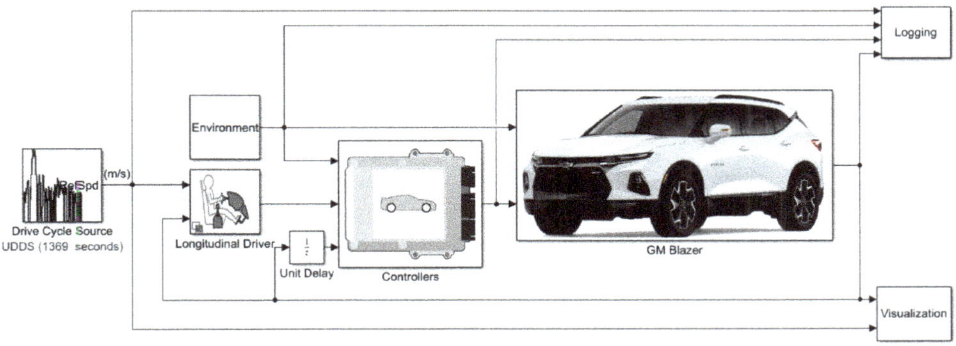

Figure 7. High-level vehicle model in Simulink.

4.1. Powertrain Components Modeling

Basic built-in blocks from the Simulink Powertrain Blockset™ were used to model each component being considered. The engine and motor blocks rely on lookup tables generated via the Model-Based Calibration Toolbox™ and simplified dynamics, while the battery block is represented by an equivalent circuit model. The built-in blocks for the components are shown in Figure 8. Linear interpolation between lookup tables breakpoints has been proven to be sufficient for standard drive cycle modeling [26].

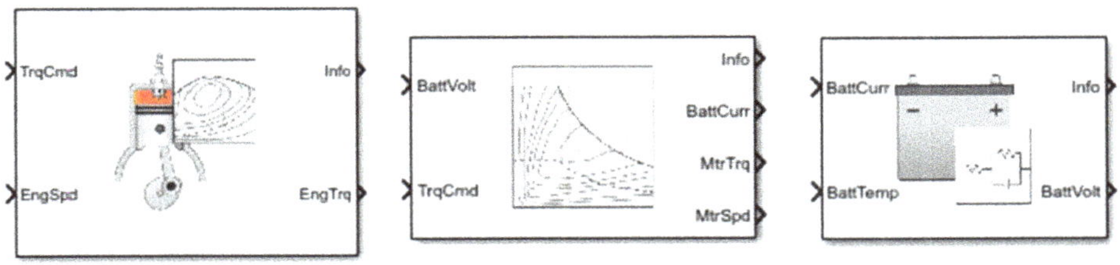

Figure 8. The engine block, motor block, and battery block in Simulink (from left to right) [27–29].

The engine model used in this study is based on the built-in "Mapped SI Engine" block [27]. This block uses a series of lookup tables to calculate the input speed and torque command. Significant lookup tables include the gas mass flow, fuel mass flow, exhaust manifold gas temperature, brake-specific fuel consumption, CO, CO_2, NO_x, and particulate matter (PM) emissions. The actuator blocks use transfer functions to determine the actuator dynamics at a given time. The emissions calculations are detailed by a set of functions that determine the level of filtrated emissions by a catalyst. These calculations assume that the operating pressure and temperature of the engine are constant and that the lookup table values are representative of practical applications.

The motor model is based on the built-in "Mapped Motor" block [28]. This block uses a lookup table to determine the motor efficiency on the basis of input torque and motor speed. Another lookup table of torque-speed data is utilized to determine maximum torque values and rotational speeds. These lookup tables were populated using data obtained directly from the motor manufacturers. The required current is calculated using the battery voltage and the mechanical power, which is obtained from speed and torque commands.

The battery model is based on the built-in "Equivalent Circuit Battery" block [29]. The parameters of the first-order equivalent circuit model were provided by the manufacturers and populated into lookup tables. The first-order equivalent circuit model has been shown to be sufficiently accurate for the purpose of battery voltage estimation [30,31].

This battery block determines the battery output voltage from the required current and operating temperature. The state of charge is calculated by coulomb counting. The temperature is assumed to be constant during operation, and the lookup tables are assumed to be representative of battery conditions. These assumptions can be made as the changes in temperature are negligible during battery operation within the standard drive cycles.

4.2. Energy Consumption Modeling

In an ICEV, only low-level controls such as individual component controllers are required. In an HEV, however, the vehicle controller needs to also determine how much power should be delivered by each of the energy sources in the vehicle [32,33]. For the UWAFT-designed vehicle, one objective is to minimize the fuel economy, which would require an effective energy management strategy. UWAFT implemented the equivalent consumption minimization strategy (ECMS), which is a method with a low computational complexity that has been used widely in hybrid vehicle applications. The ECMS is based on the concept that there is an equivalence between electrical energy and fuel energy [34]. This equivalence is evaluated by considering the average energy paths from the fuel tank to the storage of the electrical energy. In the ECMS functions, for each time t with a time step of Δt, parameters such as acceleration, speed, wheel speed, and wheel torque are measured or evaluated and used to calculate the equivalence factor. The equivalence factor determines the fuel equivalent of the electrical energy on the basis of whether the battery is being charged or discharged. This factor is used in the cost function which is minimized by adjusting the control variables. The optimal control variables then regulate the amount of torque that is provided by the electrical and fuel paths.

As previously discussed, the primary powertrain components (engine, motor, and battery) were modeled in Simulink using lookup tables provided by the component manufacturers. Without the physical components to validate these tables, it was assumed that these individual components were modeled to a sufficient degree of accuracy. There were several other assumptions that may have also caused some variance in the simulated fuel economy results. For instance, transmission shift time is a factor that can affect the fuel economy results; however, in this study, it was assumed that the transmission shift time could be neglected. The shift time used in the model was 350 ms. It has been shown that vehicle shifting tends to range from 50 ms to approximately 500 ms [35]. Therefore, different transmission shift times, from 50 to 500 ms, were run and analyzed using the stock vehicle model to see the corresponding fuel economy, as shown in Figure 9. This analysis showed a maximum difference in fuel economy of 1.9% which is small enough to be reasonably neglected in practical applications. Another assumption, not formally evaluated but inherent to the model, is environmental conditions (wind, grade, etc.) being constant. Deviations from these environmental conditions were not considered in the basic architectural design and selection. Overall, an upper estimate of the variance in the fuel economy results was ± 1.65 mpg, suggesting reasonably accurate fuel economy modeling results.

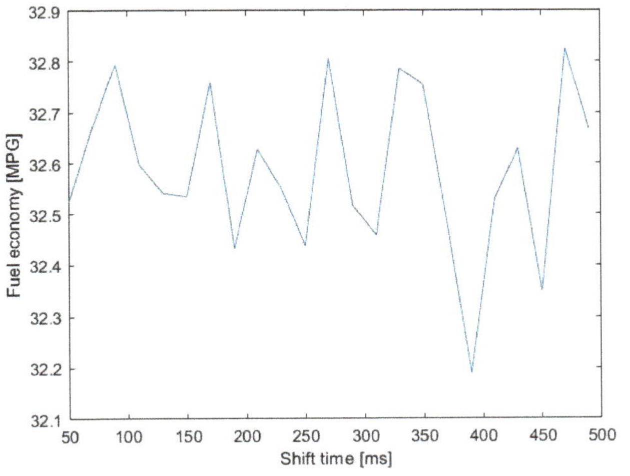

Figure 9. Fuel economy vs. transmission shift time.

5. Powertrain Simulation Results in MATLAB/Simulink

After inputting the specification parameters for the engines, motors, and batteries, and implementing the ECMS in the vehicle controller, the powertrain model was run in Simulink. Only the mechanically feasible designs were considered. For instance, only the 1.5 L engine and the Phi Power 271 s were considered for the P0 configuration because the spacing requirement for both the engine and motor in the front of the vehicle cannot accommodate any other types of engine and motor. For the P4 configuration, only the 2.5 L engine was considered for performance purposes, since, at times, the engine will have to drive the car completely by itself and, thus, will need more power. The matching electrical components were also considered. The simulation used two drive cycles to test the performance of each design, which were the urban dynamometer driving schedule (UDDS) and the highway fuel economy driving schedule (HWFET). Each of the drive cycles was run five times consecutively to obtain a longer running time. The UDDS drive cycle represents city driving and the HWFET drive cycle represents highway driving. The values shown in Table 4 are the combination of the results from running these two drive cycles, with the unsatisfactory performance metrics when considering the competition targets being labeled in red.

Table 4. Summary of performance results of various powertrain designs in Simulink. Unsatisfactory performance metrics when considering the competition targets are labeled in red.

Configuration	P0	P4	P4	P4	P4
Engine Motor Battery	GM LYX 1.5 L Phi Power 271 s HDS	GM LCV 2.5 L AAM EDU2 GM HEV4	GM LCV 2.5 L Emrax 228 GM HEV4	GM LCV 2.5 L Phi Power 271 s HDS	GM LCV 2.5 L AAM EDU4 HDS
Acceleration 0–60 mph (s)	4.85	5.77	6.13	4.95	4.78
Acceleration 50–70 mph (s)	3.97	4.51	5.05	4.01	4.15
Braking 60–0 mph (ft)	140.4	139.9	139.5	133.2	135.5
Total range (miles)	313.4	310.9	307.9	309.2	308.8
Fuel economy (mpg)	34.2	34.1	33.9	32.6	33.8
Emissions (g/mile)	159.6	17.1	15.2	87.4	43.7

From Table 4, it can be seen that, even though all five designs seem viable and satisfy most of the EcoCAR competition targets, only one design gives a fully satisfactory performance. From a high-level perspective, this design is a P4 hybrid powertrain consisting of a 2.5 L inline-four engine from GM, a 150 kW electric motor with an integrated 9.04:1 gear reduction, also known as an electronic drive unit (EDU) from AAM, and a 133 kW battery pack provided by HDS. Figure 10 summarizes the selected powertrain configuration and components. The Simulink simulation results using the vehicle model and the ECMS control strategy are shown in Table 5, in comparison with the competition targets.

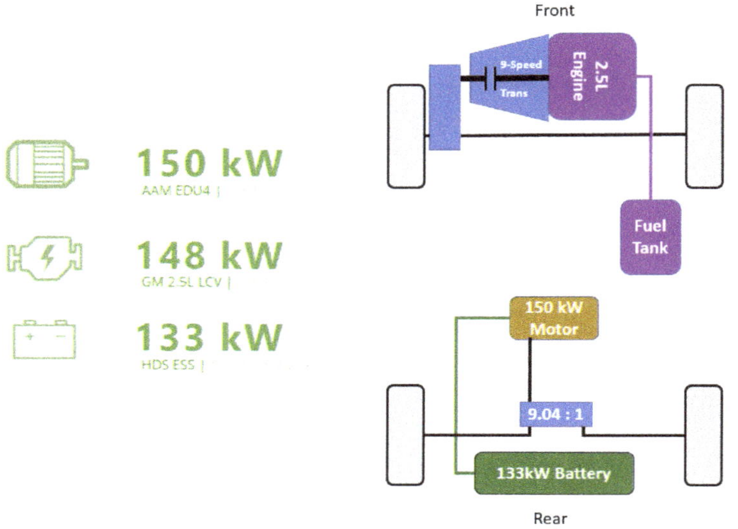

Figure 10. Selected vehicle powertrain configurations and components.

Table 5. Simulated vehicle performance of the selected design.

Criteria	Units	Competition Targets	Simulated Performance
Acceleration 0–60 mph	s	$\leq$7	4.78
Acceleration 50–70 mph	s	$\leq$6.5	4.15
Braking 60–0 mph	ft	$\leq$138.4	135.5
Total range	mi	$\geq$250	308.8
Combined fuel economy	mpg	$\geq$33.5	33.8
Total emissions	g/mi	$\leq$373	43.7

By utilizing the Simulink Powertrain Blockset™, five different powertrain designs were efficiently examined, and one was selected to be the final design to be submitted to the EcoCAR competition. The final design was shown to satisfy all of the competition targets by simulations in MATLAB/Simulink. It should be noted that this study only focused on the technical design process of the hybrid powertrain from the modeling and simulation standpoints. Other design aspects, to be addressed in our future works, such as cost analysis and mechanical/electrical risks, will also be considered before the real-life integration of the selected powertrain configuration and components into the vehicle.

6. Conclusions

This study investigated two powertrain configurations, two engines, four electric motors, and two battery packs in the process of designing a hybrid electric vehicle powertrain. The final design had to satisfy the performance requirements given by the EcoCAR Mobility Challenge competition. The model-based design methodology was utilized, specifically the first two steps, which are the definition of design requirements and software modeling and simulation. Five different designs were modeled and simulated using the Simulink Powertrain Blockset™ to obtain the performance metrics, including acceleration, braking, driving range, fuel economy, and emissions. The equivalent consumption minimization strategy was used as the energy management strategy in all examined designs. The simulation results indicated that only one design was able to meet all the given criteria. This final design was a P4 hybrid powertrain with a 2.5 L engine from GM, a 150 kW electric motor with an EDU from AAM, and a 133 kW battery pack from HDS. This study also showed that software modeling and simulation can be a good first step in the overall vehicle design process, as it can provide some general ideas of how different powertrain components such as engines, motors, and batteries work together in different powertrain architectures. MATLAB/Simulink was also validated to be a good modeling and simulation tool for powertrain design. Further research effort will focus on the next steps in the model-based design research, especially the software-in-the-loop and hardware-in-the-loop, to obtain and validate the final vehicle design. Another aspect that we will focus on in our future research is the concept of connected autonomous vehicles, where we will use technology to steer, accelerate, and brake with little to no human input.

Author Contributions: Conceptualization, M.-K.T., M.A., and R.F.; methodology, M.-K.T. and M.A.; software, M.-K.T. and M.A.; formal analysis, M.A.; writing—original draft preparation, M.-K.T.; writing—review and editing, M.A., S.P. and M.F.; visualization, M.-K.T. and M.A.; supervision, R.F., S.P. and M.F.; funding acquisition, R.F. and M.F. All authors have read and agreed to the published version of the manuscript.

Funding: This research was funded by the Natural Sciences and Engineering Research Council of Canada (NSERC) and the Collaborative Research and Development Grants—Optimizing Hybrid Electric Powertrains for Connected and Automated Vehicles (CAV) (UWAFT—ECOCAR 4—with General Motors of Canada), CRDPJ 537104-18.

Acknowledgments: This work was supported by the equipment and manpower from the University of Waterloo Alternative Fuels Team. Special thanks are given to Danielle Skeba, Vincent Leung, and Ian ShingHei Kwok for their contribution in editing the paper.

Conflicts of Interest: The authors declare no conflict of interest.

References

1. Aldhafeeri, T.; Tran, M.-K.; Vrolyk, R.; Pope, M.; Fowler, M. A Review of Methane Gas Detection Sensors: Recent Developments and Future Perspectives. *Inventions* **2020**, *5*, 28. [CrossRef]
2. Shamsi, H.; Tran, M.-K.; Akbarpour, S.; Maroufmashat, A.; Fowler, M. Macro-Level Optimization of Hydrogen Infrastructure and Supply Chain for Zero-emission Vehicles on a Canadian Corridor. *J. Clean. Prod.* **2020**, 125163. [CrossRef]
3. Taefi, T.T.; Kreutzfeldt, J.; Held, T.; Fink, A. Supporting the adoption of electric vehicles in urban road freight transport—A multi-criteria analysis of policy measures in Germany. *Transp. Res. Part A Policy Pract.* **2016**, *91*, 61–79. [CrossRef]
4. Fathabadi, H. Utilization of electric vehicles and renewable energy sources used as distributed generators for improving characteristics of electric power distribution systems. *Energy* **2015**, *90*, 1100–1110. [CrossRef]
5. Mevawalla, A.; Panchal, S.; Tran, M.-K.; Fowler, M.; Fraser, R. Mathematical Heat Transfer Modeling and Experimental Validation of Lithium-Ion Battery Considering: Tab and Surface Temperature, Separator, Electrolyte Resistance, Anode-Cathode Irreversible and Reversible Heat. *Batteries* **2020**, *6*, 61. [CrossRef]
6. Panchal, S.; Gudlanarva, K.; Tran, M.-K.; Fraser, R.; Fowler, M. High Reynold's Number Turbulent Model for Micro-Channel Cold Plate Using Reverse Engineering Approach for Water-Cooled Battery in Electric Vehicles. *Energies* **2020**, *13*, 1638. [CrossRef]
7. Garcia, R.; Gregory, J.; Freire, F. Dynamic fleet-based life-cycle greenhouse gas assessment of the introduction of electric vehicles in the Portuguese light-duty fleet. *Int. J. Life Cycle Assess* **2015**, *20*, 1287–1299. [CrossRef]
8. Reuters, Exclusive: VW, China Spearhead $300 billion Global Drive to Electrify Cars. Available online: https://www.reuters.com/article/us-autoshow-detroit-electric-exclusive-idUSKCN1P40G6 (accessed on 23 October 2020).
9. Tran, M.-K.; Fowler, M. A Review of Lithium-Ion Battery Fault Diagnostic Algorithms: Current Progress and Future Challenges. *Algorithms* **2020**, *13*, 62. [CrossRef]
10. Hajimiragha, A.; Canizares, C.A.; Fowler, M.W.; Elkamel, A. Optimal Transition to Plug-In Hybrid Electric Vehicles in Ontario, Canada, Considering the Electricity-Grid Limitations. *IEEE Trans. Ind. Electron.* **2010**, *57*, 690–701. [CrossRef]
11. Bloomberg, "Electric Cars May Be Cheaper Than Gas Guzzlers in Seven Years". Available online: https://news.bloomberglaw.com/environment-and-energy/electric-cars-may-be-cheaper-than-gas-guzzlers-in-seven-years (accessed on 23 October 2020).

12. Bonges, H.A.; Lusk, A.C. Addressing electric vehicle (EV) sales and range anxiety through parking layout, policy and regulation. *Transp. Res. Part A Policy Pract.* **2016**, *83*, 63–73. [CrossRef]
13. Sabri, M.F.M.; Danapalasingam, K.A.; Rahmat, M.F. A review on hybrid electric vehicles architecture and energy management strategies. *Renew. Sustain. Energy Rev.* **2016**, *53*, 1433–1442. [CrossRef]
14. Wu, G.; Zhang, X.; Dong, Z. Powertrain architectures of electrified vehicles: Review, classification and comparison. *J. Frankl. Inst.* **2015**, *352*, 425–448. [CrossRef]
15. Tran, M.-K.; Sherman, S.; Samadani, E.; Vrolyk, R.; Wong, D.; Lowery, M.; Fowler, M. Environmental and Economic Benefits of a Battery Electric Vehicle Powertrain with a Zinc–Air Range Extender in the Transition to Electric Vehicles. *Vehicles* **2020**, *2*, 398–412. [CrossRef]
16. Benajes, J.; García, A.; Monsalve-Serrano, J.; Martínez-Boggio, S. Optimization of the parallel and mild hybrid vehicle platforms operating under conventional and advanced combustion modes. *Energy Convers. Manag.* **2019**, *190*, 73–90. [CrossRef]
17. Di Cairano, S.; Bernardini, D.; Bemporad, A.; Kolmanovsky, I.V. Stochastic MPC with Learning for Driver-Predictive Vehicle Control and its Application to HEV Energy Management. *IEEE Trans. Control Syst. Technol.* **2014**, *22*, 1018–1031. [CrossRef]
18. Cheng, H.; Wang, L.; Xu, L.; Ge, X.; Yang, S. An Integrated Electrified Powertrain Topology With SRG and SRM for Plug-In Hybrid Electrical Vehicle. *IEEE Trans. Ind. Electron.* **2020**, *67*, 8231–8241. [CrossRef]
19. Hawkins, T.R.; Gausen, O.M.; Strømman, A.H. Environmental impacts of hybrid and electric vehicles—A review. *Int. J. Life Cycle Assess* **2012**, *17*, 997–1014. [CrossRef]
20. Dagci, O.H.; Peng, H.; Grizzle, J.W. Hybrid Electric Powertrain Design Methodology With Planetary Gear Sets for Performance and Fuel Economy. *IEEE Access* **2018**, *6*, 9585–9602. [CrossRef]
21. Kabalan, B.; Vinot, E.; Yuan, C.; Trigui, R.; Dumand, C.; Hajji, T.E. Efficiency Improvement of a Series-Parallel Hybrid Electric Powertrain by Topology Modification. *IEEE Trans. Veh. Technol.* **2019**, *68*, 11523–11531. [CrossRef]
22. Vora, A.P.; Jin, X.; Hoshing, V.; Saha, T.; Shaver, G.; Varigonda, S.; Wasynczuk, O.; Tyner, W.E. Design-space exploration of series plug-in hybrid electric vehicles for medium-duty truck applications in a total cost-of-ownership framework. *Undefined* **2017**, *202*, 662–672.
23. Lei, F.; Bai, Y.; Zhu, W.; Liu, J. A novel approach for electric powertrain optimization considering vehicle power performance, energy consumption and ride comfort. *Energy* **2019**, *167*, 1040–1050. [CrossRef]
24. Zhou, X.; Qin, D.; Hu, J. Multi-objective optimization design and performance evaluation for plug-in hybrid electric vehicle powertrains. *Appl. Energy* **2017**, *208*, 1608–1625. [CrossRef]
25. EcoCAR Mobility Challenge. Available online: https://avtcseries.org/ecocar-mobility-challenge/ (accessed on 8 September 2018).
26. Mi, C.; Masrur, M. *Hybrid Electric Vehicles: Principles and Applications with Practical Perspectives*, 2nd ed.; Wiley: West Sussex, UK, 2017.
27. MathWorks. Mapped SI Engine. Available online: https://www.mathworks.com/help/autoblks/ref/mappedsiengine.html (accessed on 2 July 2019).
28. MathWorks. Mapped Motor. Available online: https://www.mathworks.com/help/autoblks/ref/mappedmotor.html (accessed on 2 July 2019).
29. MathWorks. Equivalent Circuit Battery. Available online: https://www.mathworks.com/help/autoblks/ref/equivalentcircuitbattery.html (accessed on 2 July 2019).
30. Tran, M.-K.; Fowler, M. Sensor Fault Detection and Isolation for Degrading Lithium-Ion Batteries in Electric Vehicles Using Parameter Estimation with Recursive Least Squares. *Batteries* **2020**, *6*, 1. [CrossRef]
31. Tran, M.-K.; Mevawala, A.; Panchal, S.; Raahemifar, K.; Fowler, M.; Fraser, R. Effect of integrating the hysteresis component to the equivalent circuit model of Lithium-ion battery for dynamic and non-dynamic applications. *J. Energy Storage* **2020**, *32*, 101785. [CrossRef]
32. Zhang, H.; Wang, J. Adaptive Sliding-Mode Observer Design for a Selective Catalytic Reduction System of Ground-Vehicle Diesel Engines. *IEEE/ASME Trans. Mechatron.* **2016**, *21*, 2027–2038. [CrossRef]
33. Zhang, H.; Wang, J. Active Steering Actuator Fault Detection for An Automatically-steered Electric Ground Vehicle. *IEEE Trans. Veh. Technol.* **2016**, *66*, 3685–3702. [CrossRef]
34. Musardo, C.; Rizzoni, G.; Guezennec, Y.; Staccia, B. A-ECMS: An Adaptive Algorithm for Hybrid Electric Vehicle Energy Management. *Eur. J. Control* **2005**, *11*, 509–524. [CrossRef]
35. Asadi, B.; Vahidi, A. Predictive Cruise Control: Utilizing Upcoming Traffic Signal Information for Improving Fuel Economy and Reducing Trip Time. *IEEE Trans. Control Syst. Technol.* **2011**, *19*, 707–714. [CrossRef]

vehicles

Article

Tire Wear Reduction Based on an Extended Multibody Rear Axle Model

Jan Schütte * and Walter Sextro

Chair of Dynamics and Mechatronics, Faculty of Mechanical Engineering, Paderborn University, Warburger Str. 100, 33098 Paderborn, Germany; walter.sextro@uni-paderborn.de
* Correspondence: jan.schuette@uni-paderborn.de

Abstract: To analyze the influence of suspension kinematics on tire wear, detailed simulation models are required. In this study, a non-linear, flexible multibody model of a rear axle system is built up in the simulation software MSC Adams/View. The physical model comprises the suspension kinematics, compliance, and dynamics as well as the non-linear behavior of the tire using the FTire model. FTire is chosen because it has a separate tire tread model to compute the contact pressure and friction force distribution in the tire contact patch. To build up the simulation model, a large amount of data is needed. Bushings, spring, and damper characteristics are modeled based on measurements. For the structural components (e.g., control arms), reverse engineering techniques are used. The components are 3D-scanned, reworked, and included as a modal reduced finite element (FE)-model using component mode synthesis by Craig–Bampton. Finally, the suspension model is validated by comparing the simulated kinematic and compliance characteristics to experimental results. To investigate the interaction of suspension kinematics and tire wear, straight line driving events, such as acceleration, driving with constant velocity, and deceleration, are simulated with different setups of wheel suspension kinematics. The influence of the setups on the resulting friction work between tire and road is examined, and an exemplarily calculation of tire wear based on a validated FTire tire model is carried out. The results demonstrate, on the one hand, that the chosen concept of elasto-kinematic axle leads to a relatively good match with experimental results and, on the other hand, that there are significant possibilities to reduce tire wear by adjusting the suspension kinematics.

Keywords: elasto-kinematic; axle model; suspension kinematics; multibody simulation; tire wear

Citation: Schütte, J.; Sextro, W. Tire Wear Reduction Based on an Extended Multibody Rear Axle Model. *Vehicles* **2021**, 3, 233–256. https://doi.org/10.3390/vehicles3020015

Academic Editors: Markus Till, Ralf Stetter and Udo Pulm

Received: 8 March 2021
Accepted: 29 April 2021
Published: 18 May 2021

1. Introduction

Road traffic is a key source of dust, particulate matter, CO_2, NO_x, and other pollutant emissions [1–3]. With increasing traffic, the environmental pollution caused by cars also increases. In the face of this problem, electric mobility currently moves into the focus of politics, economy, and society. Electric vehicles do not emit any kind of exhaust emissions, so environmental pollution due to the exhaust is locally reduced. However, dust or particulate matter emissions due to tire wear remain unaffected or even increase in comparison with cars with combustion engines [4,5]. To reduce the tire wear emissions of a vehicle, other concepts are needed. Tire wear is one of the main sources of microplastics, which play a major role in water body pollution [6]. In a study by Fraunhofer Institute, it was found that tire wear constitutes the major source of microplastics in Germany [7]. As evident from Figure 1, 28% of the global releases of primary microplastics to the world oceans is tire wear [6]. The amount of tire wear released while driving a car depends on many different factors such as the rubber material, environmental conditions, and driving behavior. In particular, the interaction between suspension system and wheel has a large effect on tire wear [8]. The suspension system is the connection between the vehicle body and the road. Its main task is to guide the wheel on the road and support the associated forces. Owing to road unevenness and dynamic motion, like pitch, roll, or yaw, the tire is loaded with a combined force, comprising longitudinal, lateral, and vertical forces. On the tire side, these loads are supported in an almost postcard sized contact patch (footprint) between tire and road. On the other end of the suspension, the loads are transferred to the vehicle body. The transmission of longitudinal and lateral forces is characterized by the friction between the tire tread elements and the road surface. Hence, the positioning of the tire on the road is vital [9].

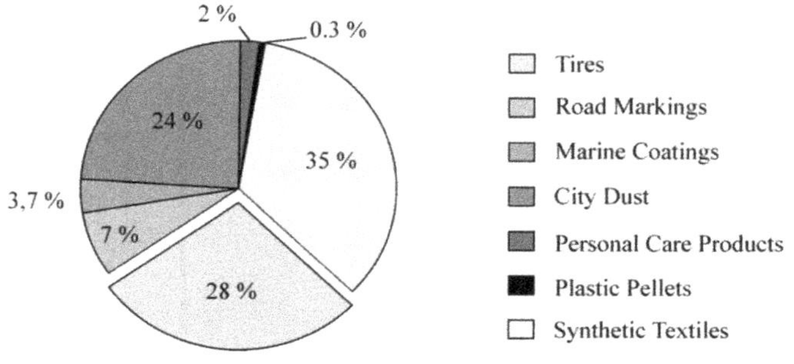

Figure 1. Global release of primary microplastics to the oceans according to [4].

Current concepts of independent suspensions consist of different linkages, which are connected to the wheel carrier on the wheel side and to the subframe on the other side. These arrangements of linkages result in a spatial wheel motion when a vertical displacement relative to the vehicle body or steering is applied. This means the wheel undergoes a motion, consisting of a translational and a rotational motion, which is defined by the number and position of the joints (kinematic points) between the suspension links [9–11]. These rigid-body motions are complemented by elastic deformations of structural components and rubber bushings. Elasticities change the kinematic behavior of the suspension system and, thereby, the wheel alignment, depending on operating forces and torques. These changes in wheel alignment, defined by the kinematics and compliance, are of particular importance for the driving behavior of a vehicle and the tire–road contact [9]. The spatial motion of the wheel affects the power transmission of the wheel, the friction process, and consequently the tire wear behavior owing to the relative sliding motion in the tire contact patch. Following this, the experimentally determined and calculated results presented in [12] show that it is possible to reduce tire wear by means of a new suspension concept.

To research tire wear caused by steady state or transient maneuvers, many papers dealing with analytical [13–17] and finite element (FE) approaches [18–20] have been published. The analytical wear models used are based on the physical phenomena of the wear process, considering the interaction between tire and road or test rig surface. These theoretical models can improve the understanding of relations between measurable tire state variables and tire wear. FE models are mainly used to optimize the tire itself. This includes adjustments of the structure of the tire, the tread pattern, or cross-section to obtain better tire characteristics, like a reduced wear rate or optimized stiffness. Furthermore, tire wear is often examined on test rigs at different quasi-static and dynamic loading conditions [21–24].

During the design process of a vehicle, the tire wear is usually not tested until a physical prototype is built up [12]. The tests are carried out either on real roads [13] or on a test track [25]. In the early design phase, simulation tools are commonly used to determine the effect of vehicle kinematics and forces on tire wear. In most cases, relatively simple vehicle models are used with detailed tire models, for example, FTire [26] or CDTire [27]. On the one hand, this approach neglects the complex nonlinear interactions between wheel and chassis, as explained in [12]. On the other hand, the changes in wheel alignment of a vehicle during driving are greatly affected by the elasticities of the suspension components, as mentioned above. Neglecting the interactions and elasticities of a suspension system leads to a significant deviation between prediction and reality. Tire wear depends not only on the magnitude of forces acting in the contact patch between tire and road, but also on the exact way they are generated [8,23]. This means the operating conditions of the tire should be simulated as realistically as possible [8], also considering the interactions with the subsystems suspension system and the road.

This paper presents an elasto-kinematic multibody rear axle modeling approach in MSC Adams/View, a widespread physics-based multibody simulation tool, to analyze the influence of the kinematics of wheel travel on friction work in the tire contact patch and tire wear. As a starting point, a non-steerable rear axle of a production car without any active elements is chosen. In passenger cars, many different axle concepts are present, depending on the major use case or price segment of the car [9]. In general, these concepts can be categorized into rigid axle, semi-rigid axle, and independent suspension systems. Rigid and semi-rigid axles do not have as many design possibilities as an independent suspension, so a multi-link independent suspension system is used.

The component models of the rear axle are parameterized by measurements and the resulting wheel suspension model is experimentally validated. To evaluate tire wear, two different methods are used in this study. First, under the assumption of a linear wear law according to Fleischer [28], the work done by friction in the tire contact patch is used to evaluate the tire wear. Secondly, the nonlinear wear law presented in [13,14] is used to determine the quantity and distribution of tire wear in the tire contact patch. Finally, the wheel suspension kinematics of the series axle are adjusted for the maximum reduction of tire wear during straight line driving at a constant velocity of 100 km/h. The adjustments stay inside the researched state-of-the-art of rear axle kinematics of wheel travel. The first results demonstrating the effect of toe and camber gradient adjustments at the design position on the overall work done by friction in the tire footprint have been published in [29]. The results were generated in a driving simulation with higher wheel load, which leads to approximately 20 mm positive wheel center displacement relative to the design position.

2. Rear Axle Model

As mentioned, the reference axle is a non-steerable independent rear suspension system—more precisely, an trapezoidal-link rear suspension of a production car. The modeling process is split into two main steps to guarantee verifiability at each step of the model design. At first, a multibody system model comprising only rigid bodies and ideal joints, which represent the kinematics, is built up and verified. Afterwards, the model is extended by additional rigid and elastic components to simulate elasto-kinematics and to consider all components of the axle system that influence the friction in the tire contact patch. Finally, the kinematics and compliance of the axle model are validated by comparing the simulated results to measured data.

2.1. Kinematic Model

The aim of the rigid body model is to verify the modeling of the kinematics of wheel travel, so that the following improvements are based on a correct model. Most of the components (toe link, camber link, wheel carrier, and subframe) can be modeled as homogenous cylinders, as shown in Figure 2. The trapezoidal link is included as a homogenous plate. The design of the components arises from the spatial positioning of the joints (kinematic points). The kinematic model consists of eleven parts, one subframe, two toe links, two camber links, two trapezoidal links, two wheel carriers, and two wheels. The wheels have no influence on the kinematics of the suspension and, at this stage, are considered as rigid dummy parts to better illustrate the changes in wheel alignment during vertical wheel center displacement on a test rig model.

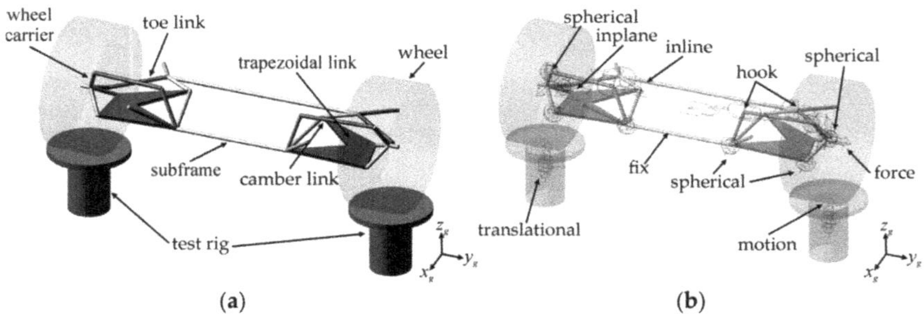

Figure 2. Kinematic model of the rear axle in Adams/View; (**a**) naming of parts and (**b**) positioning of the joints.

Subsequently, all parts are interconnected with joints to define the kinematics of the multibody system. For correct simulation of the kinematics of wheel travel, the correct selection of joint types is crucial. A single rigid body has six spatial degrees of freedom, three translational, and three rotational motions. By the joints, the degrees of freedom of the suspension system are reduced to two, the vertical wheel center displacement of both wheels. The number of degrees of freedom f is calculated by

$$f = 6n_{\mathrm{b}} - \sum_{\mathrm{i}=1}^{n_{\mathrm{j}}} (6 - f_{\mathrm{j,i}}). \tag{1}$$

In the formula, n_{b} is the number of rigid bodies, n_{j} is the number of joints, and $f_{\mathrm{j,i}}$ is the number of degrees of freedom of a specific joint i [30].

The subframe is attached to the ground by a fixed joint at the center of mass of the subframe. All other joints are positioned at the kinematic points. Toe- and camber link are attached to the wheel carrier by a spherical joint and to the subframe by a hook joint. The trapezoid link is attached to the subframe and the wheel carrier with two spherical joints at both rear kinematic points. For the second connection between trapezoidal link and wheel carrier, an inplane joint is used. This joint defines a parallel plane to the $x_{\mathrm{g}}, y_{\mathrm{g}}$-plane of the Adams ground coordinate system $(x_{\mathrm{g}}, y_{\mathrm{g}}, z_{\mathrm{g}})$. This means that the coupled points of the trapezoidal link and the wheel carrier can only move relative to each other in this plane. At the front inner side of the trapezoidal link, an inline joint is used to attach the trapezoidal link to the subframe. The inline joint only allows relative motion of the coupled bodies along a defined axis, and all rotations. The axis is oriented along the vertical direction z_{g} of the Adams ground coordinate system. The wheel dummies are fixed to the wheel carrier at the wheel center. All used joints are summarized in Table 1 and shown in Figure 2.

Table 1. Information to calculate the number of degrees of freedom of the kinematic axle model.

Number of Bodies:	$n_{\mathrm{b}} = 11$	
Number of joints:	$n_{\mathrm{j}} = 19$	
Degrees of freedom—fix joint	$f_{\mathrm{j,f}} = 0$	(3x)
Degrees of freedom—hook joint	$f_{\mathrm{j,h}} = 2$	(4x)
Degrees of freedom—spherical joint	$f_{\mathrm{j,s}} = 3$	(8x)
Degrees of freedom—inline joint	$f_{\mathrm{j,il}} = 4$	(2x)
Degrees of freedom—inplane joint	$f_{\mathrm{j,ip}} = 5$	(2x)

For analysis of the kinematics of wheel travel, the axle model is extended by a test rig model. This model consists of two rigid cylinders, one for each wheel. The test rig cylinders are attached to the ground by a translational joint in vertical direction z_{g}. This vertical degree of freedom is superimposed with a vertical motion, which can be defined individually for each wheel. The connections between the test rig and the wheels are realized by vertical springs between the test rig and the wheel center to lock only the vertical degree of freedom. During simulation, a translational sinusoidal motion is applied to the test rig and the wheel alignment is measured.

The measures of toe and camber angle and the displacement of the wheel center are implemented as described in [11]. Toe and camber angle are defined in ISO 8855 [31]. For model verification, the simulated and measured results of toe and camber angle during vertical wheel center displacement are plotted. Both curves are shown in Figure 3. The trend of both curves fits the measurements, which confirms the formal correctness of the model. At the same time, the results show that the kinematic model is not able to represent the measured kinematics of the real axle system.

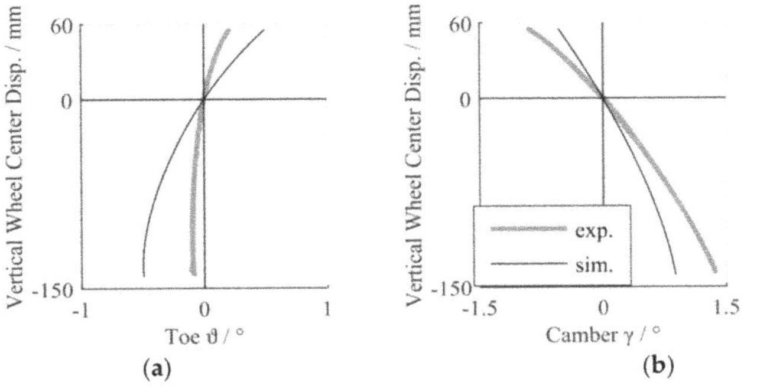

Figure 3. Comparison of experimental determined and simulated kinematics of the left wheel suspension of the kinematic model: (**a**) toe ϑ as a function of vertical wheel center displacement; (**b**) camber γ as a function of vertical wheel center displacement.

2.2. Elasto-Kinematic Model

To improve the match between the model and the measurements of the kinematics, as well as to enhance the model in view of the striven investigation, the axle model is extended by several additional components. Moreover, the elasticities of the rubber metal bushings and the structural components are integrated to realize the elasto-kinematic properties of the real axle in the model. Elasto-kinematics means the displacement of the wheels relative to the vehicle body and the changes in wheel alignment caused by longitudinal and lateral forces applied at the tire footprint or wheel center [31].

2.2.1. Spring System

At first, the spring system is integrated. The spring system consists of the suspension springs with two spring pads each, the vibration dampers with bump- and rebound stops, and the stabilizer. The spring system is relevant for ride comfort, handling, and the overall safety of a car. It has to protect the occupants from impacts due to road irregularities and undesirable vertical displacement as well as pitch and roll oscillations, and maintain an even contact between tire and road [9].

The spring of the trapezoidal rear axle is built as a coil spring and placed in front of the wheel center between two rubber spring pads, which are mounted to the vehicle body and the wheel carrier. The spring pads are modeled as a spring and damper in parallel. The spring characteristics are implemented through measured, nonlinear progressive spring characteristic curves. The normalized stiffness curve shape of both spring pads is illustrated in Figure 4a; the damping coefficient is constant in each case. The mass distribution is incorporated by rigid bodies. The coil spring can be modeled in different ways. One is to create a flexible body based on a CAD-model [12], while another way is to use a force element, as is done for the spring pads. In this study, the coil spring is modeled as a force element representing a spring with constant stiffness and no damping. The mass of the springs is represented by point masses.

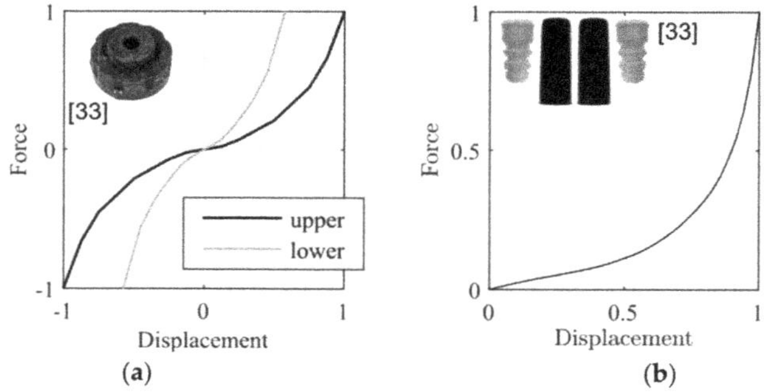

Figure 4. Normalized stiffness characteristic curves: (**a**) upper and lower spring pad; (**b**) bump stop.

In the real car, the damper is a telescopic monotube shock absorber, positioned behind the wheel center. In the axle model, the complex design of the damper is reduced to its main parts, the piston rod and main cylinder. The piston rod is mounted to the top mount, which is fixed to the vehicle body. The main cylinder is bolted to the wheel carrier by a rubber metal bushing. The piston rod and main cylinder are modeled as rigid bodies for proper mass distribution and are connected by an ideal joint. The velocity dependent damper force is nonlinear. Hence, a measured damper characteristic curve is used. The bump stop is coaxial to the piston rod and is mounted to the bottom of the top mount. It is implemented in the model as described in [11]. Its nonlinear force (Figure 4b) acts along the longitudinal direction of the damper. The rebound stop is integrated into the damper. As mentioned in [9], it is in most of the dampers. The modeling is the same as at the bump stop, but the stiffness is much more progressive. The position of the rebound stop was identified by comparing the measured overall suspension rate to the simulation results on a KnC test rig model.

A stabilizer is installed in almost every car, both at the front and the rear axle, and increases the roll stiffness without affecting the vertical spring rate [9]. When the car's body is rolling, for example, on cornering, the stabilizer twists, which results in a restoring moment about the roll axis. The center section of the stabilizer is oriented in the lateral direction of the car and is mounted to the subframe or body by two rubber elements. The stabilizer arms point in the longitudinal direction of the car and are mounted to the suspension system by links. The coupling point of the stabilizer system and the suspension system of the reference axle is at the trapezoid link. The stabilizer links have two vulcanized bushings each, in order to connect the stabilizer to the trapezoid link. To achieve a good representation of the stabilizer system in the model, its structural components are modeled as CAD-based flexible bodies. The stabilizer is tubular, which means the wall thickness defines its stiffness. The bushings are represented as general force elements in Adams.

2.2.2. Half-Shafts

Because the chosen reference axle is a driven rear axle, the drive- or half-shafts must be integrated into the model to apply the rotational motion, which normally comes from the differential. Moreover, the half-shafts affect the kinematics of wheel travel while driving [12]. The half-shafts system consists of the wheel hub, the wheel bearing, two constant velocity joints, and the half-shaft itself. The housing of the wheel bearing is bolted to the wheel carrier, which locks its relative translational degrees of freedom. The wheel hub is pressed into the wheel bearing. In the model, the housing of the wheel bearing is attached to the wheel carrier by a rotational joint, because its translational stiffness is very high and has no impact on kinematic and elasto-kinematic characteristics of the axle, as mentioned in [32]. The torsional stiffness and damping characteristics are integrated by force elements with embedded characteristic curves. The half-shafts are linked to the wheel hub and the subframe by ideal joints, so the differential side does not move relative to the subframe and the rotational speed of the half-shaft is equal to the wheel hubs. Wheel bearing, wheel hub, and half-shaft are modeled as rigid bodies. The masses of wheel hub and wheel bearing are included in the tire model. The mass and inertia of the half-shafts are specified on the basis of a CAD model.

2.2.3. Flexible Bodies

The elasticity of the structural components influences the driving behavior. As mentioned in [33], the on-load deformations of the axle components can induce displacements of the bushing positions (kinematic points) about several millimeters. This means the elasticities of the suspension links and the wheel carrier must be integrated into the model. The geometry and the properties of the material must be considered. Therefore, all suspension links and the wheel carrier are either 3D-scanned or reconstructed by SolidWorks and Catia V5. At this point, only the parts of the left wheel suspension are used to prepare CAD models, which are the basis of the flexible body models. Afterwards, the flexible body models are mirrored along the vehicle's x, z-plane of the ISO 8855 vehicle coordinate system. This is possible because all parts are symmetrical. Besides, it has the advantage of identical characteristics of the flexible bodies at the left and right wheel suspension. Components with complex geometries are 3D-scanned and rebuilt with the use of reverse engineering techniques and programs. The initial scan result is, in most cases, a faulty point cloud model composed of various, independent partial scans. This results in overlapping of the different meshes. Furthermore, each model contains additional errors like holes and the like, which are all remedied using the software tools Meshlab, Autodesk Meshmixer, Geomagic Control X, Materialise 3-Matic, and SolidWorks. After repair of the scan-based models, remeshing, and surface reconstruction, the suspension link and wheel carrier models are present as surface models. This step is exemplarily illustrated in Figure 5a,b.

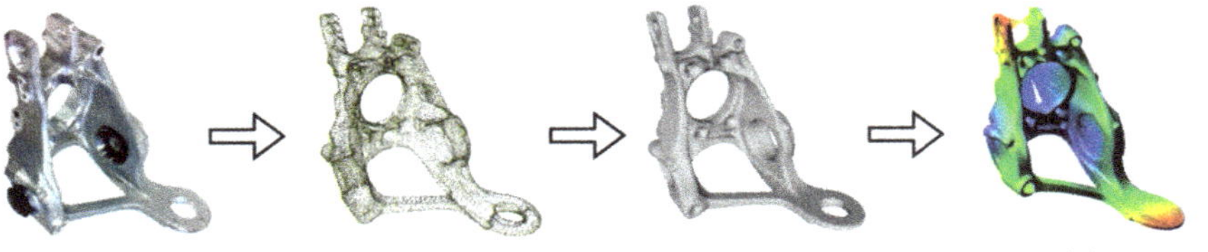

Figure 5. Exemplary process of flexible body creation of the left wheel carrier: (**a**) real part; (**b**) point cloud model; (**c**) final rigid body model; (**d**) flexible body model.

Using SolidWorks, these models are converted into solids. The bushings are cut out, and the bearing and bushing seats and screw holes are reworked to make sure the geometry of the model resembles the real parts as closely as possible (Figure 5c). Parts with simple geometry, like the stabilizer, stabilizer link, and toe link, were redesigned directly in SolidWorks.

The prepared rigid body models are imported into Adams/View and have their material properties assigned to achieve a correct mass distribution and an accurate stiffness of flexible bodies in the model. Most of the parts are made of metal. The name of the alloy used for the aluminum parts is imprinted on their surface, so density, Young's modulus, and Poisson's ratio are taken from material sheets. Stabilizers are often made of 26 MnB5 or 34 MnB5 [34], so the material data of this alloy are used for the stabilizer of the model. The stabilizer links are made of glass fiber reinforced polyamide PA 6.6 GF57, which means a glass fiber ratio of 57%. The orientation of the fibers is neglected. The used properties are listed in Table 2.

Table 2. Material properties of suspension parts.

Part	Material	Density/g/cm^3	Young's Modulus/GPa	Poisson's Ratio
Toe link	AlSi1MgMnT6	2.7	70	0.33
Camber link	AlSi1MgMnT6	2.7	70	0.33
Trapezoidal link	AlSi1MgMnT6	2.7	70	0.33
Wheel carrier	AlSi7Mg0,3	2.68	74	0.33
Stabilizer	26MnB5/34MnB5	7.85	210	0.30
Stabilizer link	PA 6.6 GF57	1.681	19.95	0.38

Following, Adams calculates mass, center of gravity, and moments of inertia of each part. Based on these rigid bodies, flexible bodies are generated using Adams/Flex, as can be seen in Figure 5d. Adams/Flex uses the FE-solver NASTRAN to generate modal reduced FE-models by the use of the Craig–Bampton method [35]. Thereby, the degrees of freedom of any finite element model are reduced to a combination of six degrees of freedom for each attachment point, six rigid body modes, and a user defined number of normal modes. For all parts, except for the trapezoidal link, tetrahedral elements with quadratic element order are used. The trapezoidal link is a hollow body, which could not be represented by the 3D-scan based CAD-model, so quadratic shell elements with a specific thickness are used. All related kinematic points of a part are defined as interface nodes, so that forces can be applied to the flexible body models. As an attachment method, compliant RBE3 is used to add no additional stiffness to the parts. After generation of the flexible bodies, all eigenfrequencies higher than 3 kHz were disabled to reduce the simulation time. The axle model with all flexible and rigid bodies is illustrated in Figure 6. The subframe and body are assembled of cylindrical rigid bodies. Mass and inertia are supplementarily defined.

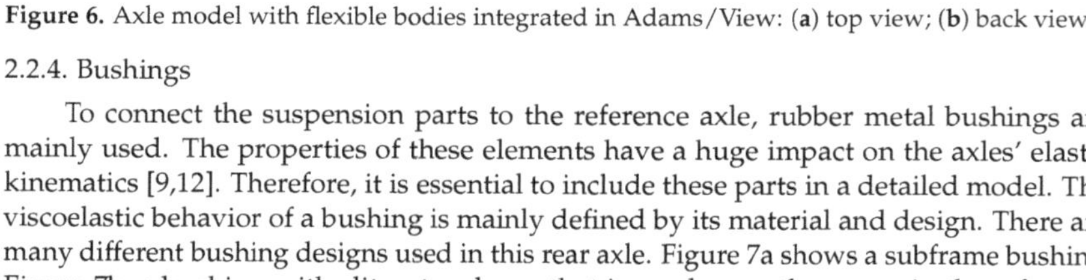

Figure 6. Axle model with flexible bodies integrated in Adams/View: (**a**) top view; (**b**) back view.

2.2.4. Bushings

To connect the suspension parts to the reference axle, rubber metal bushings are mainly used. The properties of these elements have a huge impact on the axles' elasto-kinematics [9,12]. Therefore, it is essential to include these parts in a detailed model. The viscoelastic behavior of a bushing is mainly defined by its material and design. There are many different bushing designs used in this rear axle. Figure 7a shows a subframe bushing, Figure 7b, a bushing with slit outer sleeve that is used more than once in the reference axle, and Figure 7c, the damper top mount to attach the damper to the vehicle body. In comparison with metallic components, rubber elements have some particularities thanks to their material properties.

The force–displacement curve of rubber shows a hysteresis during quasi-static deformation, which means energy dissipates due to friction between molecular chains and filler particles. Moreover, the bushing stiffness depends on the excitation frequency and amplitude, which is discussed in [37] in more detail. In this study, the viscoelastic bushing properties are modeled by Kelvin–Voigt elements, which consist of a spring in parallel to a viscous damper, as shown in Figure 8a. Kelvin–Voigt models with linear springs are the standard bushing models in Adams. Instead of the standard bushing model, a general force element is used to enable nonlinear spring characteristic curves. In Adams, general force elements are freely definable six-component force elements. In Figure 8b, the resulting frequency-dependent hysteresis of Kelvin–Voigt bushing models is shown. In the case of missing measurements for individual degrees of freedom of a bushing, properties of similar bushings are used and adjusted in the later validation process.

Figure 7. Different rubber bushing designs used on reference axle system: (**a**) subframe bushing; (**b**) bushing with slit outer sleeve; (**c**) damper top mount [36].

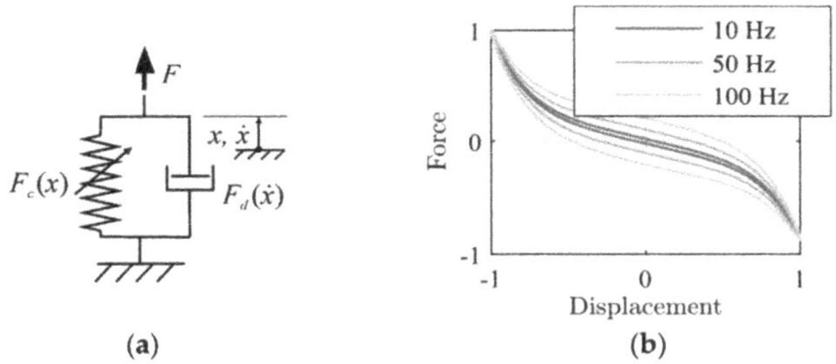

Figure 8. (**a**) Nonlinear Kelvin–Voigt element; (**b**) normalized frequency dependent hysteresis of Kelvin–Voigt element, simulated for 10, 50, and 100 Hz.

In Adams, it is not possible to directly connect flexible bodies to each other by bushings or joints. Therefore, rigid dummy bodies are used to join the flexible wheel suspension parts. These flexible body couplings are realized at each connection of a suspension link with the wheel carrier. Where it is needed, a dummy part is attached to a kinematic point of the wheel carrier with a fixed joint. The corresponding general force element realizes the connection between the dummy part and the flexible suspension link. In total, 32 general force elements are used in this model.

2.2.5. Mass and Inertia Distribution

Owing to the driving simulations envisaged in this study, the distribution of mass and inertia features is important for realistic results as well. It is essential to model the sprung and unsprung masses because they influence the dynamics of the axle system and, as a result, the dynamics at the tire contact patch. All rotating masses, except the half-shafts, such as tire, rim, brake disc, wheel bearing, hub, and bolts, are assigned to the FTire tire model. The half-shafts' mass and moments of inertia are defined as described before. The properties of the structural components are defined by the flexible body models. The bushings are weighed, and their mass is assigned to the dummy parts described in the previous section. The masses of the coil spring, brake caliper, and a proportional body mass are implemented as point masses. The moments of inertia of the subframe are derived from a CAD model and along with its mass assigned to the rigid body model. The body mass is attached to the almost massless rigid body chassis model, which can be seen in Figure 6. The point mass is positioned at the calculated height of vehicles' center of gravity and in the middle of a line between both wheel centers. The height of the center of gravity of the whole vehicle is calculated by the approximation formula from [38]. The absolute value of the body mass corresponds to the loading condition and axle load distribution of the vehicle.

2.2.6. Validation of Elasto-Kinematics

To validate the kinematics and compliance of the rear axle model, simulation results are compared to measurements, as shown in Figures 9 and 10. The results are presented for the left wheel suspension only, as the graphs are also applicable to the right side because of the axles' symmetry. To minimize the differences between the measured and simulated characteristics, finally, certain bushing parameters are adjusted. These initial deviations are normal, because measurements always underlie inaccuracies, and the properties of the rubber metal bushings scatter a lot from one another.

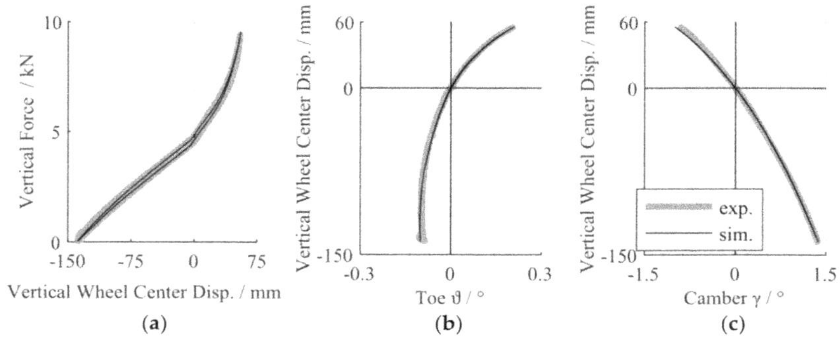

Figure 9. Comparison of experimental determined and simulated kinematics of the left wheel suspension of the elasto-kinematic model: (**a**) vertical wheel center displacement as a function of vertical force; (**b**) toe ϑ as a function of vertical wheel center displacement; (**c**) camber γ as a function of vertical wheel center displacement.

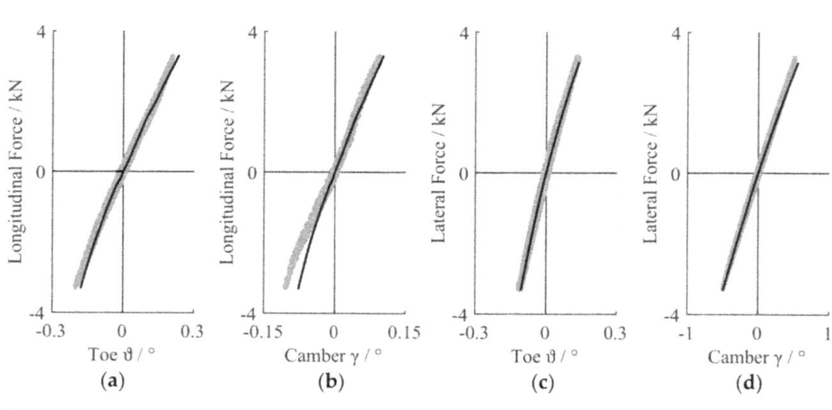

Figure 10. Comparison of experimentally determined and simulated elasto-kinematics of the elasto-kinematic model of the left wheel suspension: (**a**) toe ϑ as a function of longitudinal force; (**b**) camber γ as a function of longitudinal force; (**c**) toe ϑ as a function of lateral force; (**d**) camber γ as a function of lateral force.

Figure 9 shows the validation results of vertical wheel center displacement. In all cases, zero wheel travel is defined as the design position of the axle. In Figure 9a, the overall suspension rate is plotted. Both the linear stiffness of the suspension system at negative wheel travel and progressive stiffness at positive wheel travel, which is caused by suspension spring and bump stop in series when the bump stop hits the top of the damper tube, show a good conformity. Figure 9b,c shows measured and simulated toe and camber angle change during vertical wheel center displacement. Compared with the results of the kinematic model (Figure 3), a significantly better match is achieved.

The elasto-kinematic toe and camber angle changes, specified by the compliance of the axle, are shown in Figure 10. To determine these graphs, longitudinal and lateral forces are applied isolated in the tire contact patch and the angles of wheel alignment are measured as described in [39].

A relatively good correlation between measurements and simulation results is obtained. Only the camber change caused by a lateral force in the tire contact patch deviates for negative forces around three hundredths of a degree. This could be caused by a difference in bushing stiffness or a bushing preload that is not considered in the model. However, this small deviation should not noticeably affect the simulation results.

3. Tire and Road Model

3.1. FTire Tire Model

For this study, an experimentally parameterized and validated FTire tire model is used. FTire is a fully nonlinear 3D tire structural deformation model that covers frequencies up to 250 Hz. It is capable of covering all road irregularities, even with extremely short wavelength. The model basically comprises a structural and a brush-type dynamic tread-road contact model. Additionally, there are different model extensions like a thermal model, a tread wear model, an air volume vibration model, and a flexible and viscoplastic rim model [26,40] that are not used in this study.

The structural model of the tire is based on a structural dynamics-based tire modeling approach. It defines the tire belt as an extensible and flexible ring. The tire ring consists of a finite number of so-called belt segments with mass that are coupled to the rim on one side and to their direct neighbors by nonlinear, pressure-dependent spring-damper elements. In this context, the number of belt segments defines the level of detail of the mechanical model [40]. Normally, 90 to 360 belt segments are used [41].

The contact between tire and road is realized by a number of mass-less contact elements (tread blocks) that are assigned to every belt segment. Usually, 5 to 50 tread blocks are used per belt segment. The tread blocks are arranged in tread strips on the tire belt. This arrangement of tread strips is exemplarily shown in Figure 11.

Figure 11. Example illustration of the arrangement of brush-type contact elements (tread blocks) in lines (tread strips) on the tire belt of the FTire model.

The tread blocks carry nonlinear spring and damper characteristics in radial, tangential, and lateral directions. On calculation of the contact forces between tire and road, FTire initially checks whether or not a tread block is in contact with the road model. For tread blocks that are in contact with the road model, first an individual contact plane and then the contact forces are calculated. More information on the FTire model is given in [40–42].

A tire of size 245/40ZR18 with an inflation pressure of 270 kPa was used. The parameterization and validation of the FTire model dataset were carried out by fka mbH Aachen, Germany for this series of tire types. For FTire modeling, several tire test rigs, as shown in Figure 12, were used, and measurements such as tire footprints, tests in static conditions, tests in steady state rolling conditions, and cleat tests were performed.

Figure 12. Tire test rigs used for experimental investigation of tire properties to parameterize and validate the FTire tire model.

An overview of the FTire model parameters is given in Table 3. The model uses 88 belt segments with 44 tread blocks each. This results in 3872 tread blocks arranged in 22 lines (tread strips) along the tire circumference.

Table 3. Tire and FTire model parameters used in this study.

Size	245/40ZR18
Inflation pressure	270 kPa
Number of belt segments	88
Number of tread blocks per belt segment	44
Number of tread strips	22

3.2. Road Model

In Adams, it is possible to use several different road models that are integrated into the simulation model by a road property file. In this study, two-dimensional models with stochastic road irregularities from the Adams library are used. The roughness of these road models is defined by the spectral density of the road irregularities, which results in a good accordance between the model and real road [43].

In ISO 8608 [44], road unevenness is classified into groups from A to H. Road class A is the smoothest, with very small irregularities. A road from class H is an extremely bad road, which normally cannot be found in Europe. The waviness of all road models is 2, which is a good assumption with regard to the measurements in [45].

4. Tire Wear

The total friction force in the contact patch between tire and road comprises different individual parts. These individual forces do not need to take effect simultaneously [46] and rest on physical effects, like adhesion, hysteresis, cohesion, and viscous friction [47]. The biggest contribution to rubber friction stems from adhesion and hysteresis forces. Over the years, many different empirical wear laws were developed. More information and a deeper insight in wear processes with regard to local wear of tire tread blocks can be found in [48,49].

4.1. Linear Wear Law

In the first step of this study, Fleischer's wear law [28] is taken as a basis. This means wear is described as a result of the friction process with a proportional relation between wear volume V_{w} and friction work W_{f}. Using a proportionality factor e_{f}^{*}, the wear can be described by

$$V_{\mathrm{w}} = e_{\mathrm{f}}^{*} W_{\mathrm{f}}. \tag{2}$$

Because of this direct proportionality, the friction work W_{f} is used to describe the tire wear behavior of a tire. The friction work of each contact element (FTire tread block) of the tire model is calculated based on the results of Ftire. The contact elements of the Ftire model are arranged along lines on the tire belt with uniform distance to each other. These contact elements are not observable over time, because they do not have a unique identifier.

For each contact element i and time step n, the individual instantaneous contact forces $F_{\text{fx},i,n}$, $F_{\text{fy},i,n}$ and sliding velocities $v_{\text{sx},i,n}$, $v_{\text{sy},i,n}$ are calculated. With these parameters, the instantaneous power loss due to road friction $P_{\text{f},i,n}$ of each contact element i and time step n is calculated by

$$P_{\text{f},i,n} = F_{\text{fx},i,n} v_{\text{sx},i,n} + F_{\text{fy},i,n} v_{\text{sy},i,n}. \tag{3}$$

The total instantaneous friction power in the tire footprint at time step n is the sum of all contact elements:

$$P_{\text{f},n} = \sum_{i=1}^{k} P_{\text{f},i,n}, \tag{4}$$

provided that k is the number of contact elements that are in contact with the road model at time step n. For the calculation of the work done by friction $W_{\text{f},n}$ with a constant calculation step size Δt,

$$W_{\text{f},n} = P_{\text{f},n} \Delta t \tag{5}$$

is effective.

For evaluation of the friction work for a defined driving maneuver, the friction work W_f for a simulation time period T is defined as

$$W_\text{f} = \frac{T}{m} \sum_{n=1}^{m} P_{\text{f},n}. \tag{6}$$

In this formula, m is the number of simulation steps with a constant simulation step size.

For a better comparison and presentation of the results, the friction work W_f is normalized by the maximum of each analysis of an individual driving maneuver $W_{\text{f,max}}$. The normalized friction work is referred to as W_f^*.

4.2. Nonlinear Wear Law

In this study, a nonlinear local tire wear law based on literature data is used to calculate the amount of tire wear generated by the rear axle simulation model. The used wear law was published in [13,14] for two different unspecified tire compounds. The local wear law is derived from experimental results at 70 °C rubber bulk temperature on an abrasive surface with an equal mean micro-texture to measured data on real road tracks. It presents the relation between power loss due to road friction per unit of contact area and mass loss per unit of contact area. This relation is plotted in Figure 13.

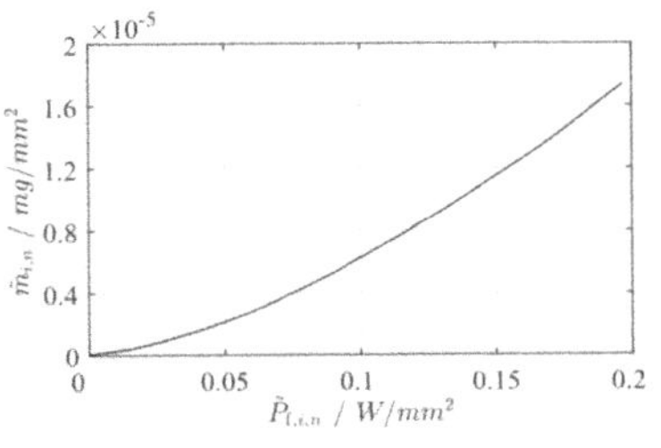

Figure 13. Local wear law of a tire tread compound at 70 °C rubber bulk temperature and a surface micro-texture having a mean wavelength of $2.3 \cdot 10^{-4}$ m according to [14].

Therefore, the instantaneous friction power $P_{\text{f},i,n}$ of an individual contact element i at time step n, calculated by the use of the FTire contact model, is divided by the present contact area $A_{i,n}$ of the contact element, which results in the area-specific instantaneous friction power:

$$\widetilde{P}_{\text{f},i,n} = \frac{P_{\text{f},i,n}}{A_{i,n}}. \tag{7}$$

The relation between instantaneous friction power per unit of contact area $\widetilde{P}_{f,i,n}$ and wear mass per unit of contact area $\widetilde{m}_{i,n}$ can be approximated as

$$\widetilde{m}_{i,n} = k_1 \widetilde{P}_{f,i,n}^{\,k_2} \tag{8}$$

as described in [13,14]. Both constants k_1 and k_2 characterise the wear behavior of the rubber compound at a specific temperature on a specific surface.

5. Simulation and Results

The final simulation model of the rear axle system is shown in Figure 14. It comprises 30 rigid bodies, 11 flexible bodies, 5 point masses, 41 ideal joints, 2 motions, and 2 FTire tire models, which results in 217 degrees of freedom. The rigid body part is connected to the ground to exclude pitch, yaw, and translational motion in the lateral direction.

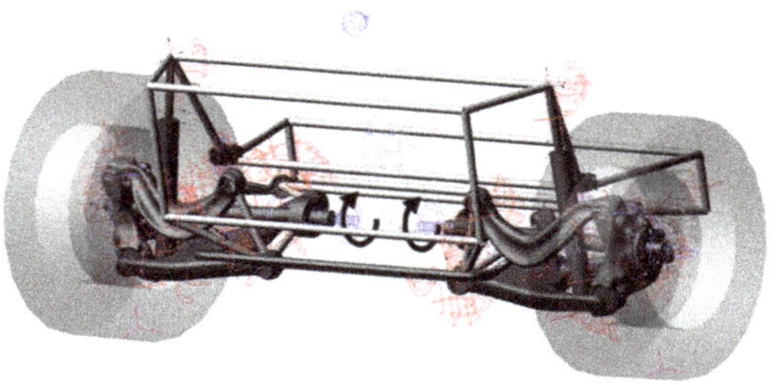

Figure 14. Final simulation model of the rear axle system in Adams/View.

For the simulations, a combination of three different driving maneuvers is defined. Because a single axle is being simulated, only straight-line driving events are considered. The first five seconds are the settle time of the model. After that, a rotational motion $\vec{\omega}$ is applied to the differential side of the half shafts, as shown in Figure 14. The acceleration phase lasts ten seconds until the predefined longitudinal velocity of 100 km/h is reached. After ten seconds of driving with a constant velocity of 100 km/h, the deceleration phase begins. During deceleration, the applied rotational motion at the drive shafts is continuously reduced. The simulation model slows down within ten seconds until it stops, and the simulation ends. This velocity curve is used for all following studies. For evaluation, the driving maneuver is divided into its three load cases→.

5.1. Variation of Initial Wheel Alignment

At first, the influence of the initial toe and camber angles on the friction work in the tire footprint on a two-dimensional road model of ISO 8608 class C is analyzed. Therefore, the initial angles are varied separately, and the simulation described above is conducted. Starting at the series initial toe angle $\vartheta_{0,\,\text{series}}$, eleven different initial toe angle settings are examined. The maximum and minimum limits are defined by

$$\vartheta_{0,\text{min}} = \vartheta_{0,\text{series}} - 3 \cdot |\vartheta_{0,\text{series}}| \tag{9}$$

$$\vartheta_{0,\text{max}} = \vartheta_{0,\text{series}} + 3 \cdot |\vartheta_{0,\text{series}}|. \tag{10}$$

The initial camber angle γ_0 is kept constant. The initial wheel alignment of each setup is listed in Table 4.

Table 4. Initial toe and camber angles at variation of the initial toe angle.

Suspension Setup	Initial Toe ϑ_0	Initial Camber γ_0
1	$-20'$	$-1°\ 20'$
2	$-14'$	$-1°\ 20'$
3	$-8'$	$-1°\ 20'$
4	$-2'$	$-1°\ 20'$
5	$4'$	$-1°\ 20'$
6 (series)	$10'$	$-1°\ 20'$
7	$16'$	$-1°\ 20'$
8	$22'$	$-1°\ 20'$
9	$28'$	$-1°\ 20'$
10	$34'$	$-1°\ 20'$
11	$40'$	$-1°\ 20'$

For evaluation of the influence of the different suspension setups on the total friction work W_f in the tire contact patch, the normalized friction work W_f^* is plotted as a function of suspension setups in Figure 15.

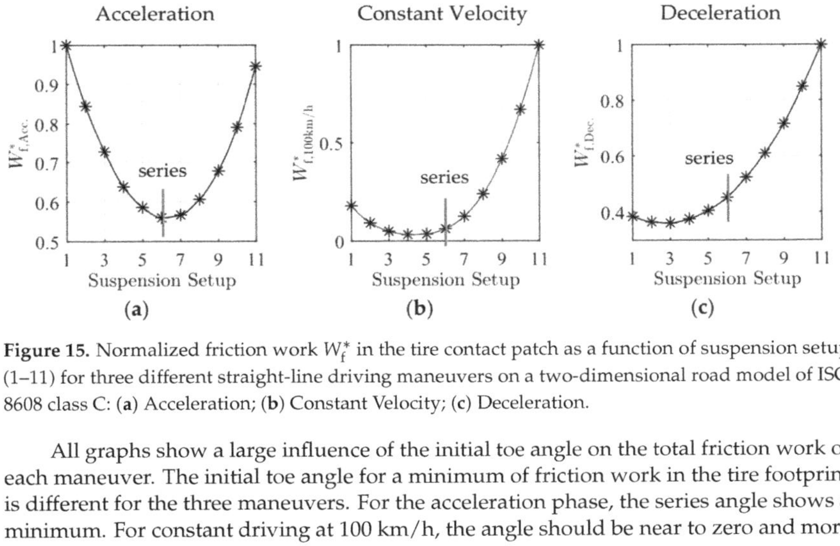

Figure 15. Normalized friction work W_f^* in the tire contact patch as a function of suspension setup (1–11) for three different straight-line driving maneuvers on a two-dimensional road model of ISO 8608 class C: (**a**) Acceleration; (**b**) Constant Velocity; (**c**) Deceleration.

All graphs show a large influence of the initial toe angle on the total friction work of each maneuver. The initial toe angle for a minimum of friction work in the tire footprint is different for the three maneuvers. For the acceleration phase, the series angle shows a minimum. For constant driving at 100 km/h, the angle should be near to zero and more negative for deceleration.

The initial camber angles are also varied. Because of the larger absolute value of the series camber angle $\gamma_{0,\text{series}}$, the angles are reduced and increased by double the absolute value, as follows:

$$\gamma_{0,\text{min}} = \gamma_{0,\text{series}} - 2 \cdot |\gamma_{0,\text{series}}| \tag{11}$$

$$\gamma_{0,\text{max}} = \gamma_{0,\text{series}} + 2 \cdot |\gamma_{0,\text{series}}|. \tag{12}$$

In total, eleven suspension setups (12–22) with equal steps in camber angle variation are simulated. The initial wheel alignment of each setup is listed in Table 5.

The simulation results are shown in Figure 16. The results are again normalized by the maximum value of overall friction work of each driving maneuver and plotted as a function of the suspension setup. The initial camber angle also has a big influence on the friction work in the tire footprint. A small absolute value of initial camber angle of the axle model leads to small friction work in all examined driving events.

Table 5. Initial toe and camber angles at variation of the initial camber angle.

Suspension Setup	Initial Toe ϑ_0	Initial Camber γ_0
12	10′	1° 20′
13	10′	48′
14	10′	16′
15	10′	−16′
16	10′	−48′
17 (series)	10′	−1° 20′
18	10′	−1° 52′
19	10′	−2° 24′
20	10′	−2° 56′
21	10′	−3° 28′
22	10′	−4°

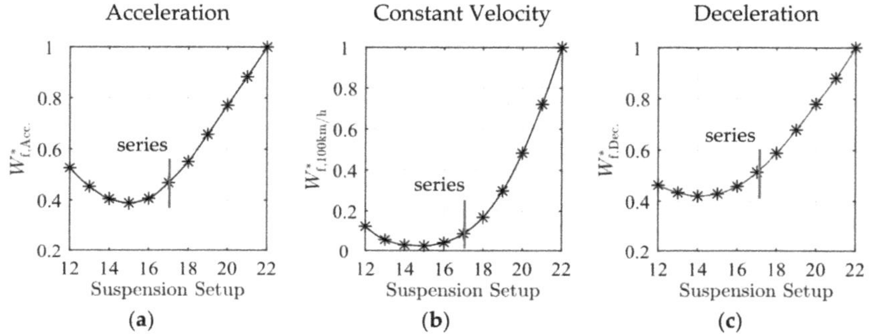

Figure 16. Normalized friction work W_f^* in the tire contact patch as a function of suspension setup (12–22) for three different straight-line driving maneuvers on a two-dimensional road model of ISO 8608 class C: (**a**) Acceleration; (**b**) Constant Velocity; (**c**) Deceleration.

5.2. Variation of Kinematics of Wheel Travel

For variation of the kinematics of wheel travel, first, a research of the state-of-the-art of the kinematics of wheel travel from actual cars is carried out and presented in Figure 17. The graphs show the change in toe and camber angle during in phase vertical wheel center displacement. There is a difference in the gradient of angle change, but the trend of the curves is the same. In both graphs, no initial angle is considered.

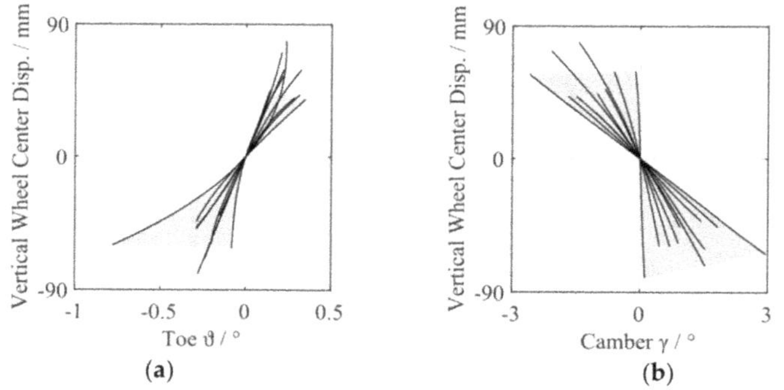

Figure 17. (**a**) Toe ϑ and (**b**) camber γ change as a function of vertical wheel center displacement during in phase vertical wheel center displacement of ten different rear axle systems according to [50–53].

For analysis of the influence of different suspension kinematics on the friction work in the tire footprint and the tire wear, the kinematics of wheel travel of the axle model are changed [29]. The changes are conducted by adjusting the position of several kinematic points. The various resulting suspension setups with different toe and camber gradients are shown in Figure 18.

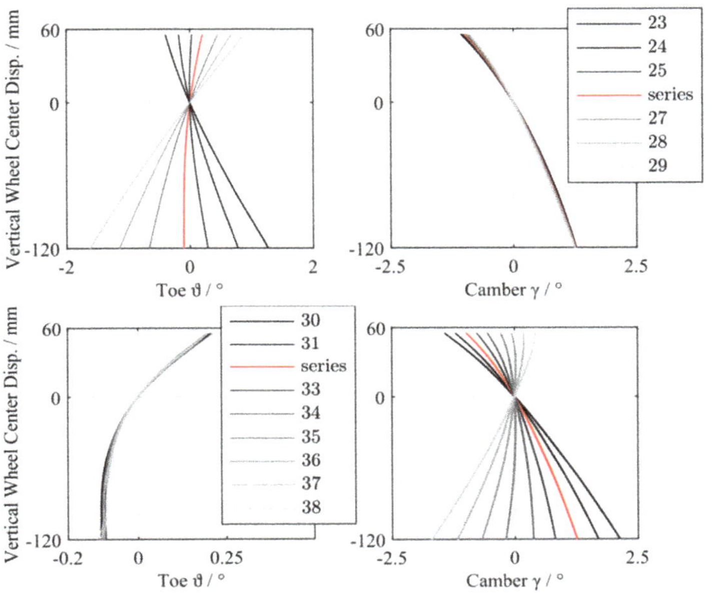

Figure 18. Fifteen different wheel suspension setups (23–38) defined by changing toe and camber gradient [29].

To get an isolated view of the impact of each angle gradient change, the changes are implemented separately. During variation of the toe angle gradient, the camber angle gradient stays almost the same. The same applies if the camber gradient is changed. For all simulations with these different suspension setups, the initial toe and camber angles of the production car were maintained. Because the mass of the vehicle body is not changed, the changes in suspension kinematics result in a different displacement of the wheel center relative to the vehicle body. This is compensated by adjusting the preload of the spring, so that the relative position of wheel center and body is equal for all simulations. All simulations are conducted on four different road classes defined by ISO 8608 [44], which results in 64 simulations.

First, the influence of the kinematic toe change in combination with a negative camber gradient (setup 23–29) on the overall friction work in the contact patch between tire and road is evaluated. The simulation results for all three maneuvers are plotted in Figures 19 and 20. For a better comparison, in Figure 19, the scaling of the axis is equal for all plots. In all plots, the resulting total friction work is normalized by the maximum value of friction work on each road.

The influence of the suspension setup is higher at a constant velocity than at the other maneuvers. Additionally, it is noticeable that, with a higher positive toe gradient, the friction work decreases. In contrast to the results of driving at a constant velocity, for acceleration and deceleration, a negative toe gradient seems to be better regarding friction work.

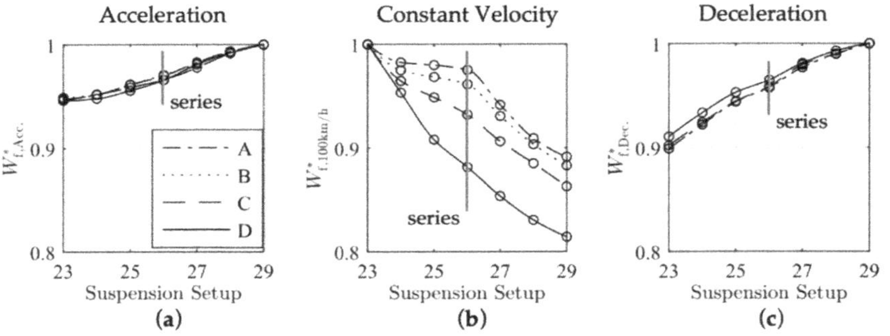

Figure 19. Normalized friction work W_f^* in the tire contact patch as a function of suspension setup (23–29) with different toe gradient for three different straight-line driving maneuvers on four different two-dimensional road models with ISO 8608 class A–D: (**a**) acceleration; (**b**) constant velocity; (**c**) deceleration.

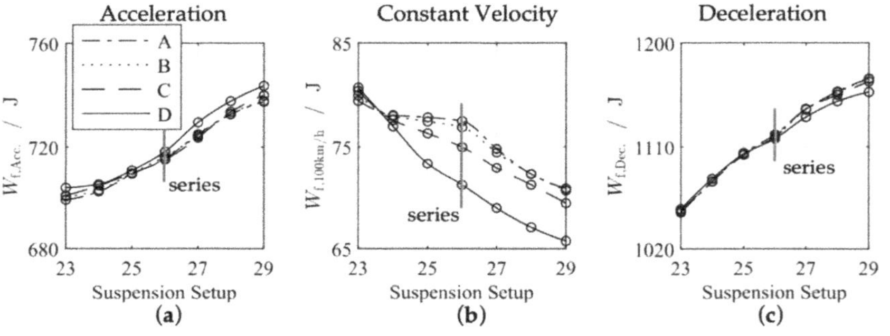

Figure 20. Friction work W_f in the tire contact patch as a function of suspension setup (23–29) with different toe gradient for three different straight-line driving maneuvers on four different two-dimensional road models with ISO 8608 class A–D: (**a**) acceleration; (**b**) constant velocity; (**c**) deceleration.

The influence of the road models with different spectral unevenness is most pronounced at the constant velocity simulation at 100 km/h. The influence at acceleration and deceleration is low. This can be explained by the different scale of the absolute values of the performed friction work in the three phases of simulation and the influence of the road (Figure 20). For constant velocity, the impact of toe gradient on the friction work increases with the unevenness of the road model starting from road class A. The amount of work done by friction significantly varies from maneuver to maneuver. The lowest friction work is done at steady state straight-line driving at 100 km/h, and the most at deceleration phase. At acceleration, the work done is almost ten times higher than at constant velocity.

Afterwards, the influence of the kinematic camber change in combination with a positive toe gradient is examined. The results are pictured in Figure 21. Again, the camber gradient produces the biggest change in friction work at a constant velocity of 100 km/h. Overall, it has to be noted that a more negative camber angle gradient in combination with a positive toe angle gradient causes a reduction of friction work in the tire footprint. During the deceleration phase, the effect is very small. The influence of the road unevenness on the results is also small during the acceleration and deceleration phases. At a constant velocity of 100 km/h, there is a significant dependence of the calculated results on the ISO 8608 road class.

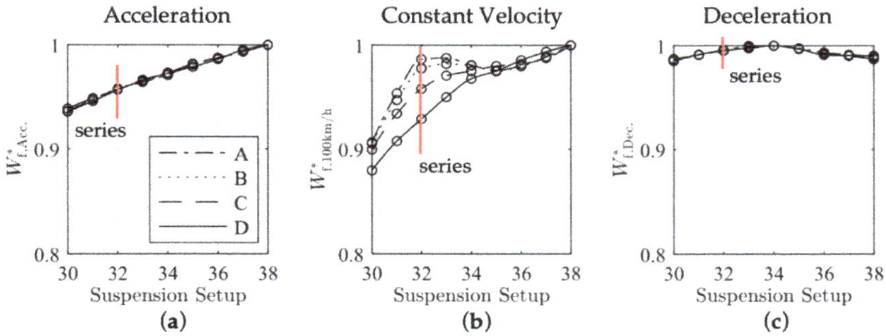

Figure 21. Normalized friction work W_f^* in the tire contact patch as a function of suspension setup (30–38) with different camber gradient for three different straight-line driving maneuvers on four different two-dimensional road models with ISO 8608 class A–D: (**a**) acceleration; (**b**) constant velocity; (**c**) deceleration.

For an exemplary calculation of the amount of tire wear, an average mileage of 40,000 km for a car tire is presumed. During its lifetime, a tire loses about 1 kg through wear on average [2]. The wear law used in [14] applies only to stationary operation conditions as present in the test rig. Regardless, this wear law is used as an example for an exemplary calculation of tire wear.

For evaluation of the saving potential, only the constant straight line driving event with 100 km/h on an ISO 8608 road model with unevenness of class C is used. The simulated results are projected to a total distance of 40,000 km to be able to make a statement regarding the lifetime of the tire. The calculation of the amount of tire wear is done with Equation (8) for all suspension setups with different kinematics (setup 23–38). In Figure 22, only the setups with minimum and maximum tire wear of the left tire are shown.

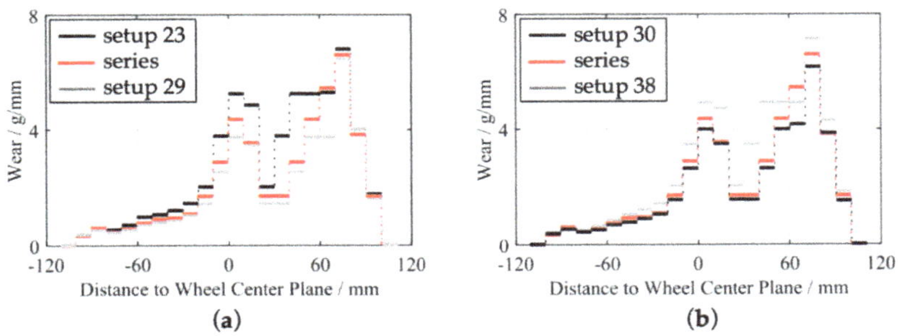

Figure 22. Comparison of tire wear distribution relative to the wheel center plane: (**a**) for three different suspension setups (23, series, and 29) at a constant velocity of 100 km/h on an ISO 8608 road model of class C; (**b**) for three different suspension setups (30, series, and 38) at a constant velocity of 100 km/h on an ISO 8608 road model of class C.

Figure 22 shows the results for two suspension setups with different toe gradients and two suspension setups with different camber gradients in comparison with the results of the series suspension setup. On the horizontal axis, the distance to the wheel center plane is used to show the tires' lateral width. The vertical axis indicates the tire wear of each strip of the FTire contact model. The horizontal lines represent the amount of tire wear in g per mm width of the tire tread. The spatial discretization of the FTire model, described in Section 3.1, results in a tread strip width of about 10.32 mm. The total tire wear is the integral over the tire width. Both plots show more wear at the side of the tire that is oriented to the center of the vehicle. This is probably caused by the initial positive toe and negative camber angles.

Using the local wear law shown in Figure 13, suspension setup 23 leads to an increased amount of tire wear of 575 g in comparison with the series setup with 470 g. Suspension setup 29 has the highest positive toe gradient and shows the minimum tire wear. The amount of tire wear is reduced by 51 g by making the toe gradient more positive, which results in a reduction of more than 10% compared with the series suspension setup. Figure 22b shows that changing the camber gradient leads to a similar distribution of tire wear as the adapted toe gradients. With 560 g, suspension setup 38 results in the highest amount of tire wear. In comparison, suspension setup 30 results in 428 g, which means a reduction of 42 g or at least 9% compared with the series setup.

For further reduction of the tire wear during the described steady state straight line driving maneuver, both setups with the highest reduction (23 and 30) are combined, which means the toe gradient of the rear axle is changed as well as the camber gradient. This results in a significantly higher reduction of 58.1% in comparison with the series setup, which means a total tire wear of 197 g. However, the toe gradient does not fit the state-of-the-art shown in Figure 17 anymore. Therefore, the toe gradient is adjusted to the positive limit of the state-of-the-art's toe gradients. The lateral distribution of tire wear is shown in Figure 23. With 204 g total tire wear, the produced quantity of tire wear of one wheel during a 40,000 km straight line ride with 100 km/h is reduced by 266 g (56.6%) in comparison with a wheel on a series axle.

The toe and camber angle changes during vertical wheel center displacement of the adjusted suspension setup are shown in Figure 24 in comparison with the state-of-the-art from Figure 17. Both curves lie inside the spanned area of the design of the kinematics of wheel travel from actual cars. Hence, it is understood that the resulting vehicle dynamics also stays inside the merchantable range.

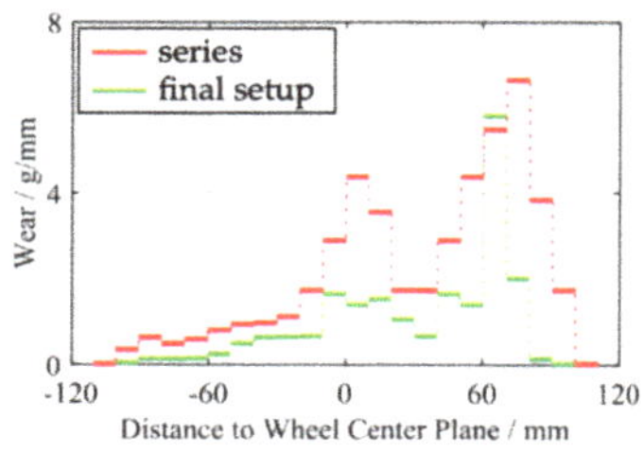

Figure 23. Comparison of tire wear distribution relative to the wheel center plane of the series and the final adjusted suspension setup at constant velocity.

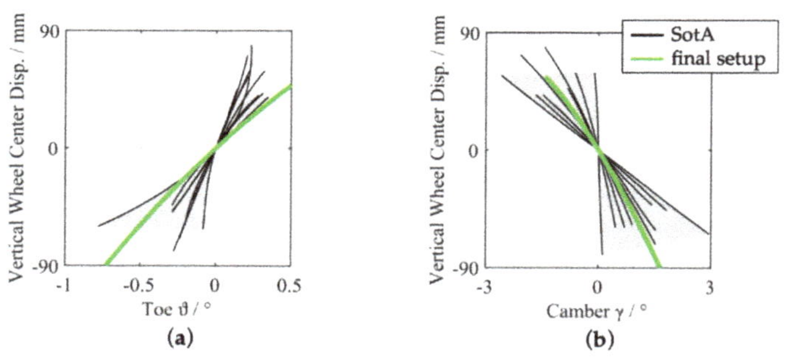

Figure 24. Kinematics of wheel travel (**a**) toe ϑ and (**b**) camber γ change as a function of vertical wheel center displacement during in phase vertical wheel center displacement of ten different rear axle systems according to [50–53] state-of-the-art (SotA) in comparison with the final setup.

6. Conclusions

The results of the exemplary tire wear calculation of setups, which leads to reduction of tire wear, are listed in Table 6. Furthermore, the tire wear reduction relative to the serial setup is shown. The final and proposed setup leads to a reduction of 56.6% tire wear while straight-line driving. This means, with the new suspension setup, the tires can be used for twice as long as on the series setup. In comparison with the experimental and simulation results of [12], it can be concluded that the identified design of suspension kinematics with the lowest tire wear is completely different. However, the reduction potential identified has equal dimensions.

Table 6. Calculated amount of tire wear for different suspension setups at straight-line driving with a constant velocity of 100 km/h.

Suspension Setup	Tire Wear/g	Tire Wear Reduction/%
Series	470	-
Setup 29	419	10.9
Setup 30	428	8.9
Final Setup	204	56.6

Based on the assumptions made and simulations done, it can be concluded that, owing to a modified suspension design process considering tire wear as the main objective, the tire wear of a car can be reduced significantly. In combination with new tire concepts with low wear, this could be a major step towards less environmental pollution.

7. Summary and Future Work

In this paper, a detailed flexible multibody simulation model of a rear axle system is built up to investigate the interactions between wheel suspension kinematics and tire wear. The modeling process was divided into two main steps, which leads to a verifiability of the model in each step. All components, which affect the dynamics in the tire contact patch, are considered, implemented, and validated on the experimental results. The elasticities of the suspension links and the wheel carrier are integrated into the model as flexible bodies. The flexible bodies are based on real geometry, derived by reengineering techniques and 3D scanning. The viscoelastic bushing characteristics are represented by Kelvin–Voigt elements with nonlinear spring stiffness. The kinematics and compliance of the final axle model are validated by measurements. For analysis of the friction process in the tire footprint while driving, the simulation model is enhanced by two experimental parameterized and validated FTire tire models and a road model with stochastic road irregularities classified by ISO 8608.

The evaluation of tire wear is conducted by the use of a linear wear law and, exemplarily, an experimentally determined nonlinear local wear law. Simulations of the three driving maneuvers, acceleration, driving with constant velocity, and deceleration, for different suspension setups were carried out. At first, the initial toe and camber angles were changed to identify their impact on the overall friction work between tire and road. The kinematics of wheel travel were changed separately, which means the toe and camber gradients were changed by adjusting the positions of kinematic points respectively without affecting each other. In this way, an isolated investigation of the influence of wheel alignment was given. Finally, the distribution of mass loss of tire tread material per unit contact width was presented for five different suspension setups for straight line driving at a constant velocity. This is the basis for a final adjustment of the suspension kinematics to achieve the largest reduction of tire wear under consideration of the compiled state-of-the-art in suspension kinematics, which leads to a significant reduction of tire wear.

It is obvious that a single axle simulation model cannot represent a vehicle as a whole. Therefore, the simulation model will be extended to a complete vehicle model. This enables pitch and yaw motion, the consideration of additional driving maneuvers like cornering or a predefined route, and a better applicability of the simulation results to the real vehicle. Especially in acceleration and deceleration, the calculated work done by friction will increase in a full vehicle simulation, because of the pitch motion, which leads to more relative motion between the wheel and chassis and, therefore, to greater changes in wheel alignment. In addition, the appropriated wear laws should be further examined.

Nevertheless, the presented results show significant potential to reduce the amount of tire wear and, in this way, environmental pollution. Up to now, this potential is not used by the automotive industry systematically. Using the proposed method, it is possible to consider tire wear as the main objective on the same level as dynamics, safety, and comfort during the design process of a suspension system.

Author Contributions: Conceptualization, methodology, validation, formal analysis, investigation, and data curation, J.S.; writing—original draft preparation, J.S.; writing—review and editing, W.S.; visualization, J.S.; supervision, W.S.; project administration, W.S.; funding acquisition, J.S. All authors have read and agreed to the published version of the manuscript.

Funding: This research was funded by Deutsche Bundesstiftung Umwelt (DBU).

sponsored by
DBU
Deutsche
Bundesstiftung Umwelt
www.dbu.de

Institutional Review Board Statement: Not applicable.

Informed Consent Statement: Not applicable.

Data Availability Statement: Not applicable.

Conflicts of Interest: The authors declare no conflict of interest.

References

1. Diegmann, V.; Pfäfflin, F.; Wiegand, G.; Wursthorn, H. *Maßnahmen Zur Reduzierung von Feinstaub Und Stickstoffdioxid*; IVU Umwelt GmbH: Dessau, Germany, 2007.
2. Willeke, H. *CO_2-Emission und Feinstaub in Verbindung mit Rollwiderstand von Reifen*; DEKRA: Essen, Germany, 2008.
3. Bundesministerium für Umwelt, Naturschutz und Nukleare Sicherheit (BMU). *Klimaschutz in Zahlen: Fakten, Trends Und Impulse Deutscher Klimapolitik*; Bundesministerium für Umwelt, Naturschutz und Nukleare Sicherheit (BMU): Berlin, Germany, 2019.
4. Timmers, V.R.; Achten, P.A. Non-exhaust PM emissions from electric vehicles. *Atmos. Environ.* **2016**, *134*, 10–17. [CrossRef]
5. OECD. *Non-Exhaust Particulate Emissions from Road Transport*; OECD: Paris, France, 2020.
6. Boucher, J.; Friot, D. *Primary Microplastics in the Oceans: A Global Evaluation of Sources*; IUCN: Gland, Switzerland, 2017; p. 43.
7. Bertling, J.; Bertling, R.; Hamann, L. *Kunststoffe in Der Umwelt: Mikro- Und Makroplastik: Ursachen, Mengen, Umweltschicksale, Wirkungen, Lösungsansätze, Empfehlungen; Kurzfassung der Konsortialstudie, Fraunhofer-Institut für Umwelt-, Sicherheits- und Energietechnik*; UMSICHT: Oberhausen, Germany, 2018. [CrossRef]
8. Pottinger, M.G. Contact Patch (Footprint) Phenomena. In *The Pneumatic Tire*; National Highway Traffic Safety Administration NHTSA: Akron, OH, USA, 2006; pp. 231–285.
9. Ersoy, M.; Gies, S. *Fahrwerkhandbuch*; J.B. Metzler: Berlin, Germany, 2017.
10. Kohl, S.; Sextro, W.; Zuber, A. Benteler Vehicle Dynamics-Fahrdynamikentwicklung Basierend Auf Einer Neuen Auslegungstheorie. In *8 Tag Fahrwerks*; RWTH Aachen University in Aachen: Aachen, Germany, 2012.
11. Blundell, M.; Harty, D. *Multibody Systems Approach to Vehicle Dynamics*; Automotive Engineering Series; Elsevier Butterworth-Heinemann: Amsterdam, The Netherlands, 2004; ISBN 978-0-7506-5112-7.
12. Kohl, S. *Analyse Der Reibleistungsverteilung Im Reifenlatsch Unter Berücksichtigung Der Fahrwerkdynamik Eines Mehrlenkerachssystems Zur Bewertung Des Reifenverschleißes*; Shaker Verlag: Aachen, Germany, 2019.
13. Lupker, H.; Cheli, F.; Braghin, F.; Gelosa, E.; Keckman, A. Numerical Prediction of Car Tire Wear. *Tire Sci. Technol.* **2004**, *32*, 164–186. [CrossRef]
14. Braghin, F.; Cheli, F.; Melzi, S.; Resta, F. Tyre Wear Model: Validation and Sensitivity Analysis. *Meccanica* **2006**, *41*, 143–156. [CrossRef]
15. Gafvert, M.; Svendenius, J. A novel semi-empirical tyre model for combined slips. *Veh. Syst. Dyn.* **2005**, *43*, 351–384. [CrossRef]

16. Sueoka, A.; Ryu, T.; Kondou, T.; Togashi, M.; Fujimoto, T. Polygonal Wear of Automobile Tire. *JSME Int. J. Ser. C* **1997**, *40*, 209–217. [CrossRef]
17. Salminen, H. Parametrizing Tyre Wear Using a Brush Tyre Model. Ph.D. Thesis, KTH R. Institute Technology, Stockholm, Sweden, December 2014.
18. Cho, J.C.; Jung, B.C. Prediction of Tread Pattern Wear by Explicit FEM. 33. *J. Automob. Eng.* **2014**, *229*, 197–213.
19. Lee, S.W.; Jeong, K.M.; Kim, K.W.; Kim, J.H. Numerical Estimation of the Uneven Wear of Passenger Car Tires. *World J. Eng. Technol.* **2018**, *6*, 780–793. [CrossRef]
20. Tong, G.; Jin, X.; Road, C. Study on the Simulation of Radial Tire Wear Characteristics. *Transp. Probl.* **2012**, *11*, 11.
21. Knisley, S. A Correlation Between Rolling Tire Contact Friction Energy and Indoor Tread Wear. *Tire Sci. Technol.* **2002**, *30*, 83–99. [CrossRef]
22. Sakai, H. Friction and Wear of Tire Tread Rubber. *Tire Sci. Technol.* **1996**, *24*, 252–275. [CrossRef]
23. Wright, C.; Pritchett, G.L.; Kuster, R.J.; Avouris, J.D. Laboratory Tire Wear Simulation Derived from Computer Modeling of Suspension Dynamics. *Tire Sci. Technol.* **1991**, *19*, 122–141. [CrossRef]
24. Stalnaker, D.; Turner, J. Vehicle and Course Characterization Process for Indoor Tire Wear Simulation. *Tire Sci. Technol.* **2002**, *30*, 100–121. [CrossRef]
25. Engel, D.; Biesse, F. Simulation based tire wear prediction by vehicle and tire model coupling. In *15. Internationale VDI-Tagung Reifen-Fahrwerk-Fahrbahn*; VDI-Berichte: Berlin, Germany, 2015; ISBN 978-3-18-092241-6.
26. Gipser, M.; Hofmann, G.F. *Tire-The Market-Leading Physics-Based Tire Model*; Esslingen University of Applied Sciences: Esslingen, Germany, 2018.
27. ITWM. *Scalable Tire Model for Full Vehicle Simulations*; CDTire: Kaiserslautern, Germany, 2017.
28. Fleischer, G. Energetische Methode Der Bestimmung Des Verschleisses. *Schmierungstechnik* **1973**, *4*, 269–274.
29. Schütte, J.; Sextro, W. Model-Based Investigation of the Influence of Wheel Suspension Characteristics on Tire Wear. In *Proceedings of the Advances in Dynamics of Vehicles on Roads and Tracks*; Klomp, M., Bruzelius, F., Nielsen, J., Hillemyr, A., Eds.; Springer International Publishing: Cham, Switzerland, 2020; pp. 1760–1770.
30. Schramm, D.; Hiller, M.; Bardini, R. *Modellbildung und Simulation der Dynamik von Kraftfahrzeugen*; Springer: Berlin/Heidelberg, Germany, 2013; ISBN 978-3-642-33887-8.
31. DIN ISO 8855:2013-11. *Straßenfahrzeuge-Fahrzeugdynamik und Fahrverhalten-Begriffe*; Beuth Verlag Gmbh: Berlin, Germany, 2013; p. 51.
32. Trzesniowski, M. *Rennwagentechnik*; Vieweg, Teubner Verlag: Wiesbaden, Germany, 2012; ISBN 978-3-8348-1779-2.
33. Braess, H.-H.; Seiffert, U. *Vieweg Handbuch Kraftfahrzeugtechnik*; Springer Vieweg: Wiesbaden, Germany, 2013; ISBN 978-3-658-01690-6.
34. Salzgitter Flachstahl GmbH Borlegierte Vergütungsstähle. Available online: https://www.salzgitter-flachstahl.de/de/produkte/warmgewalzte-produkte/stahlsorten/borlegierte-verguetungsstaehle.html (accessed on 1 August 2020).
35. Craig, R.R.; Bampton, M.C.C. Coupling of substructures for dynamic analyses. *AIAA J.* **1968**, *6*, 1313–1319. [CrossRef]
36. ZF Friedrichshafen AG WebCat-Online-Katalog-ZF Friedrichshafen AG, ZF Aftermarket. Available online: https://webcat.zf.com (accessed on 2 January 2021).
37. Popov, V.L. *Kontaktmechanik und Reibung*; Springer: Berlin/Heidelberg, Germany, 2015; ISBN 978-3-662-45974-4.
38. Burckhardt, M.; Burg, H. *Berechnung Und Rekonstruktion Des Bremsverhaltens von PKW*; Information Ambs GmbH: Kippenheim, Germany, 1988; ISBN 978-3-88550-025-4.
39. Schütte, J.; Sextro, W.; Kohl, S. Halbachsprüfstand Zur Kinematischen, Elastokinematischen Und Dynamischen Charakterisierung von Radaufhängungen. *Fachtag. Mechatronik* **2019**, *1*, 103–108.
40. Cosin Scientific Software AG. *F Tire-Flexible Structure Tire Model: Modelization and Parameter Specification*; Cosin Scientific Software AG: Munchen, Germany, 2021.
41. Gipser, M. F Tire: A Physically Based Tire Model for Handling, Ride, and Durability Part 2: Modelization. *SAE Int. J. Passeng. Cars Mech. Syst.* **2014**, *7*, 231–243.
42. Cosin Scientific Software AG Cosin Scientific Software. Available online: https://www.cosin.eu (accessed on 1 July 2020).
43. MSC Adams. *Help—Road Models*, MSC Software. 2020.
44. International Organization for Standardization ISO 8608:2016-11. *Mechanical Vibration Road Surface Profiles Reporting of Measured Data*; International Organization Standardization: Geneva, Switzerland, 2016.
45. Braun, H.; Hellenbroich, T. Messergebnisse von Strassenunebenheiten. *VDI-Berichte* **1991**, *877*, 47–80.
46. Kummer, H.W. *Unified Theory of Rubber and Tire Friction*; B-94; Pennsylvania State Univ: University Park, PA, USA, 1966.
47. Bachmann, T. *Literaturrecherche Zum Reibwert Zwischen Reifen Und Fahrbahn; Berichte aus dem Fachgebiet Fahrzeugtechnik der TH Darmstadt*; VDI-Verlag: Düsseldorf, Germany, 1996; ISBN 3-18-328612-2.
48. Moldenhauer, P. *Modellierung Und Simulation Der Dynamik Und Des Kontakts von Reifenprofilblöcken*; VDI-Verlag: Düsseldorf, Germany, 2010.
49. Sextro, W. *Dynamical Contact Problems with Friction Models, Methods, Experiments and Applications*; Springer: Berlin/Heidelberg, Germany, 2007; ISBN 978-3-540-45317-8.

50. Glaser, H.; Rossié, T.; Rüger, J.; Conrad, T.; Wagner, R. Das Fahrwerk. *ATZextra* **2010**, *15*, 32–37. [CrossRef]
51. Hudler, R.; Leitner, D.I.W.; Krome, H.; Steigerwald, A.; Fischer, D.I.S. Die Achsen des neuen Audi A4. *ATZextra* **2007**, *12*, 104–113. [CrossRef]
52. *Audi Q3: Entwicklung und Technik*; Rudolph, H.-J. (Ed.) ATZ/MTZ-Typenbuch; Springer Vieweg: Wiesbaden, Germany, 2013; ISBN 978-3-658-00852-9.
53. Hoffmann, J. *Potentiale Einer Aktiven Achskinematik zur Optimierung des Fahrverhaltens*; Schriftenreihe des Instituts für Fahrzeugtechnik, TU Braunschweig; Shaker Verlag: Aachen, Germany, 2016; ISBN 978-3-8440-4554-3.

vehicles

Article

Potential Analysis for a New Vehicle Class in the Use Case of Ride-Pooling: How New Model Developments Could Satisfy Customers and Mobility Makers

Martin Dorynek [1,*], Lisa-Theres Derle [1], Martin Fleischer [1], Alex Thanos [1], Paul Weinmann [1], Michael Schreiber [2], Sebastian Schumann [2], Tolga Tunc [2] and Klaus Bengler [1]

1 Chair of Ergonomics, TUM School of Engineering and Design, Technical University of Munich, Boltzmannstraße 15, 85748 Garching, Germany; lisa.derle@tum.de (L.-T.D.); martin.fleischer@tum.de (M.F.); alex.thanos@tum.de (A.T.); paul.weinmann@tum.de (P.W.); bengler@tum.de (K.B.)

2 Hyundai Motor Europe Technical Center GmbH, Hyundai-Platz, 65428 Rüsselsheim, Germany; mschreiber@hyundai-europe.com (M.S.); sschumann@hyundai-europe.com (S.S.); ttunc@hyundai-europe.com (T.T.)

* Correspondence: martin.dorynek@tum.de

Abstract: Due to changes in mobility and the emergence of new services, it is becoming necessary to establish new vehicle classes between conventional buses and privately owned vehicles. New mobility scenarios need concrete specifications to develop the most user-centered shuttle buses. As a result, we are looking for the requirements and needs of operators and customers. Initially, we want to determine the status quo, as there is no preliminary work in this regard. During the course of extensive literature research, expert interviews, and follow-up workshops, the respective solution space was highlighted and narrowed down. Services such as ride-pooling require adapted vehicle concepts to ensure optimal implementation of their offer. Due to its optimized processes, the automotive industry depends on producing vehicles in a certain quantity and manner. Faster changes and extensive experiments are not possible with the current production approach. Purpose-built vehicle concepts can make mobility services more attractive to customers while facilitating business operations. For instance, potential improvements can be identified in the seating concept.

Keywords: purpose-built vehicle; new mobility; ride-pooling; Mobility-as-a-Service (MaaS)

Citation: Dorynek, M.; Derle, L.-T.; Fleischer, M.; Thanos, A.; Weinmann, P.; Schreiber, M.; Schumann, S.; Tunc, T.; Bengler, K. Potential Analysis for a New Vehicle Class in the Use Case of Ride-Pooling: How New Model Developments Could Satisfy Customers and Mobility Makers. *Vehicles* **2022**, *4*, 199–218. https://doi.org/10.3390/vehicles4010013

Academic Editor: Mohammed Chadli

Received: 2 February 2022
Accepted: 28 February 2022
Published: 5 March 2022

1. Introduction

Mobility is an integral part of life. It "is a basic human need" [1] and a decisive factor for the economy [2]. Without it, many people would not be able to get to work. It makes social encounters possible in the first place. Mobility determines where we live because, for many, it is essential to live close to the nearest railway line or motorway. Most people in Germany attach particular importance to their mobility, including being flexible, independent, reliable, plannable, and safe [3].

There are a variety of influential factors that affect and change our mobility habits. Mobility is part of social change and, therefore, closely linked to it. Most trips are made in one's car, which has been the center of mobility for most people over the years. Since the Fridays for Future protests, the impact of our current mobility behavior on the climate has become a hot topic of social and political discourse. In 2018, more than 19% of Germany's climate-damaging greenhouse gas emissions (GHG) were attributable to the transportation sector [4]. The European Green Deal calls for a 90% reduction of GHG in the transportation sector by 2050 to help the EU become the first climate-neutral continent. [5].

To satisfy individual and diverse mobility needs, a variety of new offers have begun, from which the user chooses the optimal means of transportation. Users do not necessarily see optimal as optimal in terms of protecting the climate. Convincing people to give up their personally-owned cars could be one of the main obstacles to the necessary measures [6]. Public transportation needs to be more attractive to users, and new mobility concepts need to be developed.

Mobility-as-a-Service (MaaS) is a holistic system that allows users to book and pay for various means of transportation via online platforms [7]. Thus, mobility is provided as a service and is independent of vehicle ownership [8]. The service consists of having different means of transportation available from the starting point to the destination and managing them as a single source. A promising approach to an embodiment of MaaS is on-demand mobility.

On-demand mobility implies that passengers get their mobility requests fulfilled when they need it. Customers' travel requests are fulfilled flexibly and in real-time [7]. The service is enabled by an online platform that processes customer requests in real-time and notifies the driver in an available vehicle, eliminating the need for planning [9]. The term on-demand mobility covers ride-sharing (e.g., BlaBlaCar), car-sharing (e.g., ShareNow), ride-pooling (e.g., CleverShuttle), and ride-hailing (e.g., Uber) services [10]. Here, on-demand mobility can be seen as the primary means of transportation and as a feeder to the following public transportation connection point, thus covering the "first and last mile" [11]. In addition, relevant reference literature argues that on-demand mobility is the cornerstone for implementing autonomous shuttles as a complement to public transportation [10,11].

The increasing popularity is reflected in the rapid spread of the service in Germany. For example, ride-pooling is starting in more and more federal states of Germany. The following map shown in Figure 1 provides an overview of the current situation in 2021. It is particularly noticeable that services are initiated mainly in the western federal states of Germany.

User acceptance is significantly influenced by the willingness to share a mode of transportation with strangers [12]. Hall [13] specified the discomfort associated with the violation of personal privacy, which is hardly avoidable in public transportation. The perceived violation of privacy results in stress and inconvenience [14]. For this reason, it is important to adapt the seat configuration and design to privacy requirements. The switch from driver to passenger can also lead to an increased incidence of kinetosis [15]. Kinetosis describes symptoms such as nausea, headaches, and discomfort that can occur during a car trip. These symptoms occur especially during a secondary activity, such as reading or working.

On the one hand, the current concepts do not meet future customer needs and ensure increased use. On the other hand, providers could save costs with smaller and more targeted modes of conveyance. The need for conventional bus drivers with the appropriate qualifications, who are currently a scarce resource, would be minor. In theory, all parties would bene-fit.

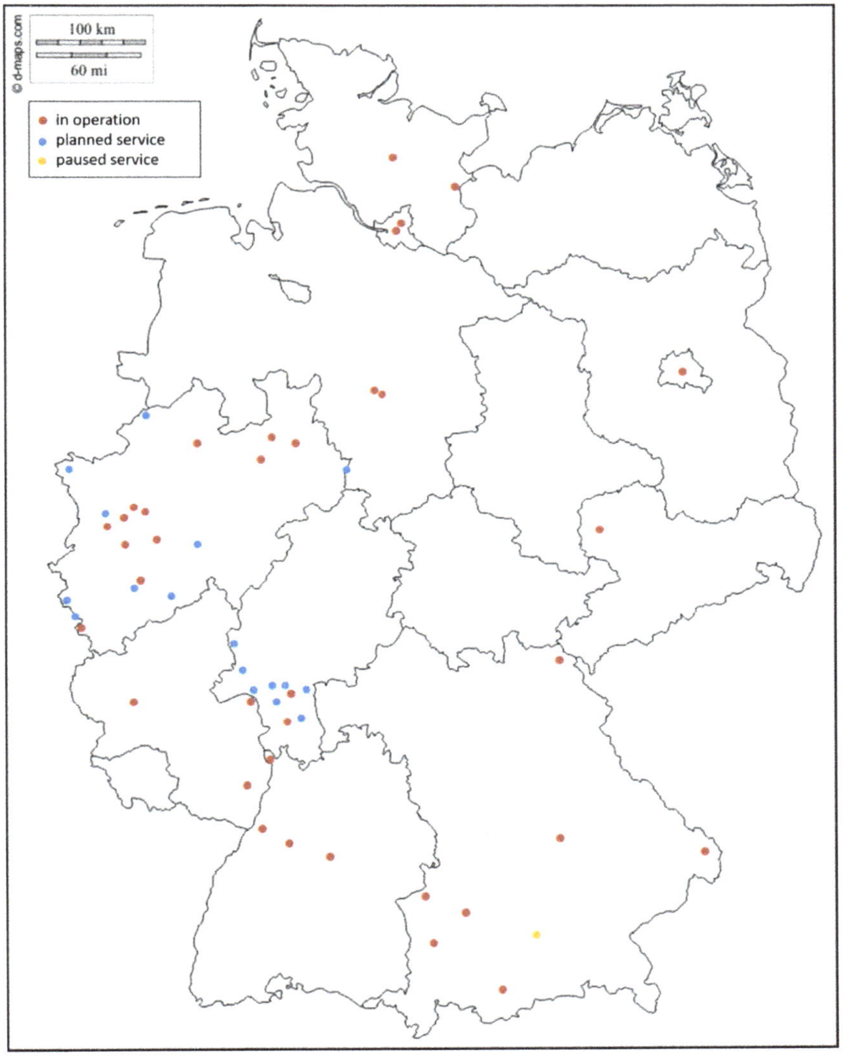

Figure 1. New mobility offers vary significantly from state to state. This map shows Germany with its 16 federal states. The data set was collected in May 2021. The red dots show the current mobility services. The yellow locations are currently suspended, while the areas marked in blue will be starting soon.

2. Theoretical Background

Many people prefer to live in cities or the surrounding suburbs. In 2017, about two-thirds of the German population lived in urban areas [16] (p. 22). More people inevitably create more traffic. This, in turn, leads to adverse effects on people and the climate, whether through traffic congestion or increased emissions [2] (p. 28).

From the user's point of view, the current public transportation system has several disadvantages that prevent it from becoming more widespread to every user group. The reasons are, e.g., [17]:

- Trips take too long compared to by car;

- There are no direct connections or distances to stops are too far;
- Comfort is restricted by overcrowding;
- There are no (convenient) options for transporting purchases;
- Passengers;
- Pollution;
- Lack of flexibility;
- Dependence on timetables;
- Costs;
- Frequency is too low;
- Delays;
- Lack of accessibility;
- An unsafe feeling.

These disadvantages can be particularly significant in certain regions such as rural areas, for specific target groups, such as people with age-related physical limitations, or particular use cases such as when having to do one's weekly shopping. Currently, public transport is struggling to bring back the travelers who stayed at home because of Corona and who will continue to commute less and work from home more often.

2.1. Position of Ride-Pooling/Different Modes of Transportation

Figure 2 compares ride-pooling to different mobility options. Offerings range from private to public transportation and are positioned vertically to show how they rate in terms of comfort and functionality. The mobility options on the side of public transportation are distinguished by reducing traffic by increasing the load factor of single transportation units.

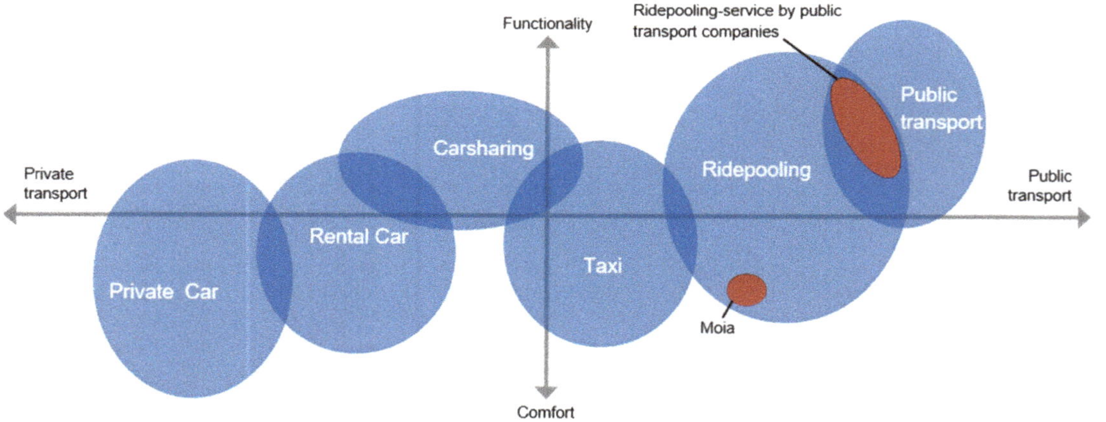

Figure 2. Position of ride-pooling compared to other mobility offerings (i.e., how it measures to the others). The situation presented and the classification made relate to the German market. Offers can differ greatly regionally due to their positioning. (Own diagram based on Gaul [18].)

Personally owned cars can be classified as individual transportation and are characterized by a high degree of driving and everyday comfort. Since rental cars are usually rented for more extended periods or at least longer distances, the aspect of convenience also dominates this concept. Both rental cars and carsharing are individual modes of transportation. While they do not really decrease the amount of traffic on roads, the space required by parked vehicles is reduced thanks to the shared use. This creates a public benefit, especially in urban areas with a high population density. When it comes to vehicle configuration and the design of the service, the focus is primarily on functionality. Trips are usually short, and customers use carsharing according to the situation. The cab is found approximately at the origin of the coordinate system. Since the vehicle is not shared, it is still a form of individual transport, but it is possible to use a cab only as part of a travel chain for which, for example, there is no public transportation alternative.

An example of this is the way home from the train station. Due to temporary mobility restrictions caused by (a lot of) luggage, the cab is preferred for such a trip. This demonstrates the functionality of cabs. In addition, the fact that passengers are driven and that other actions associated with self-driving (e.g., looking for a parking space) are not necessary also increases convenience. Public transportation is the counterpart to the privately owned car. It is public transportation that bundles different travel requests and thus reduces traffic on the road. A trip by public transportation is functional and is intended to transport customers from A to B at a low cost. The functionality aspect is also visible in the vehicles used, which do not reach the same level of convenience as, e.g., cabs, but are rather optimized for efficient use of space, inclusion, and easy cleaning.

Ride-pooling is positioned between cabs and public transportation. The idea behind pooling is to actively reduce the volume of traffic. The large area that ride-pooling occupies on the diagram illustrates its potential. Ride-pooling occurs in very different stages and forms. Depending on the design of the offer, it is possible to promote both the integration into the public transportation system as well as the convenience of the service. There are a variety of possible combinations. In addition to the design of the app (e.g., the design of the app such as MOIA) and the pricing structure, the selection of the vehicles plays a decisive role, as this is the most significant point of contact that the customers experience during the realization of the service, the transportation to the destination. Looking at the design of today's offerings, it becomes clear how enormous this design scope is.

Services provided by urban transportation companies are integrated into existing public transportation services and thus overlap with them. Vehicle concepts tend to be robust, barrier-free, and functional. The BerlKönig is an exception here. The vehicles used are slightly modified and try to be more comfortable. MOIA has advanced this optimization further with vehicles explicitly developed for the service (MOIA Pluto). With the purpose-built vehicles, the aspect of convenience has shifted more to the fore. This different design of the services makes it possible to address different customer groups (for example, leisure-related users such as students, night-owls, etc.). The mobility options of public transportation benefit from the whole range of all options. Ride-pooling has the potential to be a pull factor and accelerate the shift away from an individually owned vehicle to shared mobility. Ride-pooling adds a mobility option to public transportation that is comparably comfortable to personal transportation and can thus become a real alternative to the privately owned car.

Furthermore, there are reasons for deciding against owning a car. Driving a car is no longer a pleasure in everyday traffic and many young people therefore no longer have a driving license. The rampant lack of parking spaces and the associated search also costs time and energy. Nevertheless, ride-pooling does not reach the same level of convenience as a privately owned vehicle, since the trips are shared and comparable privacy as in individually owned cars cannot be achieved. This is partly due to the vehicle concepts in use today. However, new developments could ensure that the perceived difference at least becomes smaller.

The COVID-19 pandemic has shown that, in exceptional situations, users fall back on tried and tested means of transport, certainly also because the regulations of social distancing can be better implemented in cars and the risk of infection is reduced. However, the current trend shows that users are returning to public transport, and falling restrictions are giving shared mobility services back their former design freedom. In addition, the increasing traffic in urban areas and the challenges posed by the climate crisis do not make a return to pure individual transport possible. What remains is the increased sensitivity to hygiene, which must be taken into account in the future design of vehicles and services.

In addition to the considerable benefits that these ride-pooling PBVs (Purpose Built Vehicles) bring to users, as they are manufactured for specific needs, they offer vehicle manufacturers the opportunity to significantly reduce their vehicle-to-market costs [19]. For example, the average vehicle in the US costs just under $24,000, as it is manufactured on the premise of covering a wide range of users and use cases. In contrast, a PBV that is produced for a specific use case and a limited number of users may have a lower level of complexity, a less powerful engine, a more straightforward assembly process, shorter development times, and lower distribution costs [20]. According to Grosse-Ophoff et al. [20], all this can lead to a PBV costing on average 25% less than a conventional vehicle powered by an internal combustion engine. Bernhart et al. (2018) forecast even speaks of a cost-saving of over 50%.

The cost saving is not necessarily the case for the first generation of PBVs, such as the MOIA Pluto. This is a standard LCV (light commercial vehicle). It has been optimized for its purpose with many expensive modifications. In the future, the total cost of ownership (TCO) will be an important factor, so that even a much more expensive vehicle can be more profitable due to the high utility rate.

Furthermore, decreasing battery prices will lower the production costs of battery electric vehicles (BEVs) compared to combustion engines in the future, making electric vehicles competitive in terms of purchase price as well [21]. In addition, fuel savings allow BEVs to achieve cost parity when considering the total cost of ownership between 2022 and 2026 [22]. At the same time, the required diversity in the product range needs a platform concept, as otherwise the sub-variations would not be feasible.

2.2. Introduction and Comparison of Used Vehicles for MaaS

In a previous study, Dorynek et al. [23] already put the wide variety of different vehicles for the same mobility service to the test. Among other things, it was found that the vehicles are far too small for actual use. In the early days of ride-pooling services, various providers experimented with vehicles, including vehicles such as the VW eGolf. The available space is very limited here, making pooling hardly possible. The vehicles have only a few distinguishing features compared to taxis. With the exception of the provider MOIA, commercially available vehicles are used as in taxi operations. As a result, the vehicles hardly differ for users from those used in taxi service, especially when considering that the opportunity of isolating oneself from fellow passengers is hardly possible. This means that the service cannot be evaluated in a differentiated way. In the meantime, some companies have developed further, and small vehicles have been removed from their fleets. The trend in Germany is increasingly shifting towards minibuses. However, it is also evident that the vehicle plays an important role and must fit in order to be able to offer the service in a user-oriented way.

The vehicles are also difficult to recognize, which is especially relevant at (over-crowded) pick-up points. Although pooling vehicles are usually easy to recognize, it is sometimes difficult to identify the requested vehicle at more heavily-used virtual stops. Significant delays can occur especially in the vicinity of railway stations, as the vehicles do not have displays and do not have dedicated bays where they are supposed to wait. Thus, an obstruction of the traffic flow is possible here.

In the early years of ride-pooling, customer demands were met using diesel-powered vehicles. Mercedes V-Class were the most commonly-used vehicles. Now the focus is often placed on vehicles with an electric drive system and high passenger capacity. Climate change targets are forcing the sector towards electrification. Many operators complain due to the low supply, although the market is moving in that direction quickly.

The most frequently-used vehicles are compared in Table 1.

The left half shows vehicles that are not available worldwide or generally, such as the MOIA Shuttle or the Toyota JPN Taxi. Together with the LEVC TX5, they are all designed for the specific purpose of transporting passengers. The vehicles shown on the right, i.e., the Mercedes-Benz and Nissan minibuses, are not specifically adapted but are designed for the general public. It is obvious that the vehicles vary a lot. They can carry between four and eight passengers and rely on different body types.

Table 1. Comparison of currently available and frequently-used vehicles for on-demand mobility services in Germany. In addition, the Toyota JPN Taxi (only available in Japan) was included in the comparison because it is advertised as being particularly user-friendly.

	Purpose-Built Vehicle			Traditional Vehicle		
Vehicle Type	**MOIA**	**Toyota JPN Taxi**	**LEVC TX5**	**Nissan E-NV200 Evalia**	**Mercedes eVito/EQV Large/Extra-Large**	**MB Sprinter Mobility 23**
Length	5.99 m	4.40 m	4.86 m	4.56 m	5.14 m/5.37 m	5.93 m
Width	2.03 m	1.70 m	2.04 m	1.76 m	1.93 m	2.02 m
Height	2.59 m	1.75 m	1.89 m	1.85 m	1.91 m	2.65 m
Wheelbase	3.64 m	2.75 m	2.99 m	2.73 m	3.20 m/3.43 m	3.67 m
Boarding height	0.50 m	-	0.37 m	0.52 m	-	0.25–0.27 m
Seats	6	4	3 full + 3 tip-up	Max 7	Max 8	Max 4 wheelchairs or 8 seats
Standing passengers	0	0	0	0	0	0
Autonomous	No	No	No	No	No	No
Propulsion technology	Electric	LPG-Hybrid	Electric + range extender	Electric	Electric	Combustion engine
Range (WLTP)	300 km	≈450 km	101 km (RE 510 km)	200 km	418 km	-
Top speed	90 km/h	-	130 km/h	123 km/h	Standard 140 km/h (Max. 160 km/h)	-
Wheelchair access	No	Yes	Yes	No	No	Yes
Door opening concept	Sliding door	Conventional / opposed	Conventional / opposed	Sliding door	Sliding door	Sliding door
Seat concept	Conventional	Conventional	Face-to-face	Conventional	Conventional	Conventional

2.3. *Impact of Missing Standards*

If the OEMs do not develop and manufacture the right vehicle, the operators must improve the standard class. Figure 3 shows simple modifications made by a provider to its Mercedes-Benz Vito. To facilitate access, a grab rail was installed in the vehicle at the rear sliding door. In addition, a wooden step stool was made, which is carried in the luggage compartment. This is placed next to the vehicle when needed in order to reduce the entry height for passengers by another step. This shows that the providers are not satisfied with the vehicles in operation and that experience has been gained and used for making improvements.

Figure 3. Necessary conversions because conventional series-production vehicles do not fully meet the required specifications (image rights are held by Pikmi (VBZ)). The provider in Zurich carried out the conversion work in 2021 after it had consulted on the suitability for everyday use in an internal workshop.

3. Materials and Methods

3.1. Literature Review

During the evaluation of vehicle concepts in use, it was possible to identify major weaknesses regarding user-centered design and ergonomic vehicle design. The lack of user-centricity explains the failure of anthropometric-friendly vehicle design by (autonomous) shuttle bus manufacturers, which is made clear by the seats and seat configuration of the shuttle buses [24]. The shuttle buses have not yet been adapted to user requirements and the various user groups. This contrasts with the realization that, for in-vehicle automation and shared mobility, the overall focus is on the passenger experience and sensation [25]. To achieve this goal, ergonomics, handling, and design need to be integrated into the technical development process from the outset [26]. Ergonomic design ensures that the final product is usable and comfortable for the user groups [26]. Before adopting or redesigning the vehicle concept, the first step should be to review the relevant literature for specific recommendations. By researching recommendations for dimensions and vehicle concepts in the area of ride-pooling, it can be seen that there is still a large gap in this subject area.

Customers are likely to find that vehicle concepts become obsolete only after a few years, even though their technical components are still in good condition [19]. Market research proves the decreasing interest in vehicles when ownership changes to service. Due to a very high usage rate, the vehicles used will reach their EOL (End of Lifecycle) significantly faster than with previous use. To maintain customer enthusiasm accordingly, it must be possible to react quickly to continuously changing wishes and requirements. For this to happen cost-effectively, a modular design is needed in the development of PBVs.

Conventional vehicles must fulfill different tasks of the users, including not only short trips for shopping and going to work, but also longer trips for excursions or holiday trips. The seat design must therefore cover the entirety of the scenarios. Trips lasting several hours are also considered and must be covered to a certain extent, depending on the vehicle class.

In the case of PBVs for ride-pooling, narrow usage limits are set, and the seats must offer good support, especially on short trips. In contrast to conventional car seats, there is a significantly increased change in use at the same time. The driving profile in the ride-pooling scenario is also very different from that of a traditional passenger car; due to lower cornering speeds, less lateral support is required. Elaborate adjustment options for the seats are not necessarily required. The surfaces must be designed for heavier use compared to conventional passenger cars. Table 2 lists the different set-ups and adjustment travels for conventional seating options. The variance and resulting complexity are immense and should not be underestimated. At the same time, entries and exits are becoming more important and should be supported.

Table 2. The solution space of current and future seating concepts as a morphological box (own table based on Baumgartner [27]). Almost all available vehicles in Europe in 2021 were inspected for the analysis.

Seating concept											
Driver's workplace		Available	Not available								
Number of seats		1	2	3	4	5	6	7	8	9	
Number of rows of seats		1	2	3							
Number of seats first row		1	2	3							
Number of seats second row		1	2	3	Seat row not available						
Number of seats third row		1	2	3	Seat row not available						
Alignment first seat row	**Basic orientation**	In the direction of travel	Against the direction of travel	Parallel to the direction of travel	Variable						
	Turning possibility	Not rotatable	Small angles <15°	180° rotatable							
	Longitudinal adjustability	Adjustability within the classical seat adjustment field	High longitudinal adjustability (approx. 500 mm)	No longitudinal adjustability							
Alignment second seat row	**Basic orientation**	In the direction of travel	Against the direction of travel	Parallel to the direction of travel	Variable	Seat row not available					
	Turning possibility	Not rotatable	Small angles <15°	180° rotatable	Seat row not available						
	Longitudinal adjustability	Adjustability within the classical seat adjustment field	High longitudinal adjustability (approx. 500 mm)	No longitudinal adjustability	Seat row not available						
Alignment third seat row	**Basic orientation**	In the direction of travel	Against the direction of travel	Parallel to the direction of travel	Variable	Seat row not available					
	Turning possibility	Not rotatable	Small angles < 15°	180° rotatable	Seat row not available						
	Longitudinal adjustability	Adjustability within the classical seat adjustment field	High longitudinal adjustability (approx. 500 mm)	No longitudinal adjustability	Seat row not available						
Basic form of all seat rows		Triangle	Rectangle	Rhombus	Circular	Staggered	Asymmetrical				
Type of seat first row		Standing seat	Folding seat	Classic seat	Bench	Rotating seat	Relax seat	Reclining seat	Lounger	Different	
Type of seat second row		Standing seat	Folding seat	Classic seat	Bench	Rotating seat	Relax seat	Reclining seat	Lounger	Different	Not available
Type of seat third row		Standing seat	Folding seat	Classic seat	Bench	Rotating seat	Relax seat	Reclining seat	Lounger	Different	Not available
Armrests	**Position**	At every seat	Only at selected seats	Not available							
	Assembly	Fixed	Foldable	Concealing	Not available						
		To its own seat	To the door	In the centre console	In the seating bench	Not available					
Centre consoles		In each row of seats	In selected rows of seats	Mobile centre console	Not available						
Luggage/ Storage space	**Volume**	<200 L	200–400 L	400–600 L	>600 L						
	Integration	Integrated into the interior	Separate from the interior	Partly integrated in the interior and partly separated from the interior							

The seat is of great importance for the comfort of the passenger. A relaxed and stable position must be provided to ensure that passengers can work, read, and sleep. The view should not be restricted, and the view should be directed towards the road to prevent motion sickness. This affects the seating arrangements in the vehicle. The vehicle interior is a living space, especially due to the role switch from driver to passenger.

For this reason, there must be storage space for personal belongings, and the feeling of space must correspond to the intended use of the vehicle. The color scheme of the interior, the design, and the interior headlining influence the overall impression of the vehicle. Especially in the case of passenger transportation shuttle buses, boarding and exiting must be possible for all user groups, including users with limited mobility. Finally, every vehicle must guarantee safety. This includes, for example, seat belts that are comfortable, safe, and easy to use [28]. The ergonomic aspect of the passenger compartment is significantly influenced by the people themselves, their dimensions, freedom of movement, and accessibility limits [28].

Within the framework of the UNICARagil project for the further development of automated vehicles and their architecture, founded by the Federal Ministry of Education and Research, an autonomous taxi concept was developed, among other things. Kipp et al. [29] investigated the interior concept and the requirements for an automated driverless taxi. In a vehicle that is intended to allow the occupants to work as well as rest, the preferred viewing direction should be in the direction of travel. In addition, the seat height H30 (called by developers in packaging [30]) should be around 400 mm and adaptable to different anthropometric sizes. The seats should be equipped with armrests and tables folded into the armrests. The height of the table should also be adjustable to suit the user's needs. In addition to that, the seats should also offer at least two seating positions. Various options for stowing luggage should be considered in the design. In addition to the fixed seats, UNICARagil's auto taxi offers seats that fold in the opposite direction of travel, which can be moved across the vehicle's width. If the seats are not in use, the free space can be used, for example, as an additional luggage rack.

Another recommendation from the relevant literature concerns the vehicle at total capacity. In this case, an area for multifunctional use is required. Barrier-free access must also be guaranteed for every vehicle [31]. Mueller et al. [31] recommend a seating configuration of six seats, standing room for four persons, and additional space for luggage.

In the research conducted by Christopher Wilson et al. [32], the user experience and related interior design in autonomous vehicles were investigated, and explicit design recommendations were developed. These reduce discomfort or increase the comfort experience. Screens should be placed above the passengers so as not to influence their peripheral vision and direct it to the outside world. This reduces susceptibility to kinetosis. It is also recommended to limit the rotation of seats to 10° to avoid injuries and kinetosis. Seat belts should be built directly into the seat in order not to restrict the freedom of movement of occupants. It is recommended to pay special attention to the storage of objects during the ride in the future. Items used while the vehicle is moving also require some form of securing in case of an accident or unplanned braking of the vehicle [32].

3.2. Preparation of Expert Interviews and Workshops

In order to create a vehicle model that is specifically tailored to the needs of shared commercial passenger transportation, it is first necessary to identify those boundary conditions that differ from the development processes of conventional vehicles. To better understand the industry's necessary user demands and requirements, the current situation was investigated with the help of an expert survey. To this end, 15 experts from Germany were interviewed and asked to assess the current situation of ride-pooling in Germany.

The expert survey provided insight, experiences, and expertise from different perspectives. Among others, stakeholders and experts from operators, the public sector, OEMs, and the consulting industry were interviewed. All interviews were conducted in German and semi-structured with a guideline that contained eight questions. The focus of the interview was on the vehicles to be offered, including their equipment features. The OEMs were particularly interested in the topic so that workshops were held afterwards.

Thus, some of the interviewees were rejected to be recorded, and written notes were taken. The conversations started with a short introduction followed by an explanation of the interview procedure. The interview then moved on to general questions regarding current ride-pooling conditions, operations, and more specific questions referring to the demand for ride-pooling from a user and operator perspective.

4. Results

4.1. Expert View—Interviews for Identifying the Gap

We were provided with many interesting aspects and insight, which will be discussed one after the other in the following. Firstly, many problems and challenges were reported that occur due to unsuitable vehicles. Current vehicles are not designed for the application and are developed accordingly. That said, there are also no specifications that precisely specify what is required so that vague concepts are on the road. However, there is no longer any experimentation or new development to a large extent, as cost pressure is immense. Vehicles such as those from MOIA are the upper end of the scale and overshoot the mark. For example, a single shuttle bus costs up to €300,000, which is a lot more than that of competitors. The suppliers' ideas of what should be achieved diverge widely, even though it is the same product. Much capital is needed to turn all the ideas into reality.

Positive practical examples and experiences were also described. Since the first companies have been on the road since 2015, the first lessons learned have been derived, and the best practices were created. The output will be presented below in the five thematic blocks that concern the vehicle.

4.2. Entry and Exit

For the trip to begin at all, the entry must be suitable. The floor height still has to be lowered properly. Hand grips installed at a wide entry are desirable. Entries above the road surface higher than 400 or 500 mm are not suitable for everyday use and cannot be handled by everyone. Clever packaging coupled with technical progress can be the solution. A battery pack height of 135 mm in five years is likely and possibly even lower. The rise of other battery technologies and higher energy densities will allow for change. Another possibility would be to move the batteries from the middle floor onto the axles or under the seats.

With regard to demographic changes, the primary goal must be to guarantee people their independence, even in old age, for as long as possible. Otherwise, the health care system will collapse. This target group also wants to remain active and mobile because the desire to travel never really goes away. In addition, there is a risk of unnecessarily putting some client groups under stress because they feel that they are hindering the whole process. Unique solutions for entry are always costly, and not only for the technical enhancements themselves. It requires significant training and maintenance efforts.

4.3. Seating

The most significant point of contact in the vehicle is and remains the seat, but due to the short driving time, the focus of development shifts there as well. As a result, it is necessary to understand the use cases precisely. For example, one could easily remove the armrests and offer very spartan seats, which would be entirely comfortable. Hygiene is also a factor that should not be underestimated in shared mobility. This can lead to a complete rejection of the concept. Looking at the seat, one could also argue about other surfaces, contouring, and upholstery. That said, everything must comply with the applicable laws, which in turn depend on the application. Somebody must safeguard the product's liability. For short trips of less than 30 min, most would not adjust their seat and area precisely. After all, they would then need to be reset after the trip, and the driver certainly will not always (want to) do that.

Positioning in the vehicle is often debated, and recommendations vary widely. In (series) production vehicles, people sit too close to one another. An example is the spatial construction in minivans developed for the Asian market, where elbows easily bump together. As a result, personal space is violated, which not every guest might accept. After all, one's private space is to be maintained; otherwise, the experience resembles that of public transportation.

However, the head restraint, i.e., a wider headrest such as in MOIA vehicles or ICE, is extolled as being very positive. When transporting children or families, seat shells/child seats are essential, since no one wants to bring their own. For many, faux leather is recommended as it is more durable than velour, which is easy to slip on. Some respondents argued for modern fabric technologies that are highly hygienic.

4.4. Loading and Unloading

The perception is quite different when taking luggage with you or transporting objects. On the one hand, one's car is frequently used to transport stuff or run many kinds of errands. Whether shopping at the supermarket, DIY, or furniture store, people need to stow their purchases somewhere. On the other hand, shopping online is also flourishing. More start-ups and traditional market players deliver almost everything in a few minutes. There are also other means of transport, so not everything has to be served by ride-pooling. However, it is necessary to provide enough space for walkers, strollers, and hand luggage as well, since these things are an integral part of everyday life. Ideally, a boot and tailgate can be dispensed with, as this entails additional operating steps. This poses a risk of injury or safety in everyday use. It can be another cause of delay and lateness. After all, one's own luggage should not be that accessible to third parties.

4.5. Complexity Reduction and Scalability

The more diverse the shuttles, the more difficult it becomes. Complex vehicle systems have to be understood and operated by temporary staff. Most drivers are students or senior citizens and mainly work to earn extra money or voluntarily. Vehicle problems are compensated or compounded by the workforce. Sometimes drivers shy away from these additional tasks because they find them annoying and unpleasant. After all, they are only paid for driving.

A good vehicle concept creates acceptance among users (can help win new customer groups) and among driving personnel, as this is scarce. Based on accounts provided by inclusion taxi operators, we know that lifts are more accepted than ramps. That said, one should not underestimate the complexity of the technology in day-to-day operations, as has already been confirmed by the study conducted by Dorynek et al. [33].

What does the application of small, divided, commercial passenger transportation look like? Vehicle body manufacturers can respond to individual wishes and thus offer batch size one. A common industry standard, as it is for (large) city buses, could help, but there is no need for groundbreaking innovations because the individual parts are already mastered. The possible scaling of a purpose-built vehicle is otherwise difficult. Anything can be built. More handles are likely but already exist as well.

4.6. Other

Due to the other use, the driver's seat has now become a workplace. With changing driving personnel, i.e., several people using one vehicle, time is lost due to the need to regularly readjust the seat. Between trips and shifts, drivers want and need to take a break. They cannot always return to the depot. In addition to that, drivers also want to pass the time during loading breaks. For instance, they many want to listen to their own music/radio station without disturbing passengers.

Overall, the appearance must fit into the mobility concept. Further variations with larger fleets can experiment with higher/more variable prices and comfort seats. A catalog of ergonomic requirements alone is of no use if it cannot be fulfilled. The focus is clearly on the benefits of each passenger and their needs.

Vandalism is not a problem since drivers are always present, and users are registered and not anonymous. The feeling of safety has not been negatively noticed so far.

The employees in charge often have no corresponding knowledge or previous experience in the sector. They sometimes come up with inconceivable ideas so that the gap between what is possible and what is not is sometimes huge.

Due to the rapid developments in entertainment and communication electronics, OEMs should not compete. Otherwise, the vehicles will quickly appear outdated or old-fashioned. A good interface with a fast-charging function should be the goal instead. The possibility of playing advertisements for customers during the entire trip and using the free spaces should be further considered.

4.7. Concluding Remark

The London Cab is a good starting point but would have to be modified for the application. The following changes would be necessary first: it should be 1 m longer, 20 cm higher, and offer two more seats. Vehicles such as the Mid-City Buses, which are currently in use, are designed as cargo variations for logistics or pallet transport. For example, the sliding door is 120 cm wide, whereas 80 cm would be sufficient. Other novel door concepts fail either because of crash and safety requirements or sealing problems. There may be problems with liquids entering the vehicle or the seals freezing in cold conditions (or any other issue).

A particular buffer capacity would be ideal for rush-hour. It should not be forgotten that public transportation does not stand for luxury. Functional bells and whistles are unnecessary and only complicate the vehicles or make them more susceptible to repair. Uptime, i.e., the time such cars are on the road, transporting customers and thus generating revenue, must be maximized. The focus must be on robustness, easy cleaning, and repair. After all, these drain financial resources over the long-term and are decisive differentiators vis-à-vis the clientele (no one wants to be on the road in dirty or old-fashioned vehicles; rental companies renew their fleets constantly for good reasons). Since the COVID-19 pandemic, we know that surfaces must also be able to withstand all disinfectants.

Overall, all experts agreed that a new mobility service such as MoD/ride-pooling requires new and different vehicles.

5. Discussion

5.1. Benefits from a Purpose-Built Vehicle

Conventional vehicles intended for private use are developed on the premise that they can be used for various purposes. This means that they are not really optimized, however, for any of these purposes. For example, a privately-owned conventional vehicle is used to do the shopping, commute to work, and go on family trips, among other things. However, the requirements are not optimally met for any of these purposes. With the introduction of PBVs, the focus in development is shifting to conceptualizing vehicles manufactured for a specific purpose. This enables concept optimization, as the developed PBV is specifically tailored to the requirements derived for the particular purpose. Thus, the precise definition of the purpose is the fundamental prerequisite in the development of PBVs.

5.1.1. Shifting from a Driver-Centered to a Passenger-Centered Experience

The future mobility shift from private vehicle ownership to shared mobility concepts requires a rethinking in the development of PBV concepts. As a result, the focal point of vehicle design is shifting from the driver to the passenger. Formal requirements relating to the driver's cab are moving to the background and will eventually disappear with the future introduction of autonomous systems. This, in turn, will lead to a greater need to address the requirements of the passenger. Accordingly, the development of PBVs in the future requires a passenger-centered (user-centered) approach. This is the most important constraint in development, as the main objective should be to ensure maximum fulfillment of requirements. However, the exact definition is still pending or has not yet been made.

5.1.2. Focusing on Functionality

Conventional vehicle concepts focus primarily on drivers, as they are in most cases the prospective buyers or customers. In contrast, the focus of PBVs for shared mobility is on the functions offered to the passenger and for fulfilling the operator requirements. This includes functionalities regarding the use of space, especially for passengers with limited mobility and more effortless operation in the overall context of the service. The added value is thus no longer defined by driving characteristics but by new passenger-oriented functions [34]. Characteristic features and key figures such as top speed and acceleration, on the other hand, only play a subordinate role.

5.1.3. Development Focus on Interior Conceptualization

Compared to the development process for conventional vehicles for private ownership, the development of PBVs for shared mobility should shift the focus more to interior design. The exterior design becomes less critical as fleet operators or service providers own the vehicles. Thus, in contrast to the conventional development process, where the interior is determined independent of the exterior design, the interior is designed first, and the exterior is adapted to it. This leads to the development process being reversed to a certain extent and running from the inside to the outside.

5.1.4. Modular Design

As purpose-built vehicles will be part of a digital mobility service, they will have to evolve at a similar pace similar to consumer electronics compared to traditional vehicles. Customers are likely to find that vehicle concepts are obsolete after only a few years, even though their technical components are still in good condition [19]. To maintain customer enthusiasm accordingly, it must be possible to react quickly to continuously changing wishes and requirements. To do this cost-effectively, PBVs require a modular design. This means that the individual assemblies, such as seats or displays, can be quickly and easily replaced with modified versions. This can be ensured at various levels, such as the entire vehicle through the platform strategy, in which the whole vehicle consists of two modules. On the one hand, there is the driving platform, which includes all drive modules, and on the other hand, there is the vehicle cabin. So, the interior can be easily replaced for a new vehicle model, but the driveline can be taken over from the old model. This ensures significant cost savings.

5.1.5. Increasing System Attractiveness and Leveraging Synergies

With the many changes in our daily lives, existing mobility services need to be better connected and timed in users' interest. The COVID-19 crisis accelerated the shift to teleworking, and during the period of mobility restrictions, companies and employees had to perform their work online wherever possible. It is predicted that employees will pursue a hybrid form of work in the future, i.e., work partly from the home office and partly in the office. Thus, the individual mobility requirement for trips to work is likely to decrease slightly. The same applies to business trips. For example, the pandemic has shown that various business trips can be avoided and are more efficient and less expensive online. In summary, it can be said that one or the other route will be eliminated as a result of new possibilities offered by digitization, which will further reduce the occupancy rate of existing public transport vehicles. By introducing new vehicle concepts, an increase in interest in such mobility services is to be expected, since previous disadvantages will be eliminated. In times of MaaS, an on-demand shuttle is considered a guarantor for multimodality. To enable the mobility of people in Germany, there is a dense network of different options and offers from which people can choose. This offer structure differs in detail locally very much. For example, new mobility services in particular are generally found in metropolitan areas and are not available in rural areas. Thus, residents of these areas cannot include such an option in their choice. This kind of new vessel would be quickly ready for use and could be scaled to the respective place of application accordingly. In this regard, there are already many considerations. For example, a flexible shuttle could replace weak scheduled services or connect new development areas with existing public transportation. Even the fine distribution from the station on the first and last mile would be solvable herewith.

5.2. Possible Adaptations in the Seating Environment

A comparison will be made between conventional seats and the possible seating concept of an on-demand service. Considering the large number of different vehicle seats, many comparisons would be conceivable. Since this is an investigation during the concept phase, we restricted the solution space to a more abstract level. During the course of gathering information, experts in seat development were asked about the possibilities and boundary conditions of such a development.

Based on these findings, the possibilities were outlined and sorted into categories. This resulted in the framework in Table 3. It compares the various requirements in four main categories and divides them into requirement (R), desire (D), or not required (NR). Due to indispensability, the categories with requirements are highlighted in red. The possible savings potential is shown in the next column with up to three € signs. In addition, further comments can be found on the far right. Finally, the central technical properties are indicated in the lower section. However, no details were given here for a purpose-built vehicle.

Table 3. Potential analysis in seat development for PBVs compared to existing seat concepts. For comparison, a conventional vehicle seat is used, as it is installed in most vehicles in Europe today. The potentials are estimated based on five main categories previously addressed by the experts.

Category		**Requirement**	**Passenger Car**	**PBV**	**Savings Potential [€] with Elimination**	**Remarks**
			(Passenger Seat)	**(Passenger Seat)**		
			Requirement [R]/Desire [D]/ Not Required [NR]	**Requirement [R]/Desire [D]/ Not Required [NR]**		
The requirements for adjusters and comfort functions differ considerably between passenger cars and PBVs. Only a few points are congruent.						
Seat adjusters and Comfort	1	Electric seat adjustment	D	NR	€€€	Nice to have
	2	Seat belt height adjustment	D	D	€	Useful for comfort & safety when passengers do not conform to the norm.
	3	Longitudinal adjustment	R	NR	€	Necessary for passenger cars, also to be able to adjust the legroom in the second row. Rather not desired for PBVs.
	4	Height adjustment	D	NR	€€	Nice to have for co-drivers in cars, not absolutely necessary in PBVs due to short distances and transport times.
	5	Backrest angle adjustment	R	D	€€	

Table 3. *Cont.*

Category		Requirement	Passenger Car (Passenger Seat) Requirement [R]/Desire [D]/ Not Required [NR]	PBV (Passenger Seat) Requirement [R]/Desire [D]/ Not Required [NR]	Savings Potential [€] with Elimination	Remarks
	6	Headrest: Minimum in 2 ways adjustable (if not integrated)	R	D	€	Important safety feature in passenger cars, similarly important in PBVs but tends to be impractical because passengers may not take the time to adjust the headrest to themselves.
	7	Seat cushion tilt adjustment	D	NR	€€	Not needed for PBV. In some cases considerable savings potential in terms of costs and development time. If necessary, seat heating can be considered, as it is easy to implement and low cost - in addition, it is already almost part of the standard equipment in most vehicles.
	8	Seat cushion extension	D	NR	€€	
	9	Side bolster adjustment	D	NR	€€€	
	10	Lordosis support (min. 2-way)	D	NR	€€	
	11	Memoryfunction	D	NR	€€	
	12	Massage function	D	NR	€€€	
	13	Seat heating	D	NR	€	
	14	Seat ventilation	D	NR	€€	
	15	Angle adjuster in y/ rotating seat rail	NR	D	€€	For PBVs, possibly interesting to increase privacy.
	16	Relax/recline function	D	NR	€€€	An absolute luxury feature in modern passenger cars. Not necessary or not desired in PBVs.
Environment	17	USB port	D	D	€	Nice to have
	18	Grab handle	NR	R	€	Necessary in the PBVs to facilitate access for all passengers.
	19	Armrest	D	D	€€€	Can have a significant impact on comfort, but can also be a cost generator - depending on the solution & effort involved
	20	Storage area	R	R	€	Suitable for both vehicle types
	21	Storage compartment	D	NR	€€€	Storage compartment for PBV not necessary: Danger of things being forgotten in it
	22	Cupholder	D	D	€	Practical and can be implemented relatively inexpensively

Table 3. *Cont.*

Category		Requirement	Passenger Car (Passenger Seat) Requirement [R]/Desire [D]/ Not Required [NR]	PBV (Passenger Seat) Requirement [R]/Desire [D]/ Not Required [NR]	Savings Potential [€] with Elimination	Remarks
Ergonomics	23	Seat height: Even small people equipped with good all-round visibility	R	NR	-	In the PBV, privacy may take priority over good all-round visibility.
	24	Headroom: Tall people should be able to sit upright	R	R		Absolute necessity
	25	Good entry & exit	D	R		Much more important due to frequent entries & exits in PBV.
	26	Good to very good lateral support	R	D		With PBV good entry & exit > Side support (height of side bolsters)
Material	27	Easy to clean / washable	D	R		Today's hygiene standards require a maximum of cleanliness and, if necessary, the possibility of disinfection, especially for ridesharing and ridepooling concepts.
	28	Seat cover with high coefficient of friction	D	D		Can significantly support the lateral support. Especially for seats of PBVs where the side bolsters should not be so pronounced to ensure good entry and exit.
	29	High abrasion resistance	R	NR		Significantly lower number of entrances & exits in the passenger car, therefore lower requirements than in the PBV
	30	Very high abrasion resistance	D	R		
Technical data	31	Seat cushion hardness	6 ± 0.5 KPA *	?		
	32	Backrest hardness	6 ± 0.5 KPA *	?		
	33	Height side bolsters seat cushion	20.8 mm (For Reference) *	?		
	34	Height side bolsters backrest	40.4 mm (For Reference) *	?		
	35	H point height	421.5 mm *	?		

* values of the Mercedes-Benz V-Class/e-Vito in the second and third row of seats.

In the context of seat development, it is unrealistic to initiate entirely new trends rapidly. The number of pieces and tools are designed and optimized for other processes. Otherwise, it becomes costly. As soon as a significant or critical change is made, the entire seat must be cleared again. The requirements for homologation and other approvals should not be underestimated. High crash requirements must also be met, which cannot be circumvented. All extras, such as lumbar support or electric adjustment, show higher equipment costs.

Simple seat concepts still predominate, or the take-rate is considerably high in the lower-priced segment. A viable option: based on an existing seat, one could be tempted to adopt it. If the carrier and basic frame are not changed, no new extensive testing needs to be carried out. Foams and covers can still be changed. Seats in airplanes or train seats have much lower complexity. The focus here was elsewhere, namely maximum lightweight construction vs. suitability for everyday use or traveling comfort. Both are significantly more resistant and easier to care for. However, the application in the vehicle must be explicitly examined, as the boundary conditions are entirely different and, for example, vibrations play a role again, which was not taken into account before. With new kits, extra equipment can be considered early so that it is easy to react to later adaptations and costs remain low. Some solutions are already pre-developed. The focus for innovations and new developments is on Generation Z.

Reference is made to the incredible complexity of seat development. There are many people involved and lengthy processes that cannot be easily changed. Through iterations, development and production are now highly optimized. Therefore, this use case should instead be located in the commercial vehicle industry. There, small quantities and particular adaptations are easier to implement. This was also the case with Volkswagen's MOIA Pluto. With the many expected passenger changes, the seat concept should support easy and fast boarding and exiting.

On the one hand, this can be taken into account in the package through a higher arrangement. On the other hand, active systems would be conceivable and easy to implement but are more expensive. Of course, this can be expanded further and can lead to changes in the entire vehicle concept. Many user-friendly attributes known from family vans or residential properties are not necessarily allowed. For example, the seats may not be rotated while driving, and folding tables may only be used when stationary. Anything else is invalid according to the operating permit. There is a lot of company-internal and -specific knowledge that is not easily shared or published. Thus, superficial knowledge can be found in a few standard works. New technologies are already being used on a global scale, such as dirt-resistant and antibacterial covers. Conventional car seats will not stand up to the strain over time and will need rehashing. However, they are not designed for this, i.e., you cannot simply change the cover, foams, etc. They are assembled in the factory once in their lifetime. Replacement is theoretically possible but not common in practice and cannot be done by one person alone. Standards and specifications are completely different depending on vehicle class. A simple example is that seats in trains must be inflammable, but in buses, everything is allowed. A new vehicle class or registration category could create clarity here and simplify many things.

6. Conclusions

On-demand mobility services should support different customers in their everyday mobility and make public transportation more attractive. To achieve this, vehicle concepts must be analyzed, optimized, and tested from the passengers' point of view. The increasing shift from privately owned motorized transportation towards shared mobility is creating entirely new market segments. OEMs and other participants from the mobility industry must position themselves at an early stage. One such market segment is the so-called "purpose-built vehicle" for ride-pooling services. Companies may manage to establish themselves early in this emerging market segment, win customers, and gain valuable early experience. In that case, they will already be well prepared for the next step towards autonomous driving services.

Overall, PBVs could significantly reduce their vehicle-to-market costs and further customize and fully integrate the user experiences. Our work revealed weak spots in currently used ride-pooling vehicles, and possible improvements for future (autonomous) shuttle buses were identified. For this purpose, an expert survey and workshops were conducted to determine user-specific requirements for ride-pooling vehicles. Based on the work undertaken, it can be clearly stated that new needs and requirements are placed for this vehicle type.

Since every provider will have special requests, this can hardly be taken into account in vehicle production. Automobile production lives on other quantities. Everything else is a unique construction and/or the subject of body manufacturers/converters. As a result, prices automatically go up.

So far, the ride-pooling fleet in Germany is still very small, which is why both the level of awareness and the share of the modal split are low. Larger fleets could increase public awareness and make the development of PBVs economically attractive. This would subsequently lead to an improvement in the quality of service.

The design of ride-pooling vehicles offers much room for further research. Comfort is, therefore, a subjective interpretation of an interplay of different sensations based on experiences, expectations, and preferences [35]. New mobility is a very young field and thus enables ergonomics to be integrated into the vehicle development process from the beginning.

Author Contributions: M.D.: conceptualization, methodology, formal analysis, investigation, interpretation, writing—original draft. P.W.: conceptualization, visualization, investigation, writing—original draft. L.-T.D.: investigation, writing—original draft. A.T.: investigation, writing—original draft. M.F.: methodology, formal analysis, writing—review and editing. K.B.: supervision; writing—review and editing. M.S.: interpretation, specification requirements, writing—review and editing. S.S.: specification requirements, visualization, writing—review and editing. T.T.: specification requirements, visualization, writing—review and editing. All authors have read and agreed to the published version of the manuscript.

Funding: This publication was funded by the Open Access publication fund of the Technical University of Munich.

Institutional Review Board Statement: Not applicable.

Informed Consent Statement: Informed consent was obtained from all subjects involved in the study.

Data Availability Statement: The data sets generated and/or analyzed during the current study contain personal information and as a result are not publicly available due to data protection reasons but are available from the corresponding author on reasonable request.

Acknowledgments: We would like to thank all participating experts for their insights, stories and information.

Conflicts of Interest: The authors declare no conflict of interest.

References

1. Baumann, S.; Püschner, M. Nutzungsszenarien I. In *Smart Mobility: Trends, Konzepte, Best Practices für die Intelligente Mobilität*; Flügge, B., Ed.; Springer Fachmedien Wiesbaden: Wiesbaden, Germany, 2016; pp. 91–98. ISBN 978-3-658-14371-8.
2. Flügge, B. (Ed.) *Smart Mobility: Trends, Konzepte, Best Practices für die Intelligente Mobilität*; Springer Vieweg: Wiesbaden, Germany, 2016; ISBN 978-3-658-14370-1.
3. Kahle, H.N.; Yunus, M. *Mobilität in Bewegung: Wie Soziale Innovationen Unsere Mobile Zukunft Revolutionieren*; GABAL: Offenbach, Deutschland, 2021; ISBN 9783967390605.
4. Federal Ministry for the Environment, Nature Conservation and Nuclear Safety—Federal Environment Agency. Emissionsquellen. Available online: https://www.umweltbundesamt.de/themen/klima-energie/treibhausgas-emissionen/emissionsquellen#energie-stationar (accessed on 16 November 2021).
5. European Commission. Verkehrsbedingte Emissionen Sollen bis 2050 Um 90 Prozent Sinken. 2020. Available online: https://www.eiz-niedersachsen.de/verkehrsbedingte-emissionen-sollen-bis-2050-um-90-prozent-sinken/ (accessed on 16 November 2021).
6. Canzler, W.; Radtke, J. Der Weg ist das Ziel: Verkehrswende als Kulturwende. Oder: Zur schwierigen Entwöhnung vom Auto. *Politik und Zeitgesch.* **2019**, *69*, 33–38.
7. Knie, A.; Canzler, W.; Ruhrort, L. Autonomes Fahren Im Öffentlichen Verkehr—Chancen, Risiken und Politischer Handlungsbedarf. 2019. Available online: https://www.gruene-hamburg.de/wp-content/uploads/2019/04/Autonomes_Fahren_Gutachten_030419.pdf (accessed on 27 August 2021).
8. Wong, Y.Z.; Hensher, D.A.; Mulley, C. Mobility as a service (MaaS): Charting a future context. *Transp. Res. Part A Policy Pract.* **2020**, *131*, 5–19. [CrossRef]
9. Thaithatkul, P.; Seo, T.; Kusakabe, T.; AAsakura, Y. A Passengers Matching Problem in Ridesharing Systems by Considering User Preference. *J. East. Asia Soc. Transp. Stud.* **2015**, *11*, 1416–1432. [CrossRef]
10. Greenblatt, J.B.; Shaheen, S. Automated Vehicles, On-Demand Mobility, and Environmental Impacts. *Curr. Sustain. Renew. Energy Rep.* **2015**, *2*, 74–81. [CrossRef]

11. Canzler, W.; Knie, A. Autonom und Öffentlich: Automatisierte Shuttles für mehr Mobilität mit weniger Verkehr. *böll.brief Grüne Ordnungspolitik Nr.13:. Heinrich-Böll-Stiftung e.V.: Berlin, Deutschland, Oktober 2019*. Available online: https://www.boell.de/de/2019/10/24/autonom-und-oeffentlich (accessed on 1 February 2022). [CrossRef]
12. Merat, N.; Madigan, R.; Nordhoff, S. *Human Factors, User Requirements, and User Acceptance of Ride-Sharing In Automated Vehicles*; International Transport Forum: Paris, France, 2017; ISBN 2223-439X.
13. Hall, E.T. *The Hidden Dimension, Reprinted*; Anchor Books: New York, NY, USA, 1990; ISBN 9780385084765.
14. Evans, G.W.; Wener, R.E. Crowding and personal space invasion on the train: Please don't make me sit in the middle. *J. Environ. Psychol.* **2007**, *27*, 90–94. [CrossRef]
15. Bohrmann, D. Probandenstudie—Vom Fahrer zum Passagier. *ATZ Extra* **2019**, *24*, 36–39. [CrossRef]
16. Nobis, C.; Kuhnimhof, T.; Follmer, R.; Bäumer, M. *Mobilität in Deutschland—Zeitreihenbericht 2002–2008–2017: Studie von infas, DLR, IVT und Infas 360 im Auftrag des Bundesministeriums für Verkehr und Digitale Infrastruktur (BMVI) (FE-Nr. 70.904/15)*; BMVI: Berlin, Germany, 2019; Available online: http://www.mobilitaet-in-deutschland.de/pdf/MiD2017_Zeitreihenbericht_2002_2008_2017.pdf (accessed on 27 August 2021).
17. Suder, E.; Pfaffenbach, C. Alltagsmobilität in Kommunen zwischen Niederrhein und Ruhrgebiet. Aus welchen Gründen wird der ÖPNV nicht häufiger genutzt? *Standort* **2021**, *45*, 31–37. [CrossRef]
18. Gaul, M. Neue Mobilität: Evolution oder Revolution? Available online: https://www.dekra-solutions.com/2020/09/neue-mobilitaet-evolution-oder-revolution/ (accessed on 16 November 2021).
19. Wolfgang, B.; Jan-Philipp, H.; Johan, K.; Marc, W. *A New Breed of Cars: Purpose-Built Electric Vehicles for Mobility on Demand*; Springer Fachmedien Wiesbaden: Wiebaden, Deutschland, 2018.
20. Grosse-Ophoff, A.; Hausler, S.; Heineke, K.; Möller, T. How Shared Mobility will Change the Automotive Industry; McKinsey&Company. 2017. Available online: https://www.mckinsey.com/industries/automotive-and-assembly/our-insights/how-shared-mobility-will-change-the-automotive-industry (accessed on 16 November 2021).
21. Hagman, J.; Ritzén, S.; Stier, J.J.; Susilo, Y. Total cost of ownership and its potential implications for battery electric vehicle diffusion. *Res. Transp. Bus. Manag.* **2016**, *18*, 11–17. [CrossRef]
22. Lutsey, N.; Nicholas, M. Update on Electric Vehicle Costs in the United States through 2030. 2019. Available online: https://theicct.org/sites/default/files/publications/EV_cost_2020_2030_20190401.pdf (accessed on 17 November 2021).
23. Dorynek, M.; Weinmann, P.; Bengler, K. Warum die Veränderung der Mobilität noch länger dauern könnte, als viele gehofft hatten. Bewertung der eingesetzten Fahrzeuge auf dem deutschen Ridepooling-Markt. In *GfA 67. Frühjahrkongress. Arbeitswissenschaftlichen Kongress*; Gesellschaft für Arbeitswissenschaft e. V.; GfA-Press: Dortmund, Germany, 2021.
24. Wedler, D.; Vietor, T. Potentials of modular autonomous vehicles for variable scenarios of public transport. In *19. Internationales Stuttgarter Symposium, Wiesbaden, 2019*; Bargende, M., Reuss, H.-C., Wagner, A., Wiedemann, J., Eds.; Springer Fachmedien Wiesbaden: Wiesbaden, Germany, 2019; pp. 557–570. ISBN 978-3-658-25939-6.
25. Diels, C.; Erol, T.; Kukova, M.; Wasser, J.; Bos, J.; Cieslak, M.; Payre, W.; Miglani, A.; Mansfield, N.J.; Hodder, S. Designing for Comfort in Shared and Automated Vehicles (SAV): A Conceptual Framework. In Proceedings of the Presented at the 1st International Comfort Congress (ICC2017), Salerno, Italy, 7–8 June 2017.
26. Frohriep, S.; Schneider, F. People on the Move: Collaboration of Ergonomics, Usability and Design in Vehicle Interior Development. In Proceedings of the International Conferences, Aachener Kolloquium 2020. Aachen, Deutschland, October 2020.
27. Baumgartner, M. Entwicklung von Lösungsansätzen für Innenraumkonzepte in vollautomatisierten zukünftigen Fahrzeugkonzepten. Master's Thesis, Technische Universität München, München, Germany, 2018.
28. Bubb, H.; Bengler, K.; Grünen, R.E.; Vollrath, M. *Automobilergonomie*, 1st ed.; Springer Fachmedien Wiesbaden GmbH: Wiesbaden, Germany, 2015; ISBN 9783834822970.
29. Kipp, M.; Bubb, I.; Schwiebacher, J.; Schockenhoff, F.; Koenig, A.; Bengler, K. Requirements for an Autonomous Taxi and a Resulting Interior Concept. In *HCI International 2020—Posters*; Stephanidis, C., Antona, M., Eds.; Springer International Publishing: Cham, Switzerland, 2020; pp. 374–381. ISBN 978-3-030-50732-9.
30. Human Accom and Design Devices Stds Comm. J2732_200806-Motor Vehicle Dimensions. SAE International; 23.06.2008. Available online: https://saemobilus.sae.org/content/j2732_200806 (accessed on 1 February 2022).
31. Mueller, A.; Beyer, S.; Kopp, G.; Deisser, O. User-Centered Development of a Public Transportation Vehicle Operated in a Demand-Responsive Environment. In *Advances in Human Factors of Transportation*; Stanton, N., Ed.; Springer International Publishing: Cham, Switzerland, 2020; pp. 545–555. ISBN 978-3-030-20503-4.
32. Christopher, W.; Diane, G.; Andrew, M. Re-Inventing the Journey Experience—A Multifaceted Framework to Comfort in Autonomous Vehicles. In Proceedings of the 2nd International Comfort Congress, Delft, The Netherlands, 29–30 August 2019.
33. Dorynek, M.; Guthardt, A.; Bengler, K. Developing a Standard One-Fits-All Boarding Assistance System as a Universal Accessibility Solution. In *Congress of the International Ergonomics Association*; Springer: Cham, Switzerland, 2021; Volume 220, pp. 229–238. [CrossRef]
34. Schütte, G. Die Zukunft des Entwicklungsprozesses in der Automobilindustrie. 2019. Available online: https://www.goetzpartners.com/fileadmin/user_upload/Ergebnisbericht_Entwicklung_in_der_Automobilindustrie.pdf (accessed on 17 November 2021).
35. Ulherr, A.D.J. Bewertung des aktuellen Vorgehens bei Diskomfortuntersuchungen im Sitzen. Ph.D. Thesis, Technische Universität München, München, Germany, 2019.

vehicles

MDPI

Article

Numerical Study of Longitudinal Inter-Distance and Operational Characteristics for High-Speed Capsular Train Systems

Bruce W. Jo

Advanced Dynamics Mechatronics and Collaborative Robotics (ADAMS) Laboratory, Department of Mechanical Engineering, State University of New York (SUNY), Stony Brook University, Incheon 21985, Korea; bruce.jo@stonybrook.edu

Abstract: High-speed capsular vehicles are firstly suggested as an idea by Elon Musk of Tesla Company. Unlike conventional high-speed trains, capsular vehicles are individual vessels carrying passengers and freight with the expected maximum speed of near 1200 [km/h] in a near-vacuum tunnel. More individual vehicle speed, dispatch, and position control in the operational aspect are expected over connected trains. This numerical study and investigation evaluate and analyze inter-distance control and their characteristics for high-speed capsular vehicles and their operational aspects. Among many aspects of operation, the inter-distance of multiple vehicles is critical toward passenger/freight flow rate and infrastructural investment. In this paper, the system's equation, equation of the motion, and various characteristics of the system are introduced, and in particular control design parameters for inter-distance control and actuation are numerically shown. As a conclusion, (1) Inter-distance between vehicles is a function of error rate and second car start time, the magnitude range is determined by second car start time, (2) Inter-distance fluctuation rate is a function of error rate and second car start time, however; it can be minimized by choosing the correct second car start time, and (3) If the second car start time is chosen an integer number of push-down cycle time at specific velocity error rate, the inter-distance fluctuation can be zero.

Keywords: hyperloop actuation; capsular vehicles; inter-distance control; magnetic levitation; control and actuation; numerical studies; longitudinal distance

Citation: Jo, B.W. Numerical Study of Longitudinal Inter-Distance and Operational Characteristics for High-Speed Capsular Train Systems. *Vehicles* **2022**, *4*, 30–41. https://doi.org/10.3390/vehicles4010002

Academic Editor: Mohammed Chadli

Received: 31 October 2021
Accepted: 31 December 2021
Published: 5 January 2022

1. Introduction

The conventional modes of transportation of people consist of four unique types and that are rail, road, water, and air. As transportation technology has been advanced in recent years, people are seeking much faster and more efficient ground transportation ways all over the world. Many countries have developed high-speed railway transportation. The most famous one is the Chinese bullet train- China Railway High-speed(CRH) train series as shown in Figure 1. with a normal speed of 350 [km/h], and now the Shanghai Maglev can reach a speed of 430 [km/h] during daily operation [1]. Other countries like France has SNCF TGV series train with a speed of 350 [km/h], Japan has Shinkansen series train with a speed of 320 [km/h], Germany has ICE with 320 [km/h], Spain AVE with 310 [km/h], South Korea KTX train with 305 [km/h], etc. [2–5].

Figure 1. An exemplary picture of high-speed connected vehicles (image courtesy by China Railway High-speed).

2. High-Speed Capsular Vehicles and Related Works

High-speed capsular systems are vehicles that travel within a vacuumed tunnel, which is designed to travel a long distance within a very short time. Due to reduced aerodynamic resistance, the vehicles normally can travel at ultra-high speed, more than 1000 [km/h]. The transportation system departures with a passenger-vehicle capsule pod as depicted in Figure 2. Hyperloop is one of the high-speed capsular vehicles, which was first published by Elon Musk of Tesla Company in August 2013 [6–10]. It is a kind of vehicle system incorporating reduced-pressure tubes in which pressurized capsules ride on air bearings driven by linear induction motors and axial compressors [11–13]. Reduced air resistance could permit vactrains to travel at very high speeds with relatively little power. Unlike conventional high-speed trains, capsular vehicles are individual vessels carrying passengers and freight with the expected maximum speed of near 1200 [km/h] in a near-vacuum tunnel and CFD (Computational fluid dynamics) works have been performed to validate feasibilities [14–16]. Capsular vehicles have a lot of advantages over traditional on-ground transportations like high speed, capacity, fuel efficiency, operational cost, environmentally friendly, etc. [17–19]. Among various subject areas for realization including communication [20,21], software systems and their designs [22], performance modeling and simulations [23], energy management [24] and eco-friendly effect [25], train control management and operation [26–28], this paper focuses on the inter-distance control aspect in operation since its inter-distance is directly related to the number of passengers, freight, and the overall efficacy of hyperloop system operation.

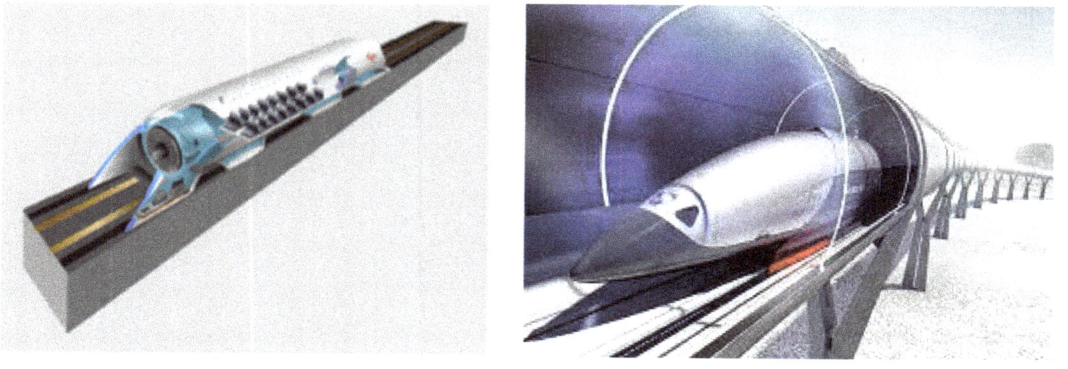

Figure 2. Conceptual rendering of capsular vehicles (image from KRRI).

M. Overton et al. focused on and described the capsule hazards and mitigations along with specialized evacuation processes appropriate for Hyperloop [29]. C. Taylor et al. did a high-level Hyperloop commercial feasibility analysis overview about system costs, market, safety, capacity, etc. [30]. K. Goeverden et al. did Hyperloop performance analysis and comparison with high-speed rail and air passenger transport system about the operation, finance, and environment, found out that Hyperloop is better for the environment with long-distance travel, but not for operation and finance [31]. Mitropoulos and et al. did a construction design about the Hyperloop transportation system. They proposed the model of the tube, capsule, compressor, suspension, and propulsion [32]. The MIT Hyperloop Team and some academic institutes competed in the SpaceX Hyperloop competition from June 2015 to January 2017 [17,33]. They did a detailed explanation of the design of the pod, the construction of the pod, and the tests that were performed–including an analysis of the competition run.

R. Janzen proposed a high-speed vehicle model–Transpod. It is designed to carry passengers at speeds exceeding 1000 [km/h], presented concerning dynamics considerations. The system is based on electromagnetic propulsion of vehicles within a protected tube guideway, whose air pressure is reduced and controlled for improved performance at high speed. The tube environment is designed for levitation systems, stability systems, and safety support systems, to permit multiple TransPod vehicles to run simultaneously with high-frequency departures [12].

3. Systems Dynamic Model

In this paper, a specific system model and its specification has been used as shown in Table 1. Traffic dynamics modeling is an essential task for longitudinal control. For single-lane traffic modeling, considering constant space platoon, the dynamics function will be defined:

$$X_i = X_{i-1} - D_i \tag{1}$$

where X_{i-1} and X_i are the x (longitudinal axis) co-ordinates of the preceding vehicle and the vehicle under the consideration and D_i is a constant distance between adjacent two vehicles. For single-vehicle dynamic modeling,

$$m\ddot{X} = \sum F(t) = F_{input}(t) - F_d - F_{EDS} \tag{2}$$

Table 1. Given System Specifications.

Given System Specifications	
Pressure	1/1000 [atm]
Max speed	1220 [km/h] (339 [m/s])
Weight	24,000 [kg] (No passenger)
	26,000 [kg] (with full passenger)
Dimension	1.25 × 1.25 × 26 [m^3]
Levitation gap	over 100 [mm]
Levitation	Superconductor EDS
Propulsion	LSM
Acceleration	(+,−) 0.5 g [m/(s^2)]

Aero resistance force F_d is

$$F_d = C_D \times \frac{1}{2} \times \rho \times V^2 \times A(N) \tag{3}$$

where:

$$\rho = 0.00116\ \text{kg/m}^3,$$

$$C_D = 6.5,\ A = 1.41\ \mathrm{m}^2,\ Fd = 0.0053\ \mathrm{V}^2.$$

EDS resistance force is a resistance force caused by magnetic levitation. We obtained experimental data of the real-time EDS force corresponding to a series of velocities from 0 [km/h] to 1220 [km/h]. Using the discrete data, we generated the continuous equation to simulate the data set, which is given by Equation (4). To better illustrate the proximity of the equation, we draw the trend line of the data set as shown in Figure 3 below.

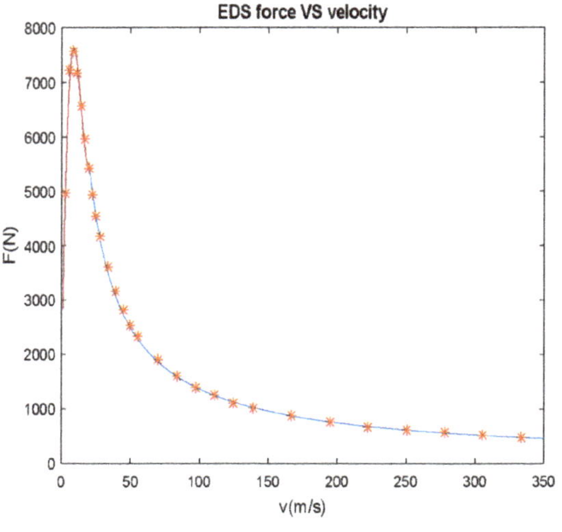

Figure 3. EDS force simulation with velocity.

In this system, the aerodynamic drag force and EDS resistance force are both functions of velocity. Thus, the system equation can become:

$$m\dot{V} = F_{input}(t) - F_d(V) - F_{EDS}(V) \tag{4}$$

Furthermore, we have:

$$\dot{V} = F_{input}(t)/m - F_d(V)/m - F_{EDS}(V)/m \tag{5}$$

If the input force is given always proportional to the mass, then the system becomes a first-ordinary differential equation about velocity no matter how much the system masses are. Thus, to solve this system is to solve the velocity of this system.

4. Numerical Analysis

In our simulation, the mass is given by two states: 24,000 [kg] with no load and 26,000 [kg] with a full load. We assume in this system two vehicles are departing at a certain time, first with no load and second with a full load. At first, to make the system simple and efficient, we give a pulse force denoted in Figure 4a. Then the vehicles' acceleration output response is shown in Figure 4b. For the first vehicle, if we give the initial input force to push the vehicle to maximum speed, then stop pushing, let the vehicle drive freely in the tunnel, after a certain time, give the same negative force to decelerate, and we name these three regions as acceleration region, free region, and deceleration region. As you can see, for the corresponding velocity figure, the acceleration region, the speed is increasing rapidly with an acceleration of about 4.9 [m/s^2], for the free drop region, the speed is slowly dropping, due to there is not a lot of resistance, the deceleration is small, for the deceleration region, the speed is decreasing sharply. The second car starts 50 [s] later and is given the same force pattern. The acceleration output response of the second vehicle is the same as the first one.

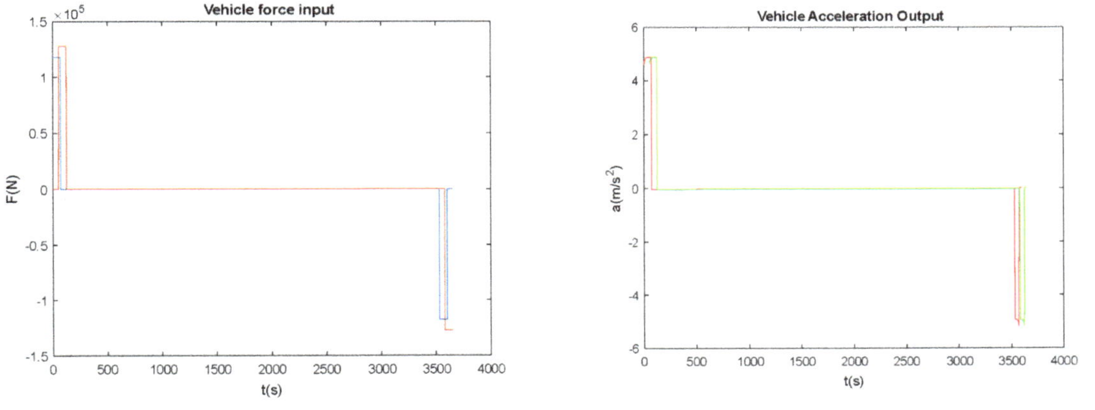

(**a**) (**b**)

Figure 4. (**a**) Input force and (**b**) vehicle accleration output when (m1 = 24,000 [kg] (blue), m2 = 26,000 [kg] (red), t_scndstart = 50 [s]) in the duration of 3600 [s] of operation, (red) and (green) respectively.

As shown in Figure 4b, two vehicles' acceleration outputs are having the same pattern—acceleration-slow dropping-deceleration. Figure 5a shows the acceleration region, as the velocity goes up before 19.82 [m/s], the acceleration is decreasing a little, and after the velocity passes the 19.82 [m/s], the EDS force goes down, the acceleration goes up to the about 4.9 [m/s^2]. Figure 5b shows for the constant but deceleration region due to aerodynamic drag and EDS resistance, the velocity is dropping at an acceleration of −0.04 [m/s^2]. For the deceleration region shown in Figure 6a, the absolute value of deceleration is increasing as the velocity goes down. Figure 6b shows the corresponding velocity response of two vehicles where 50 [s] is apart from the start time.

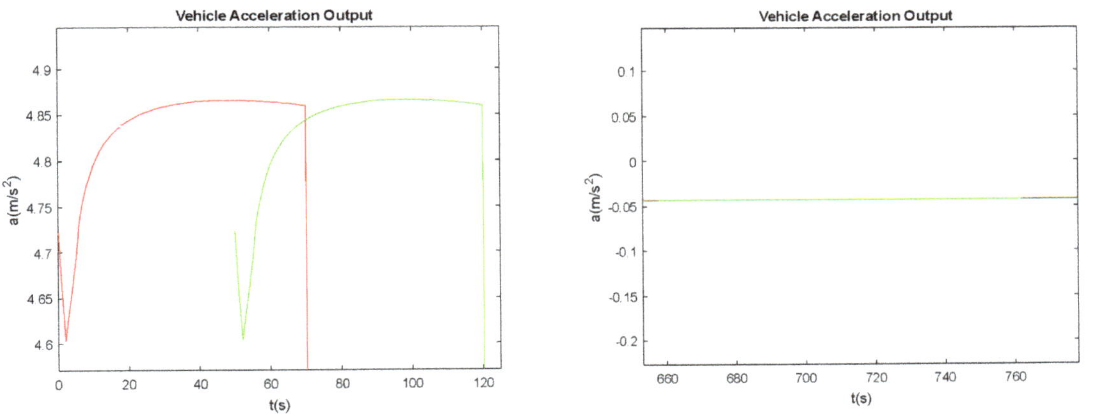

(**a**) (**b**)

Figure 5. (**a**) Acceleration response in the initial stage (before the velocity reaches 19.82 [m/s] or initial 120 [s] (red and then green), and (**b**) while the vehicle is cruising when (m1 = 24,000 [kg] (red), m2 = 26,000 [kg] (green), t_scndstart = 50 [s]).

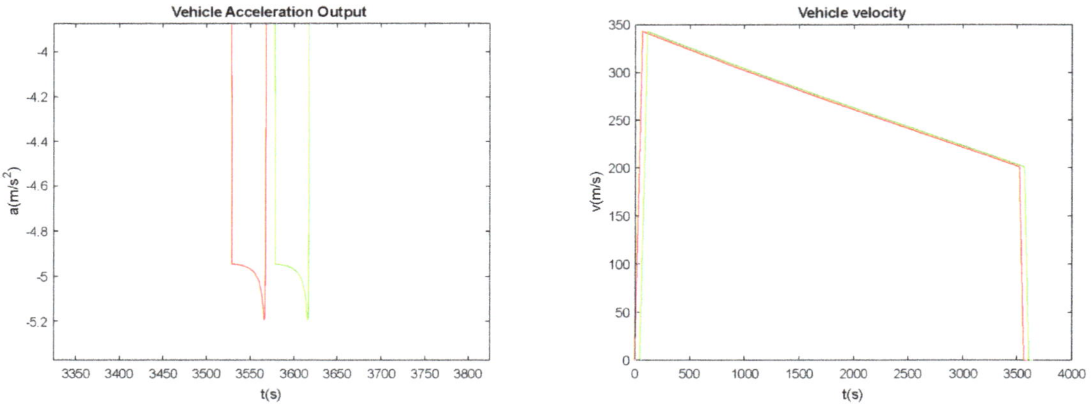

(a) (b)

Figure 6. (**a**) Acceleration output in the final state near 3600 [s] when (m1 = 24,000 [kg] (red), m2 = 26,000 [kg] (green), t_v2 = 50 [s]) and (**b**) velocity response corresponding to the acceleration output Figure 4b and input force as shown in Figure 4a.

Another important aspect of the capsular vehicle's operation is its inter-distance or displacement. The displacement of two vehicles is shown in Figure 7a. They are having the same nonlinear pattern with a time difference of 50 [s]. The inter-distance between two vehicles is changing as the figure shows in Figure 7b. Inter-distance is a nonlinear pattern. Inter-distance is going up quickly up to maximum at first 120 [s], then it starts dropping gradually because the speed of the first vehicle is smaller than the second one, finally, it drops rapidly to zero at the end.

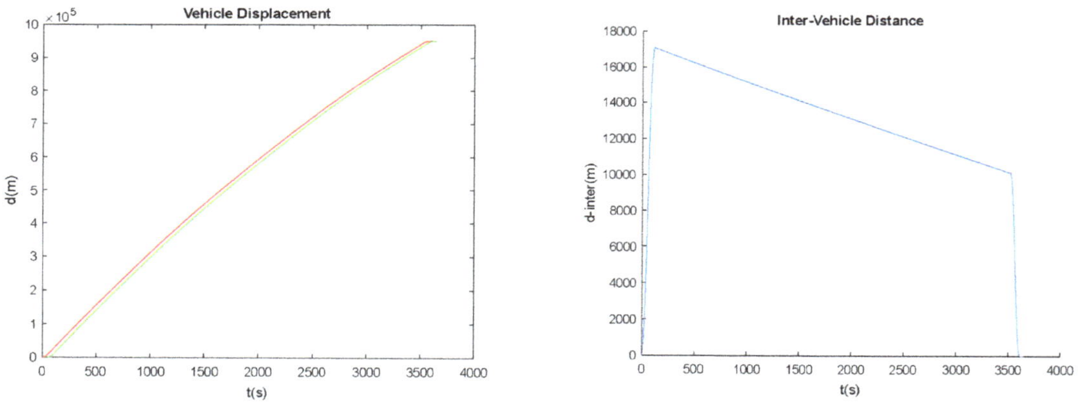

(a) (b)

Figure 7. (**a**) Each vehicle's displacement (m1 = 24,000 [kg] (red), m2 = 26,000 [kg] (green), t_scndstart = 50 [s]), and (**b**) inter-distance between two vehicles when when (m1 = 24,000 [kg], m2 = 26,000 [kg], t_scndstart = 50 [s]) in the duration of 3600 [s] of operation (blue).

The inter-distance is changing a lot in the middle region. However, from a system stability view, this is not good, thus we need to control the inter-distance to a proper range.

5. Vehicle's Actuation and Operation Aspects

To control the inter-distance, we need to adjust the two vehicles' displacement. Then we will need to regulate the vehicle velocity. We have two choices—either we control the single-vehicle velocity pattern or adjust the second car start time. To make our system more stable, we are trying to give the input during the free drop region for a single vehicle. We introduced the velocity drop error, which is defined by:

$$Error = (Vmax - Vmin)/Vmax \quad (6)$$

where $Vmax$ is the designed maximum speed, $Vmin$ is the free drop minimum speed allowance, and $Error$ is the speed drop allowance;

From this design, first, we push the vehicle to the maximum. When the velocity dropped to a certain speed, we give one input to push the vehicle speed back to maximum and then let it go freely. When the speed drops to the allowed minimum again and we give the push again, then keep doing the same loop until it gets to the final deceleration region.

Figure 8a shows a velocity response with an error of 0.1. In this case, the maximum speed is 339 [m/s], then the minimum speed allowance is 305.1 [m/s]. Whenever the velocity dropped down to 305.1 [m/s], the vehicle will get a push to 339 [m/s]. Then the designed input force for this scenario is shown in Figure 8b. For two vehicles with different masses, the force impulses are 118 [KN] and 128 [KN] separately.

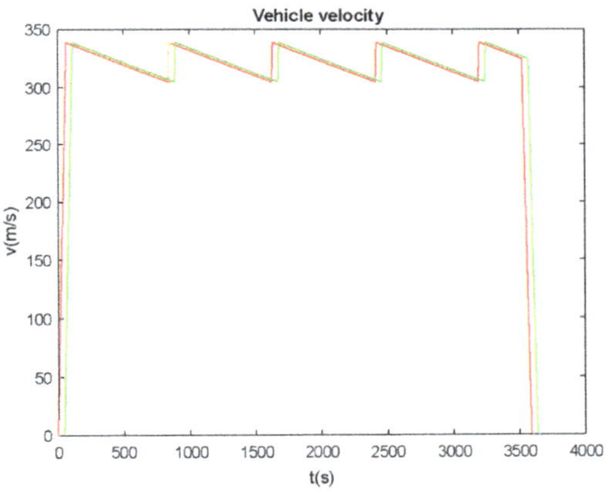

(a)

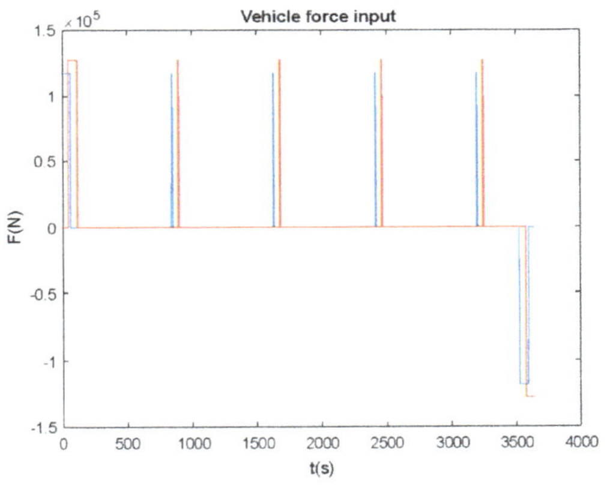

(b)

Figure 8. (**a**) Regulated velocity response with error = 0.1 rate (red for 24,000 [kg], green for 26,000 [kg], and (**b**) Corresponding force input to generate the regulated velocity as shown in Figure 8a when (m1,m2) = (24,000 (blue),26,000 (red) [kg], t_scnd = 50 [s]) respectively.

Then the vehicle output acceleration is shown in Figure 9a. For the acceleration and deceleration region, the acceleration output is the same as without regulation. For the middle region, the acceleration gets the same positive pulses when the input force has functioned.

The vehicle's displacement is shown in Figure 10a. The trend is the same as before. Figure 10a shows the inter-distance after regulation, as we see, the inter-distance is get controlled in one range, which means the velocity control is effective. However, it is getting multi up-down regions. Moreover, various error rates were applied as shown in Figure 10b.

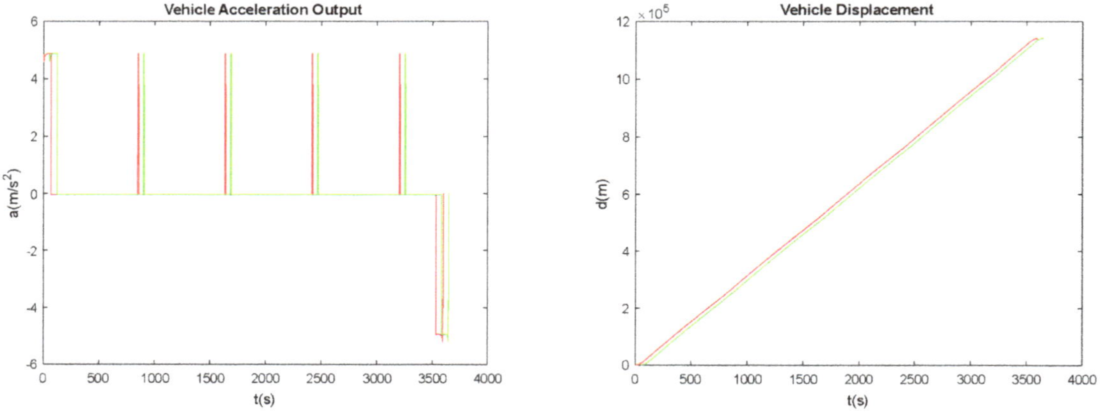

(a) (b)

Figure 9. (**a**) Acceleration output when regulated velocity with error = 0.1 rate, (red for 24,000 [kg], green for 26,000 [kg], and (**b**) Vehicles' displancement when regulated velocity with error = 0.1 rate (m1,m2) = (24,000, 26,000 [kg], t_scnd = 50 [s]) respectively.

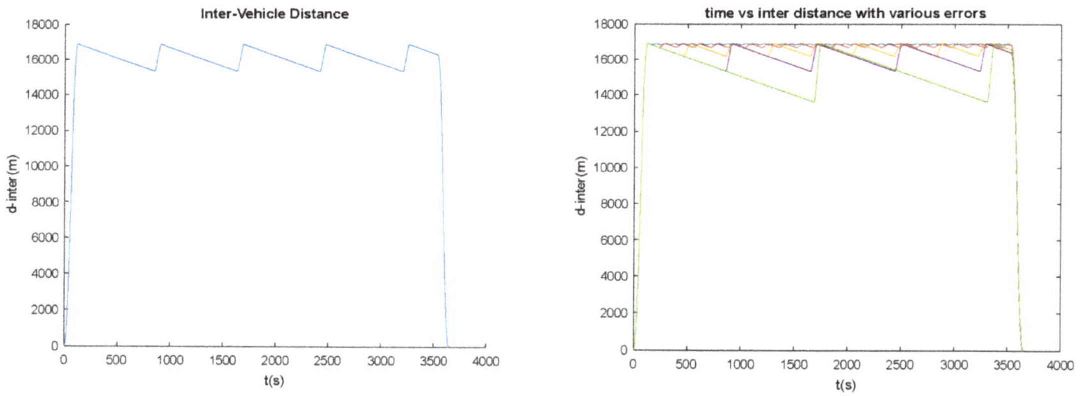

(a) (b)

Figure 10. (**a**) Inter-distance between two vehicles when regulated velocity with error = 0.1, and (**b**) Inter-distnace with various error rates (error = 1%, 2%, 5%, 10%, 20%, t_scndstart = 50 [s], m1 = 24,000 [kg], m2 = 26,000 [kg]), green, purple, yellow, orange, blue respectively.

5.1. Minimizing the Flunctuation Distance

To minimize the inter-distance fluctuation range, we are going to try two ways: (1) control the velocity error, and (2) control the second car start time. First, we fix the second vehicle start time at 50 [s] and vary the velocity error. Figure 10b shows the inter-distance with various errors at 1%, 2%, 5%, 10%, 20%. Inter-distance fluctuation is increasing when the error increases.

For one up-down region, we name it a node. As we see in Figure 11a, the nodes number is decreasing exponentially when the error increases. Moreover, the pushing time for pushing vehicles back to maximum speed from minimum is increasing linearly when error increases, see Figure 11b. The inter-distance fluctuation (df), which refers to the difference of maximum value and minimum value of fluctuation region, is also increasing linearly when error increases, also shown in Figure 11b.

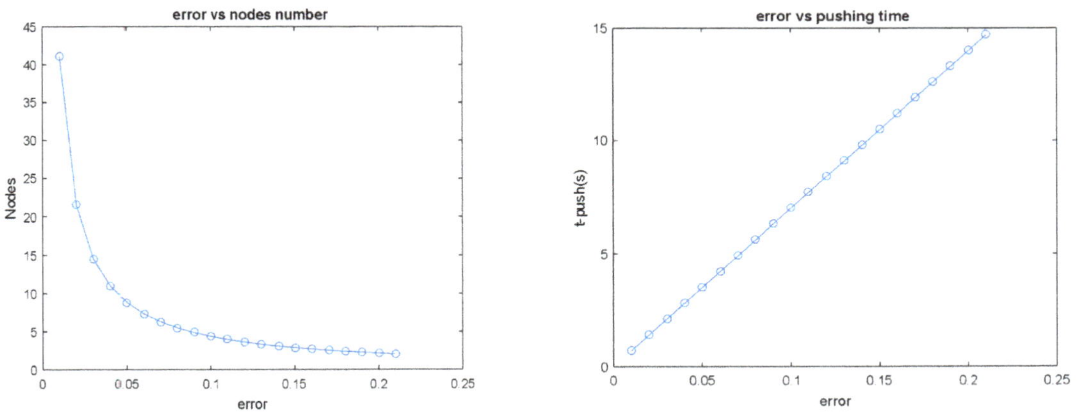

(**a**) (**b**)

Figure 11. (**a**) Various error rates vs. number of nodes, and (**b**) various error rates vs. pushing time when (error = 1%, 2%, 3%, ... , 20%, t_scndstart = 50 [s], m1 = 24,000 [kg], m2 = 26,000 [kg]).

5.2. Second Vehicle Start Time Actuation and Characteristics

After studying the velocity error effect, we find out, when the error is smaller, the inter-distance fluctuation is smaller. Now we fix the error at 1% and vary the second vehicle start time. Figure 12a shows the inter-distance with various second vehicle start time at 5 [s], 10 [s], 20 [s], 50 [s], 100 [s]. Inter-distance fluctuation normal value(d_normal), which refers to the average of maximum value and minimum value of fluctuation region, is increasing when the second vehicle start time increases, see Figure 12b. The nodes number is kept the same when the error increases, which means the nodes number is the only function of error. Moreover, the pushing time is also kept the same when error increases, which means the pushing time is also only a function of error.

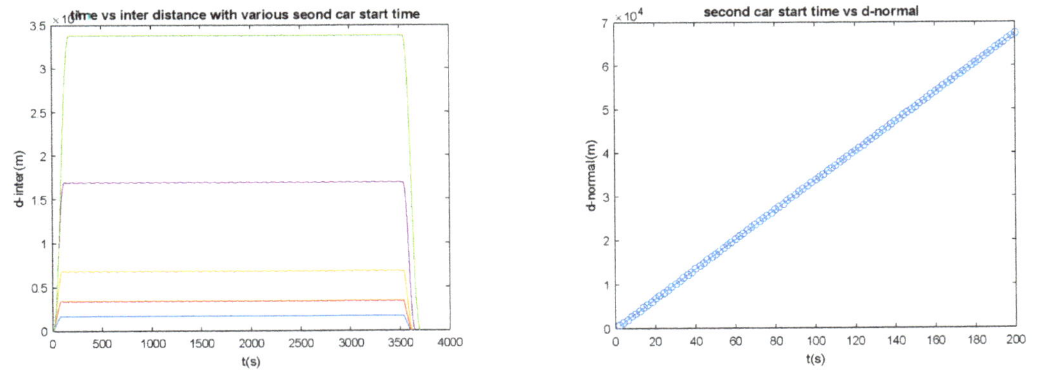

(**a**) (**b**)

Figure 12. (**a**) Inter-distance with various second vehicle start time and (**b**) second car's start time vs. inter-distance flunctuation normal distance when (error = 1%, t_scndstart = 5, 10, 20, 50, 100 s, m1 = 24,000 [kg], m2 = 26,000 [kg]) green, purple, yellow, orange, blue respectively.

Interestingly, the inter-distance fluctuation is showing a periodical pattern when error increases, shown in Figure 13a. From the figure, we can see, at a certain point, the fluctuation can be maximum, and at a certain point, the fluctuation can be zero, which means, at some second vehicle start time, there is no fluctuation for inter-distance. If we take the first zero fluctuation point, which is 76.64 [s], the velocity simulation is shown in Figure 13b, we can see that the two vehicles' velocity figures become overlapped after the first vehicle's velocity drops down to a minimum. Then the first pushing region of the first vehicle coincides with the second vehicle acceleration. Moreover, the corresponding inter-distance is shown in Figure 14a. After the inter-distance reaches to maximum, then the inter-distance is keeping the same value, there is no fluctuation anymore.

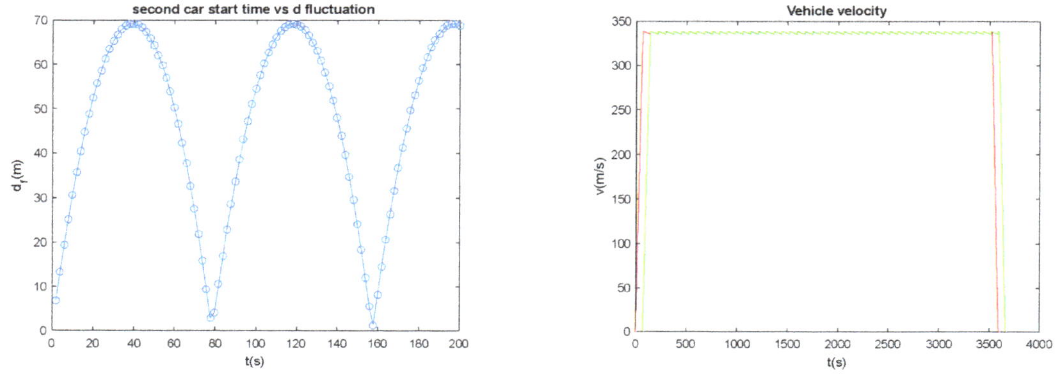

(a) (b)

Figure 13. (**a**) Inter-distance fluctuation with various second vehicle start time, and (**b**) velocity response (error = 1%, t_scndstart = 5 [s], 6 [s], 7 [s], . . . , 100 [s], m1 = 24,000 [kg] (red), m2 = 26,000 [kg] (green)).

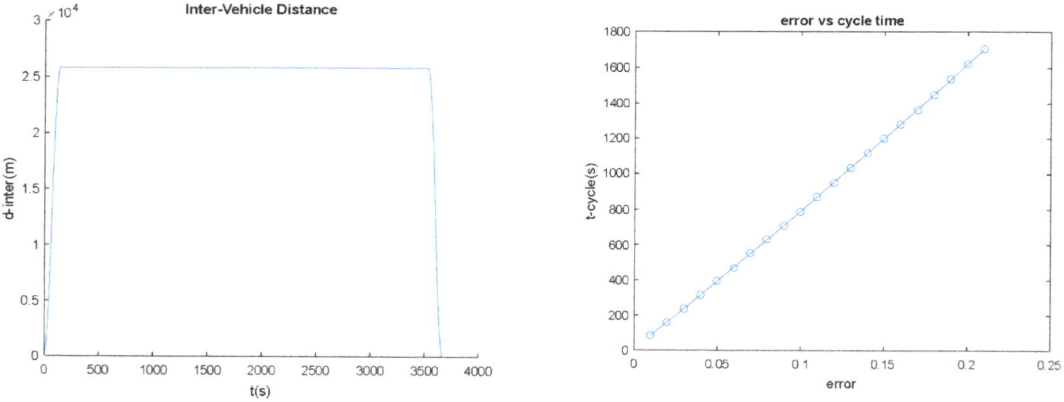

(a) (b)

Figure 14. (**a**) Inter-distance at error = 1%, t_scndstart = 76.64 [s], and (**b**) various error rates vs. push-down cycle time when (m1 = 24,000 [kg], m2 = 26,000 [kg]).

The second vehicle start period for zero inter-distance fluctuation is a push-down cycle time. Figure 14b shows the cycle time for different velocity errors. For specific velocity error rates, when the second car starts at an integral number of cycle times, the inter-distance fluctuation can always be minimized.

6. Conclusions

In this paper, operational control of high-speed capsular vehicles, numerical analysis, and evaluation has been performed with multiple control parameters. The goal is to find the best scenario to minimize the inter-distance and its fluctuation. The major accomplishment of this effort is that we generated a dynamic motion model for our unique system and proceeded with a numerical analysis toward operational control of high-speed vehicles. After we studied the numerical analysis of the system, we found out:

1. Inter-distance is a function of error rate and second car start time, the magnitude range is determined by second car start time.
2. Inter-distance fluctuation rate is a function of error rate and second car start time. However, it can be minimized by choosing the correct second car start time.
3. If the second car start time is chosen as an integer number of push-down cycle times at a specific velocity error rate, the inter-distance fluctuation can be zero.

Through this important finding based on the numerical evaluation, authors can conclude that we can design a control law to minimize the inter-distance and fluctuation rate by setting error rate and second car start time aiming for maximized passenger or freight rate with given time for future application.

Funding: This work is sponsored by KRRI (Korea Railroad Research Institute).

Institutional Review Board Statement: Not applicable.

Informed Consent Statement: Not applicable.

Data Availability Statement: Not applicable.

Acknowledgments: The author greatly appreciates Yuanyuan Zhou at the University of Maryland at College Park (former M.S. student at Tennessee State University) for her data curation.

Conflicts of Interest: The authors declare no conflict of interest.

References

1. Xiangming, W. Experience in operation and maintenance of Shanghai Maglev demonstration line and further application of Maglev in China. In Proceedings of the Maglev' 2006 Germany Proceedings, Dresden, Germany, 13–15 September 2006; Volume 1, pp. 17–19.
2. Leheis, S. High-speed train planning in France: Lessons from the Mediterranean TGV-line. *Transp. Policy* **2012**, *21*, 37–44. [CrossRef]
3. Hood, C. *Shinkansen: From Bullet Train to Symbol of Modern Japan*; Routledge: London, UK, 2006.
4. Sunduck, D.; Keun-Yul, Y.A.N.G.; Jae-Hoon, L.E.E.; Byung-Min, A.H.N. Effects of Korean Train Express (KTX) operation on the national transport system. *Proc. East. Asia Soc. Transp. Stud.* **2005**, *5*, 175–189.
5. Song, D.; Yang, C.H.; Hong, S.K.; Kim, S.B.; Woo, M.S.; Sung, T.H. Study on application of piezoelectricity to Korea Train eXpress (KTX). *Ferroelectrics* **2013**, *449*, 11–23. [CrossRef]
6. Abdelrahman, A.S.; Sayeed, J.; Youssef, M.Z. Hyperloop Transportation System: Analysis, Design, Control, and Implementation. *IEEE Trans. Ind. Electron.* **2017**, *65*, 7427–7436. [CrossRef]
7. Nar, P.; Narkar, B.; Subhash, S.; Palekar, L.S.; Ramchandra, Y. Review of Hyperloop Technology: A New Mode of Transportation. *Int. J. Sci. Res. Dev.* **2018**, *6*, 73–75.
8. Nemchenko, A.; Boyarskaya, A. The Hyperloop High-Speed Train. 2017. Available online: https://rep.bntu.by/bitstream/handle/data/33629/The%20Hyperloop%20High-Speed%20Train.pdf?sequence=1 (accessed on 2 May 2019).
9. Ross, P.E. Hyperloop: No Pressure. *IEEE Spectr.* **2015**, *53*, 51–54. [CrossRef]
10. Singh, A.; Mahajan, A.; Mahajan, N.; Gaikwad, A. Hyperloop Transportation System. *Int. Res. J. Eng. Technol.* **2019**, *6*, 2554–2558.
11. Dhandapani, C. A Review: Hyperloop Transportation System. *Cikitusi J. Multidiscip. Res.* **2019**, *6*, 5.
12. Janzen, R. Transpod ultra-high-speed tube transportation: Dynamics of vehicles and infrastructure. *Procedia Eng.* **2017**, *199*, 8–17. [CrossRef]
13. Ji, W.Y.; Jeong, G.; Park, C.B.; Jo, I.H.; Lee, H.W. A study of non-symmetric double-sided linear induction motor for Hyperloop All-In-One System (propulsion, levitation, and guidance). *IEEE Trans. Magn.* **2018**, *54*, 1–4. [CrossRef]

14. Lluesma-Rodríguez, F.; González, T.; Hoyas, S. CFD simulation of a hyperloop capsule inside a closed environment. *Results Eng.* **2021**, *9*, 100196. [CrossRef]
15. Oh, J.S.; Kang, T.; Ham, S.; Lee, K.S.; Jang, Y.J.; Ryou, H.S.; Ryu, J. Numerical analysis of aerodynamic characteristics of hyperloop system. *Energies* **2019**, *12*, 518. [CrossRef]
16. Yang, Y.; Wang, H.; Benedict, M.; Coleman, D. Aerodynamic simulation of high-speed capsule in the Hyperloop system. In Proceedings of the 35th AIAA Applied Aerodynamics Conference, Denver, CO, USA, 5–9 June 2017.
17. Opgenoord, M.M.J.; Merian, C.; Mayo, J.; Kirschen, P.; O'Rourke, C.; Izatt, G. *MIT Hyperloop Final Report*; Massachusetts Institute of Technology: Cambridge, MA, USA, 2017.
18. Dudnikov, E. Advantages of a new Hyperloop transport technology. In Proceedings of the 2017 Tenth International Conference Management of Large-Scale System Development (MLSD), Moscow, Russia, 2–4 October 2017.
19. Almujibah, H.; Kaduk, S.I.; Preston, J. Hyperloop—Prediction of Social and Physiological Costs. *Transp. Syst. Technol.* **2020**, *6*, 43–59. [CrossRef]
20. Zhang, J.; Liu, L.; Han, B.; Li, Z.; Zhou, T.; Wang, K.; Wang, D.; Ai, B. Concepts on Train-to-Ground Wireless Communication System for Hyperloop: Channel, Network Architecture, and Resource Management. *Energies* **2020**, *13*, 4309. [CrossRef]
21. Zhang, J.; Liu, L.; Wang, K.; Han, B.; Piao, Z.; Wang, D. Analysis of the Effective Scatters for Hyperloop Wireless Communications Using the Geometry-Based Model. In *International Conference on Machine Learning for Cyber Security*; Springer: Berlin/Heidelberg, Germany, 2020.
22. Nikolaev, R.; Idiatuallin, R.; Nikolaeva, D. Software system in Hyperloop pod. *Procedia Comput. Sci.* **2018**, *126*, 878–890. [CrossRef]
23. Van Goeverden, K.; Milakis, D.; Janic, M.; Konings, R. Analysis and modelling of performances of the HL (Hyperloop) transport system. *Eur. Transp. Res. Rev.* **2018**, *10*, 41. [CrossRef]
24. Lafoz, M.; Navarro, G.; Torres, J.; Santiago, Á.; Nájera, J.; Santos-Herran, M.; Blanco, M. Power supply solution for ultrahigh speed hyperloop trains. *Smart Cities* **2020**, *3*, 642–656. [CrossRef]
25. Janić, M. Estimation of direct energy consumption and CO_2 emission by high speed rail, transrapid maglev and hyperloop passenger transport systems. *Int. J. Sustain. Transp.* **2021**, *15*, 696–717. [CrossRef]
26. Ahmadi, E.; Alexander, N.A.; Kashani, M.M. Lateral dynamic bridge deck–pier interaction for ultra-high-speed Hyperloop train loading. In *Proceedings of the Institution of Civil Engineers-Bridge Engineering*; Thomas Telford Ltd.: London, UK, 2020; Volume 13, pp. 198–206.
27. Bose, A.; Viswanathan, V.K. Mitigating the Piston Effect in High-Speed Hyperloop Transportation: A Study on the Use of Aerofoils. *Energies* **2021**, *14*, 464. [CrossRef]
28. Eichelberger, M.; Geiter, D.T.; Schmid, R.; Wattenhofer, R. High-Throughput and Low-Latency Hyperloop. In Proceedings of the 2020 IEEE 23rd International Conference on Intelligent Transportation Systems (ITSC), Rhodes, Greece, 20–23 September 2020.
29. Sarin, M.; Overton, M. Complex Hyperloop Capsule Safety Requirements and Risk Mitigations. In Proceedings of the 36th International System Safety Conference, Phoenix, AZ, USA, 13–17 August 2018.
30. Hyde, D.J.; Barr, L.C.; Taylor, C. *Hyperloop Commercial Feasibility Analysis: High Level Overview*; John, A., Ed.; Volpe National Transportation Systems Center (US): Cambridge, MA, USA, 2016.
31. Alexander, N.A.; Kashani, M.M. Exploring Bridge Dynamics for Ultra-High-Speed, Hyperloop, Trains. *Structures* **2018**, *14*, 69–74.
32. Mitropoulos, L.; Kortsari, A.; Koliatos, A.; Ayfantopoulou, G. The Hyperloop System and Stakeholders: A Review and Future Directions. *Sustainability* **2021**, *13*, 8430. [CrossRef]
33. Gkoumas, K. Hyperloop Academic Research: A Systematic Review and a Taxonomy of Issues. *Appl. Sci.* **2021**, *11*, 5951. [CrossRef]

vehicles

MDPI

Article

Analysis and Research on the Comprehensive Performance of Vehicle Magnetorheological Regenerative Suspension

Jinhui Huang, Enrong Wang and Hailong Zhang *

Vibration Control Lab, School of Electrical and Automation Engineering, Nanjing Normal University, Nanjing 210097, China; 181835014@njnu.edu.cn (J.H.); erwang@njnu.edu.cn (E.W.)
* Correspondence: 61204@njnu.edu.cn

Received: 17 September 2020; Accepted: 13 October 2020; Published: 22 October 2020

Abstract: Magnetorheological (MR) regenerative suspension system can not only achieve excellent comprehensive suspension performance but also effectively recover and utilize vibration potential energy, which has been a research hotspot in the field of vehicle engineering. In this paper, for the 1/4 vehicle's MR regenerative suspension system parallel with a tubular permanent magnet linear motor (TPMLM), the dynamic model of the MR semi-active suspension system and the TPMLM finite element model are established separately to form a joint simulation platform. The simulation analysis of the comprehensive suspension performance and regeneration performance under different road excitations is performed. The results show that installing TPMLM does not change the natural resonance frequency of the suspension system, which ensures good driving comfort and handling stability. At the same time, it has considerable regeneration power. This research can provide a reference for the stability analysis and popularization of the vehicle's MR regenerative suspension system.

Keywords: magnetorheological suspension; regenerative energy; joint simulation; suspension performance

1. Introduction

The suspension is an important assembly and functional part of the vehicle. Magnetorheological (MR) semi-active suspension system has become a research hotspot in the field of vehicle suspension due to its simple structure, low energy consumption, and adaptability to various road conditions. In addition, with the increasingly prominent worldwide energy problems, energy conservation and utilization have become another important topic in the field of vehicle engineering. Research shows that in the daily use of cars, 14–26% of the fuel energy is used to overcome road friction and air resistance to drive the car [1]. The main way is to dissipate the heat energy through suspension damping damper. This part of the energy has not been fully utilized yet and has considerable recovery value.

Karnopp [2] and Browne [3] et al. carried out the researches on the recovery and utilization of vibration energy of vehicle suspension for the first time in the 1980s, indicating that suspension vibration energy has a great potential utilization value. Huang et al. [4] established a vehicle regenerative suspension model for the electrodynamic active regenerative suspension of permanent magnet direct current brushless motor combined with a ball screw structure and analyzed its mechanical properties and regenerative characteristics. The conclusion that its ride comfort needs to be improved was found. Yu et al. [5] reviewed and summarized the researches on regenerative suspension in recent years. By comparing the mechanical regenerative suspension with the electromagnetic regenerative suspension, it was found that the electromagnetic regenerative suspension is convenient for energy conversion and convenient for storage and reuse. Dong et al. [6] used a ball screw combined with a rotating motor as the power generation device and analyzed the corresponding relationship between the performance of regenerative generation and the external excitation of the two working modes of power generation device and magnetorheological damper (MRD) connected in series and parallel. The idea that vibration generation storage and suspension vibration reduction completed at the same time was proposed. Kou et al. [7] proposed a parallel ball screw semi-active suspension actuator structure, which is transformed into a unidirectional rotation power generation of a direct current brushless motor through a ball screw mechanism and a parallel gearbox structure. Zhu et al. [8] carried out research on a vehicle suspension regenerative power generation device integrated with tubular permanent magnet linear motor (TPMLM) and MRD, to directly supply power to the MRD through vibration power generation, without additional driving power. Kou et al. [9] designed a novel regenerative suspension power generation device integrating TPMLM and MRD, using skyhook semi-active control. This research showed that the suspension could realize energy self-supply. Lv et al. [10] made a comprehensive analysis of the hydraulic regenerative suspension system and analyzed the shock-absorbing and regenerative characteristics of the system, but the authors did not make a comprehensive analysis of the MR regenerative suspension in parallel with TPMLM.

The above work shows that the vibration energy of vehicle suspension has considerable recycling value. On the one hand, it can supply power for MRD, and on the other hand, it can recover and store energy to increase the cruising range of electric vehicles. However, the addition of the regenerative device may cause the suspension resonance point to shift, thus changing the inherent suspension characteristics of the vehicle, which has not been effectively improved. To solve this problem, based on the joint simulation, this paper establishes the MR regenerative suspension system simulation model in parallel with TPMLM, to analyze the system's comprehensive suspension performance and regenerative characteristics by observing the inherent suspension characteristics under different road excitations.

2. 1/4 MR Regenerative Suspension System

Figure 1 shows the structure diagram of the 1/4 MR regenerative suspension system, mainly composed of carriage, suspension spring, MRD, regenerative motor, and wheels. Wherein, m_s is the sprung mass, m_u is the unsprung mass. K_s is the stiffness coefficient of the suspension spring. F_d is the output damping force of the damper. F_m is the electromagnetic force of the linear motor. K_t is the equivalent stiffness coefficient of the tire. C_t is the tire equivalent damping coefficient. x_i, x_s, and x_u are the displacement of the road, the vertical relative displacement of the carriage, and the vertical relative displacement of the chassis, respectively.

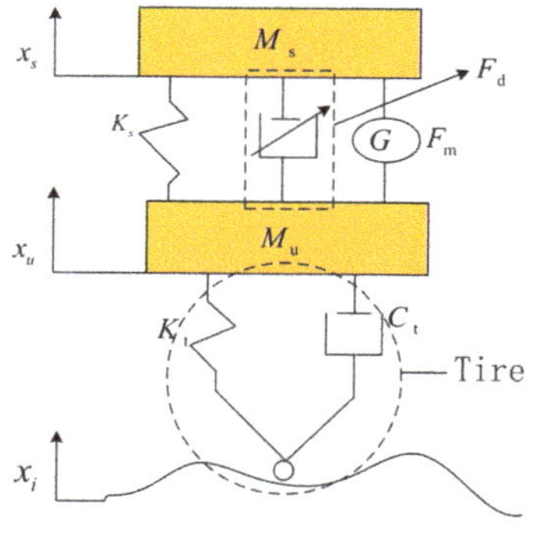

Figure 1. The structure of the 1/4 magnetorheological (MR) regenerative suspension system. The explanations of m_s, m_u, K_s , K_t, F_d, F_m, C_t, x_i, x_u, x_s are listed in the main-text.

2.1. MR Semi-Active Vehicle Suspension

Assuming that the sprung mass and the unsprung mass are rigid bodies, the motion is only perpendicular to the ground direction. The suspension system has a linear stiffness coefficient and the damping coefficient. The effective stroke of the damper and the linear motor is not exceeded in the process of motion. Furthermore, the dynamic model of the regenerative suspension system shown in Equation (1) is established.

$$\left.\begin{array}{r} m_s\ddot{x}_s = -K_s(x_s - x_u) - (F_d + F_m) \\ m_u\ddot{x}_u = K_s(x_s - x_u) + F_d + F_m \\ -K_t(x_u - x_i) - c_t(\dot{x}_u - \dot{x}_i) \end{array}\right\} \tag{1}$$

The output damping force of MRD selects the modified Bouc–Wen phenomenon model [11], of which driving current modulation is separated from the hysteresis operator.

$$F_d(i_d, x_r) = c(i_d)F_h(x_r) \tag{2}$$

$$c(i_d) = 1 + \frac{k_2}{1 + e^{-a_0(i_d+I_0)}} - \frac{k_2}{1 + e^{-a_0 I_0}} \tag{3}$$

$$F_h(x_r) = c_1\dot{y} + k_1(x_r - x_0) \tag{4}$$

$$\left.\begin{array}{r} \dot{y} = \frac{1}{c_0+c_1}\left[\alpha z + c_0\dot{x}_r + k_0(x_r - y)\right] \\ \dot{z} = -\gamma\left|\dot{x}_r - \dot{y}\right|z|z|^{n-1} - \\ (\beta|z|^n - A)\left(\dot{x}_r - \dot{y}\right) \end{array}\right\} \tag{5}$$

where i_d is the drive current. $c(i_d)$ is the drive current modulation function. $F_h(x_r)$ is the Bouc–Wen hysteresis operator. y is the internal piston displacement of the Bouc–Wen model. I_0 is the current saturation adjustment parameter, and x_0 is the initial displacement. z is a unitless progressive variable, and $k_1, k_2, a_0, c_0, c_1, \alpha, \beta, \gamma, A$ are constants of model parameters.

The improved skyhook semi-active controller is applied. Namely, when the acceleration of vehicle sprung mass is inconsistent with the relative velocity of damper, the MRD output damping force is increased. On the contrary, the MRD output damping force is reduced to suppress the resonance amplitude of sprung mass. The MRD semi-active control current i_d is as follows.

$$\left.\begin{array}{r} k_{sky}\left|\dot{x}_s\right|^2, \dot{x}_s\left(\dot{x}_s - \dot{x}_u\right) \geq 0 \\ 0, \dot{x}_s\left(\dot{x}_s - \dot{x}_u\right) < 0 \end{array}\right\} \tag{6}$$

where k_{sky} is the ceiling control coefficient.

2.2. Design of Regenerative TPMLM

Moving-coil type TPMLM is selected in this paper. Compared with the same size moving-magnetic one, it has a higher energy density, the smaller mass of moving magnetic parts, and higher electromechanical conversion efficiency. The structure diagram is shown in Figure 2, which is divided into I-V according to the material medium. The five areas are core area, winding area, air gap area, permanent magnet area, and back iron area. The magnetic tile adopts the Halbach magnetization arrangement method. This method has the highest average magnetic field energy density. The axial magnetic field distribution accounts for a high proportion of the fundamental wave component, thus reducing the electromagnetic conversion output harmonic content [12].

The regenerative suspension mainly uses TPMLM to recover electric energy. The power can be stored or used directly after conversion. The side effect can be ignored in the analysis. The principle and equation of the rotating motor are directly applied to calculate and design the linear motor and perform park transformation. Turn the motor model from $a-b-c$ stationary coordinate system to a rotating $d-q-o$ coordinate system.

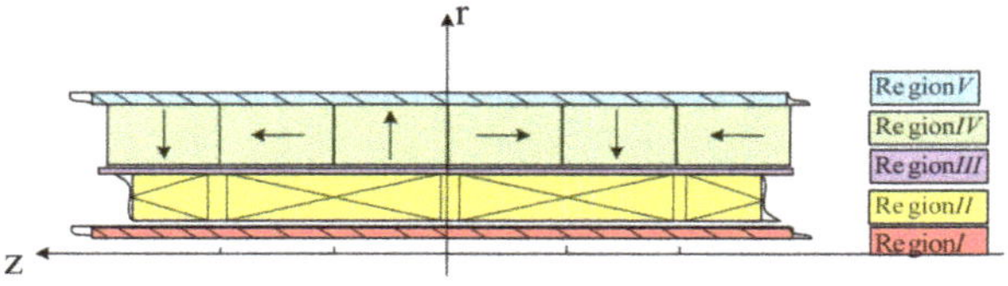

Figure 2. Structure of moving-coil tubular permanent magnet linear motor (TPMLM).

In $d-q$ coordinate system, the induced electromotive force is as follows.

$$\left.\begin{array}{l} u_{sd} = \frac{d\psi_{sd}}{dt} - \omega\psi_{sq} + R_s i_{sd} \\ u_{sq} = \frac{d\psi_{sq}}{dt} + \omega\psi_{sd} - R_s i_{sq} \end{array}\right\} \tag{7}$$

where u_{sd} and u_{sq} are the components of stator d and q coordinate axis. i_{sd} and i_{sq} are the d and q coordinate axis components of stator current. R_s is stator resistance. ψ_{sd} and ψ_{sq} are stator flux linkage.

The flux linkage equation in the $d-q$ coordinate system is as follows.

$$\left.\begin{array}{l} \psi_{sd} = L_s i_{sd} + \psi_f \\ \psi_{sq} = L_s i_{sq} \end{array}\right\} \tag{8}$$

where L_s is the stator inductance, ψ_f is the flux linkage of the permanent magnet poles.

The motion equation of the mover is as follows.

$$m_d \frac{d_v}{d_t} = F_p - F_{gen} - B \cdot v \tag{9}$$

where m_d is the mass of the mover. F_{gen} is the electromagnetic force on the mover. F_p is the suspension pressure of the vehicle on the mover, and B is the mechanical damping coefficient.

The expression of F_{gen} is as follows.

$$F_{gen} = \frac{3\pi}{2\tau} p\left(\psi_{sd} i_{sq} - \psi_{sq} i_{sd}\right) = \frac{3\pi}{2\tau} p\psi_f i_{sq} \tag{10}$$

where p is the number of pole pairs.

3. Establishment of a Joint Simulation Platform

The TPMLM finite element model is established in Ansys/Maxwell, and the vehicle MR semi-active suspension system dynamics model is established in Matlab/Simulink to build the joint operation on the Simplorer platform. Figure 3 is the schematic diagram of the simulation platform.

In Maxwell 2D simulation, the solving domain boundary conditions are set as balloon boundary. The left boundary of the iron core and the right boundary of back iron are set as vector potential boundaries. A winding is set, and six coils are added to it. The moving type in motion setup is set to the positive z-axis direction. The simulation time is from 0 to 1 s, and the time step size is 1 ms. The permanent magnet material uses NdFeB30.

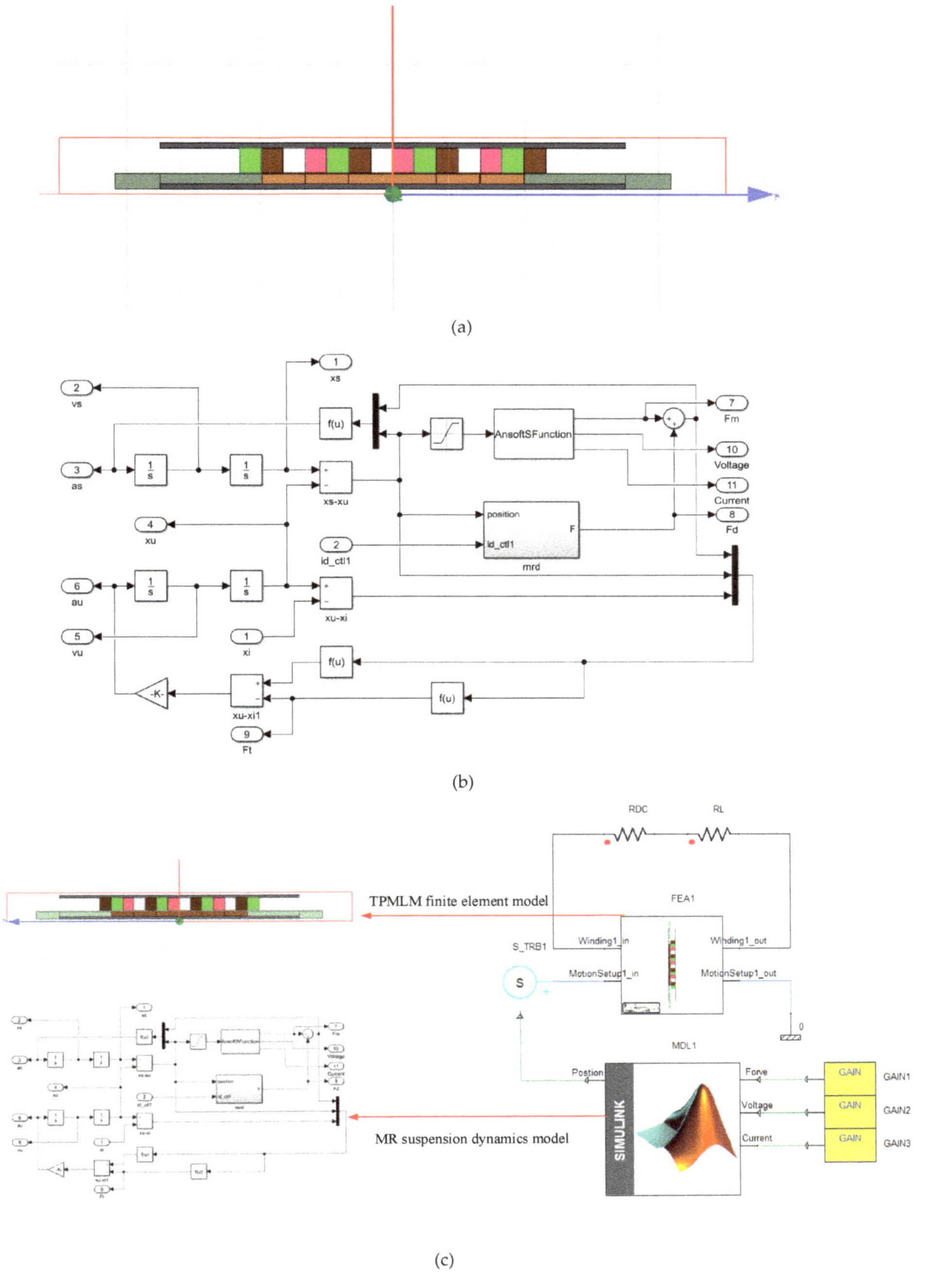

Figure 3. Schematic diagram of the MR regenerative suspension system simulation platform. (**a**) A finite element simulation model, (**b**) Suspension dynamics simulation model, (**c**) Joint simulation model.

By running the MR suspension dynamic model in Matlab/Simulink under different road conditions x_i, a displacement $x_{position}$ is obtained, which is the output of the position interface in Figure 3c. Meanwhile, $x_{position}$ is input to excite the finite element model of the TPMLM through the Simplorer (here named S_TRB1) module. Then, the joint simulation is implanted. After that, the comprehensive performance analysis of the vehicle MR regenerative suspension system is carried out based on the data obtained by the joint simulation.

The TPMLM structure parameters used in the simulation are shown in Table 1. $R_0, R_I, R_{II}, R_{III}, R_{IV}, R_V$ are the radius of the cross-sectional circle of each area of the motor model from the inside to the outside. $\mu_{rI}, \mu_{rII}, \mu_{rIII}, \mu_{rIV}, \mu_{rV}$ are the relative permeability of the material in area I-V, and τ_p is the permanent magnet pole pitch. B_{rem} is the permanent magnet remanence. m_N is the number of winding slots. N_w is the number of winding turns. K_c is the winding full-slot ratio. R_{dc} is the winding direct current resistance. ρ_c is the winding wire conductivity. S_w is the winding single-slot cut area. MRD is the actual identification parameter; $k_0 = 185.954$, $k_1 = 2750.7$, $k_2 = 9.991$, $a_0 = 7.601$, $I_0 = 0.0683$, $\alpha = 19970$, $\beta = 179900$, $\gamma = 8901.5$, $c_0 = 1367.4$, $c_1 = 6204.1$, $n = 2$, $A = 24.518$, $x_0 = -0.004$.

Table 1. Data parameters of 1/4 magnetorheological (MR) regenerative suspension [13].

Structural Parameters		Value	Unit
Cross-sectional radius	R_0	4	mm
	R_I	8	
	R_{II}	16	
	R_{III}	17	
	R_{IV}	36	
	R_V	40	
Relative permeability	μ_{rI}	400	H/m
	μ_{rII}	1	
	μ_{rIII}	1	
	μ_{rIV}	1.02	
	μ_{rV}	400	
Permanent magnet polar distance	τ_p	33	mm
Permanent magnet remanence	B_{rem}	1.23	T
Winding slots number	m_N	6	
Winding turns number	N_w	100	
Winding full slot rate	K_c	0.9	
Winding direct current resistance	R_{dc}	0.42	Ω
The conductivity of the winding wire	ρ_c	17.5	nS/m
Winding single slot cross-sectional area	S_w	264	mm^2
Regenerative suspension parameters	m_s	562.5	Kg
	m_u	90	Kg
	k_s	57,000	N/m
	c_s	2500	$N/(m \cdot s^{-1})$
	k_t	285,000	N/m
	c_t	100	$N/(m \cdot s^{-1})$

4. Comprehensive Performance Analysis of Regenerative Suspension

4.1. Suspension Performance

Firstly, the frequency response of the MR regenerative suspension system is analyzed. For nonlinear systems, constant amplitude harmonic excitation will cause a large acceleration excitation signal at high frequencies and saturation of suspension components. Therefore, the segmented harmonic excitation signal shown in Equation (11) is used to meet the requirement of limiting the amplitude of the displacement harmonic signal at high frequencies.

$$\left.\begin{array}{c} a_m \sin(2\pi f t)\ f \le f_T \\ a_m\left(\frac{f_T}{f}\right)\sin(2\pi f t)\ f > f_T \end{array}\right\} \quad (11)$$

where a_m is the single-frequency harmonic amplitude. f_T is the turning frequency. The natural frequency band of the car body and wheels is covered, setting $a_m = 2$ cm, $f_T = 2.1$ Hz.

Using the system frequency domain response about sprung mass acceleration transmission rate T_{as}, unsprung mass acceleration transmission rate T_{au}, suspension dynamic travel transmission rate T_{dr}, and tire's dynamic load coefficient (DLC), the overall suspension performance of vehicle suspension ride comfort and operational stability is evaluated, including resonance suppression, vibration isolation, suspension travel limit, and tire grip [11]. It can be seen from Figure 4 that the first resonance point of T_{as} is 1.5 Hz, which meets the requirements of passenger car stability. T_{au} is slightly distorted due to simulation reasons. Both T_{dr} and DLC show that the TPMLM does not significantly change the handling performance of MR suspension.

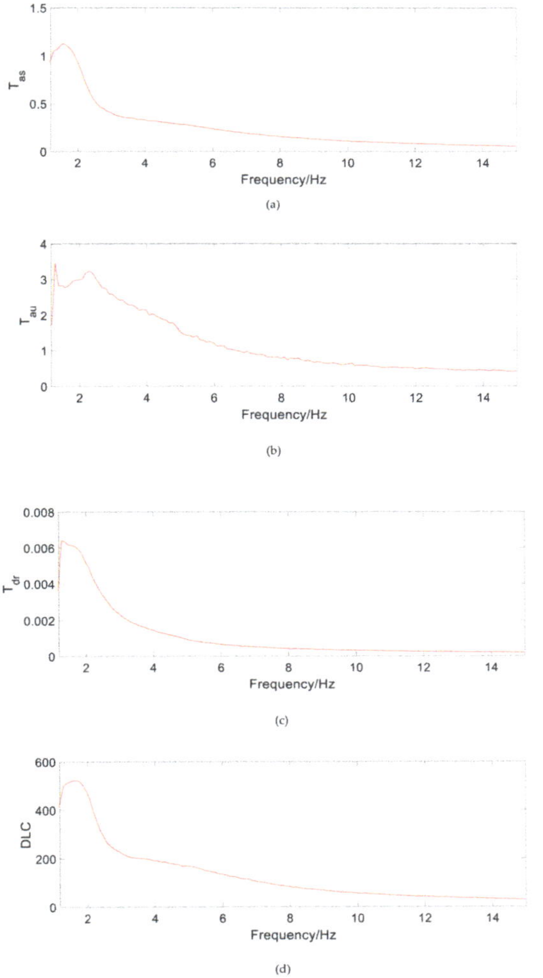

Figure 4. The frequency response of MR regenerative suspension. (**a**) Transmission rate of the sprung mass acceleration, (**b**) Transmission rate of the unsprung mass acceleration, (**c**) Transmission rate of the suspension dynamic travel, (**d**) Dynamic load coefficient of the tire.

Vehicles often encounter various impact roads during driving, which is a short time and high intensity. The smooth pulse signal excitation shown in Equation (12) is used [14].

$$x_i = 0.25a_m e^2 (\mu\omega_0 t)^2 e^{-\mu\omega_0 t}; \mu = 1, 2, 3 \tag{12}$$

In the equation, a_m is the amplitude of the smooth pulse signal, ω_0 is the fundamental frequency, and μ is the pulse stiffness coefficient. A larger μ value will cause the excitation to produce larger shock and acceleration signals.

Figure 5 shows the time-domain transient response of a_s, a_u, x_r, and F_t under smooth pulse excitation. The stabilization time of each response adjustment of the system is within 1 s. Due to the addition of TPMLM, small-scale oscillations appear in the initial stage of impulsive excitation, especially the sprung mass acceleration a_s, but the response amplitude does not increase significantly.

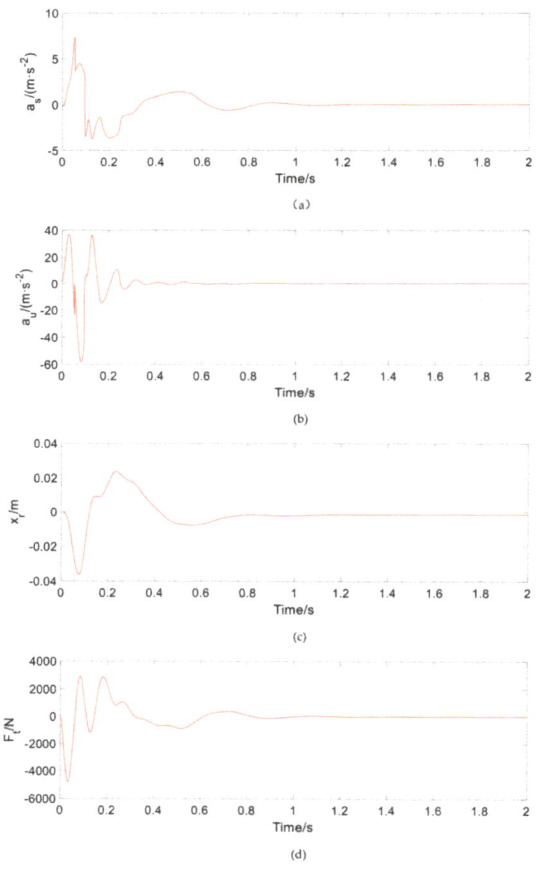

Figure 5. Dynamic characteristics under smooth pulse excitation. (**a**) Sprung mass acceleration under smooth pulse, (**b**) Free spring-mass acceleration under smooth pulse, (**c**) Suspension dynamic travel under smooth pulse, (**d**) Tire dynamic force under smooth pulse.

The random road is established based on the filtered white noise method, which is used to simulate the wheel excitation input when the vehicle is running on the actual road and gives a general analysis of the performance of the suspension system. The differential equation of road roughness is

$$\dot{q}(t) + 2\pi n_{00} u q(t) = 2\pi n_0 \sqrt{G_q(n_0) u W(t)} \tag{13}$$

where $q(t)$ is the function of road roughness. $W(t)$ is the Gaussian white noise signal with power spectral density (PSD) of 1. n_{00} is the cutoff frequency of the road spatial frequency, $n_{00} = 0.011\ \mathrm{m}^{-1}$, n_0 is the reference spatial frequency of the road. $n_0 = 0.1\ \mathrm{m}^{-1}$. $G_q(n_0)$ is the PSD of the road under the reference spatial frequency n_0, related to pavement classification and speed u in steady-state. This experiment selects $u = 20\ \mathrm{m/s}$, $G_q(n_0) = 64 \times 10^{-6}\ \mathrm{m}^3$.

Figure 6 shows a_u, x_r, and F_t under random road excitation. It can be seen that the first resonance point of the MR regenerative suspension system is 1.5 Hz, and the second resonance point is about 9.5 Hz, which is consistent with the transmission rate under the above variable amplitude harmonic excitation. It is also indicated that the addition of TPMLM does not significantly change the system stability.

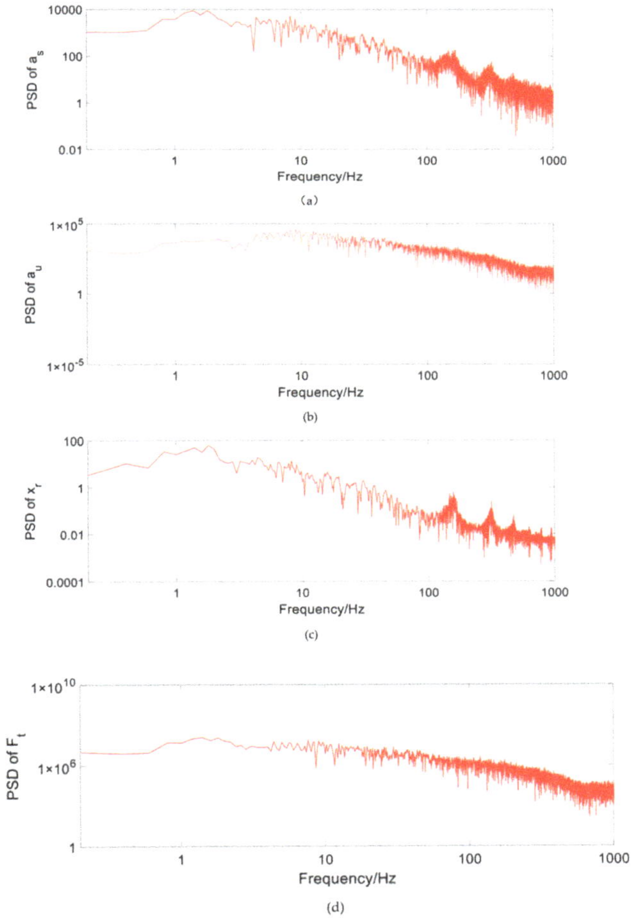

Figure 6. Power spectral density (PSD) under random road excitation. (**a**) PSD of the sprung mass acceleration, (**b**) PSD of the unsprung mass acceleration, (**c**) PSD of the suspension dynamic stroke, (**d**) PSD of the tire dynamic force.

4.2. Regenerative Characteristics

Under the single-frequency harmonic road excitation in Equation (14), above smooth road excitation, and random road excitation, the output regeneration characteristics are observed. The average output power $\overline{P}_m$ and the effective power value P are employed as the evaluation indicators, as shown in Equations (15) and (16).

$$x_i = a_m \sin(2\pi f t) \tag{14}$$

where a_m is excitation amplitude, f is frequency. The typical frequencies observed here are 1.5 Hz near the resonant frequency of the sprung mass, 6 Hz at the intermediate frequency, and 15 Hz near the resonant frequency of the unsprung mass.

The average output power of the regenerative motor

$$\overline{P}_m = \frac{1}{T}\int_0^T u(t)i(t)dt \tag{15}$$

Power effective value

$$P = \sqrt[2]{\frac{\sum_0^T {P_m}^2}{T}} \tag{16}$$

The instantaneous output power is shown in Figures 7–9. Table 2 lists the corresponding average output power and power effective value. It can be seen that the effective value of the output power at 1.5 Hz under harmonic excitation reaches about 500 W. As the frequency is increased, the regenerated power is greatly increased. On the impact road, the regenerative time is very short, but the effective value of output power is also about 50 W. On a general road, the recovered power is only 1.5 W on a flat A-class road. With the increase of road unevenness, the output power increases observably, which reaches about 46 W on a C-class road. The above results show that under the excitation of typical frequency harmonics, the MR regenerative suspension has a considerable performance of power regeneration.

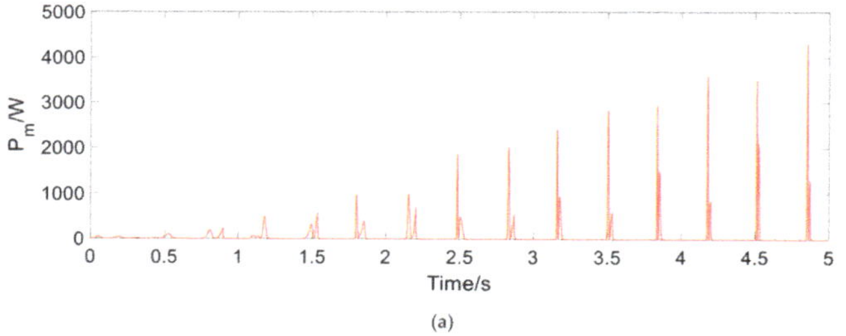

(a)

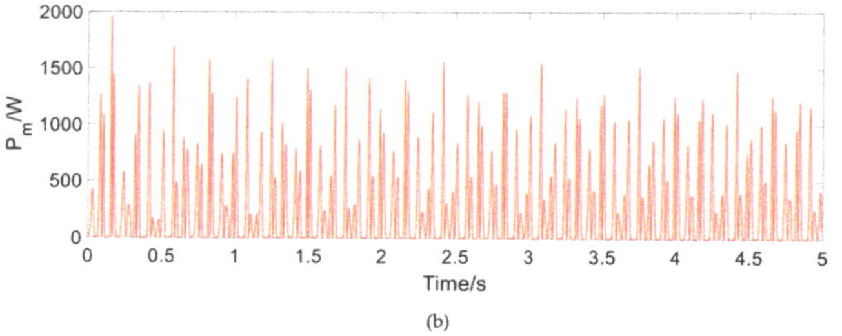

(b)

Figure 7. *Cont.*

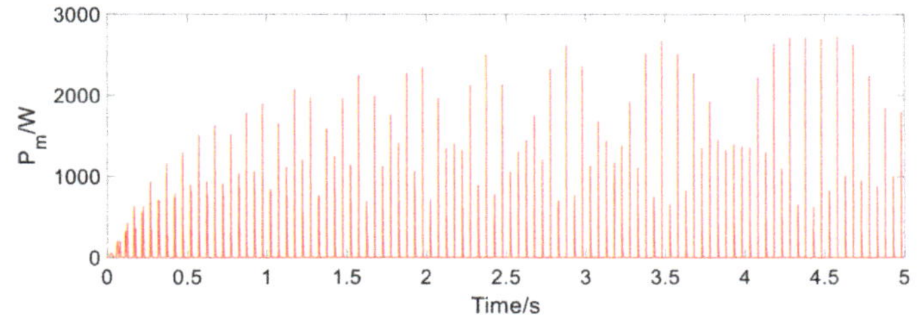

(c)

Figure 7. The instantaneous output power of regenerative motor under sinusoidal excitation. (**a**) Output power at 1.5 Hz, (**b**) Output power at 6 Hz, (**c**) Output power at 10 Hz.

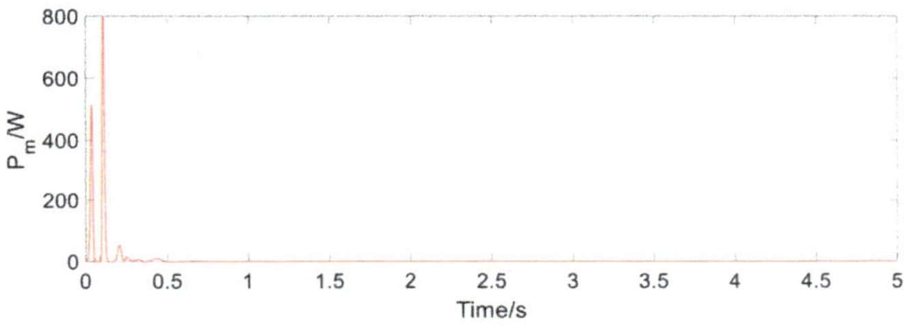

Figure 8. The instantaneous output power of regenerative motor under a smooth pulse.

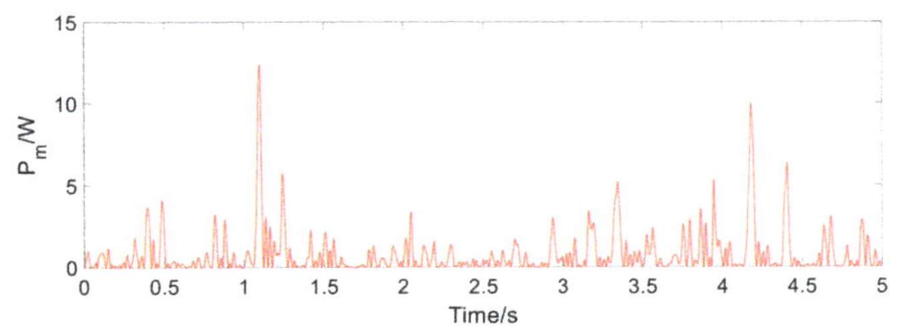

(a)

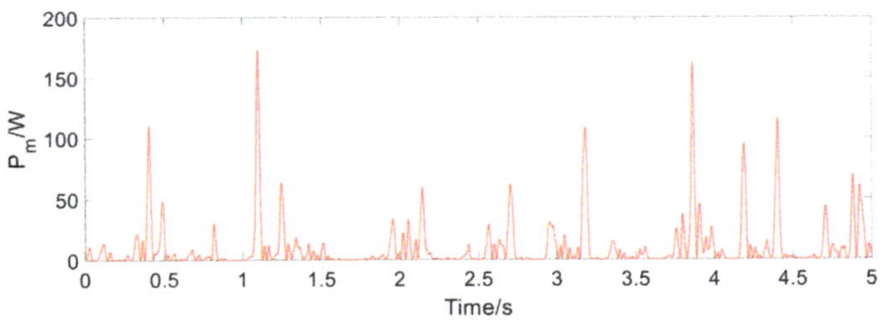

(b)

Figure 9. *Cont.*

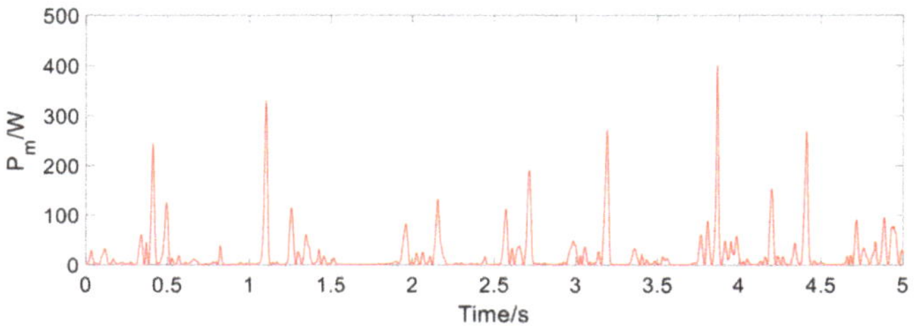

(c)

Figure 9. The instantaneous output power of regenerative motor under the random road. (**a**) Output power under A-class road, (**b**) Output power under B-class road, (**c**) Output power under C-class road.

Table 2. The average and effective output power of the regenerative motor.

Types of Road Excitation		$\bar{P}_m$	P
Sinusoidal	Frequency (Hz)		
	1.5	142.2825	497.2003
	6	308.8829	493.1127
	15	389.2749	771.0291
Smooth pulse		5.8977	52.2608
Random road	Road class		
	A	0.7540	1.4778
	B	9.7138	22.2265
	C	18.6516	46.0925

5. Conclusions

(1) For the MR regenerative suspension system in parallel with TPMLM, a stability analysis method based on the comprehensive suspension performance under different road excitation is proposed, especially observing whether the resonance point shifts after installing TPMLM;

(2) Dynamic model of the MR suspension system and a finite element model of TPMLM are established, respectively. Matlab/Simulink and Ansys software are used for joint simulation to obtain the comprehensive suspension performance and regenerative characteristics under the road excitation of harmonic, impact, and random pavement.

(3) The simulation results show that the MR regenerative suspension system with improved skyhook control meets the requirements of the natural resonance frequency of the vehicle suspension and effectively improves the driving comfort. At the same time, it has considerable regeneration power.

Author Contributions: Conceptualization, J.H. and H.Z.; methodology, J.H. and E.W.; software, J.H.; formal analysis, J.H. and H.Z.; writing—original draft preparation, J.H.; writing—review and editing, E.W. and H.Z.; All authors have read and agreed to the published version of the manuscript.

Funding: This research received no external funding.

Conflicts of Interest: The authors declare no conflict of interest.

References

1. Where the Energy Goes: Gasoline Vehicles. Available online: https://www.fueleconomy.gov/FEG/atv.shtml (accessed on 6 May 2019).
2. Karnopp, D. Power requirements for traversing uneven roadways. *Veh. Syst. Dyn.* **1978**, *7*, 135–152. [CrossRef]
3. Browne, A.L.; Hamburg, J.A. On-road measurement of the energy dissipated in automotive shock absorbers. In *Symposium on Simulation and Control of Ground Vehicles and Transportation Systems*; American Society of Mechanical Engineers: New York, NY, USA, 1986; Volume 80, pp. 1–10.

4. Huang, K.; Zhang, Y.C.; Yu, F. Coordinated optimization of comprehensive performance of electric active energy feedback suspension. *J. Shanghai Jiaotong Uni.* **2009**, *43*, 226–230.
5. Yu, F.; Zhang, Y.C. Active suspension technology of energy feedback vehicle. *Trans. Chin. Soc. Agric. Mach.* **2010**, *41*, 1–6.
6. Dong, X.M.; Peng, S.J.; Yu, J.Q. Study on characteristics of self powered vehicle magnetorheological damper. *J. Mech. Eng.* **2016**, *52*, 83–91. [CrossRef]
7. Kou, F.R.; Liang, J.; Wei, D.D. Research on a parallel semi-active suspension actuator. *China Mech. Eng.* **2017**, *28*, 2318–2324.
8. Zhu, Z.Y.; Zheng, L.; Wang, K. Energy feedback characteristics and experimental study of self powered rheological damper. In Proceedings of the 2014 Annual Meeting of China Society of Automotive Engineering, Shanghai, China, 22 October 2014; pp. 1210–1216.
9. Kou, F.R.; Chen, L.; Zhang, W. Study on characteristics of self powered magnetorheological semi-active suspension. *Hydraul. Pneum.* **2016**, *11*, 10–14.
10. Lv, X.Y.; Ji, Y.J.; Zhao, H.Y.; Zhang, J.; Zhang, G.; Zhang, L. Research Review of a Vehicle Energy-Regenerative Suspension System. *Energies* **2020**, *13*, 441. [CrossRef]
11. Zhang, H.L.; Wang, E.R.; Min, F.H.; Subash, R.; Su, C. Skyhook-based Semi-active Control of Full-vehicle Suspension with Magneto-rheological Dampers. *Chin. J. Mech. Eng.* **2013**, *26*, 498–505. [CrossRef]
12. Zhu, Z.Q.; Howe, D. Halbach Permanent Magnet Machines and Applications: A Review. *IEE Proc. Electr. Power Appl.* **2001**, *148*, 299–308. [CrossRef]
13. Lu, D.S.; Zhang, H.L.; Liu, J.; Wang, E. Structural Parameter Optimization of a Tubular Permanent-Magnet Linear Machine for Regenerative Suspension. *Shock Vib.* **2019**, *2019*, 11. [CrossRef]
14. Young, K.D.; Utkin, V.I.; Ozguner, U. A control engineer's guide to sliding mode control. *IEEE Trans. Control Syst. Technol.* **1999**, *7*, 328–342. [CrossRef]

vehicles

MDPI

Article

Investigation into Multiaxial Character of Thermomechanical Fatigue Damage on High-Speed Railway Brake Disc

Chun Lu [1,2], Jiliang Mo [1,2,*], Ruixue Sun [1,2], Yuanke Wu [1,2] and Zhiyong Fan [1,2]

1 School of Mechanical Engineering, Southwest Jiaotong University, Chengdu 610031, China; clu@swjtu.edu.cn (C.L.); sunruixue@my.swjtu.edu.cn (R.S.); wykhappy@my.swjtu.edu.cn (Y.W.); zhiyongfan@swjtu.edu.cn (Z.F.)

2 Technology and Equipment of Rail Transit Operation and Maintenance Key Laboratory of Sichuan Province, Chengdu 610031, China

* Correspondence: jlmo@swjtu.cn

Abstract: The multiaxial character of high-speed railway brake disc thermomechanical fatigue damage is studied in this work. Although the amplitudes and distributions of temperature, strain and stress are similar with uniform and rotating loading methods, the multiaxial behavior and out-of-phase failure status can only be revealed by the latter one. With the help of a multiaxial fatigue model, fatigue damage evaluation and fatigue life prediction are implemented, the contribution of a uniaxial fatigue parameter, multiaxial fatigue parameter and out-of-phase failure parameter to the total damage is discussed, and it is found that using the amplitude and distribution of temperature, stress and strain for fatigue evaluation will lead to an underestimation of brake disc thermomechanical fatigue damage. The results indicate that the brake disc thermomechanical fatigue damage belongs to a type of multiaxial fatigue. Using a uniaxial fatigue parameter causes around 14% underestimation of fatigue damage, while employing a multiaxial fatigue parameter without the consideration of out-of-phase failure will lead to an underestimation of about 5%. This work explains the importance of studying the thermomechanical fatigue damage of the brake disc from the perspective of multiaxial fatigue.

Keywords: thermomechanical fatigue; multiaxial fatigue; out-of-phase failure; brake disc; life prediction; high-speed railway; damage mechanism

Citation: Lu, C.; Mo, J.; Sun, R.; Wu, Y.; Fan, Z. Investigation into Multiaxial Character of Thermomechanical Fatigue Damage on High-Speed Railway Brake Disc. *Vehicles* **2021**, *3*, 287–299. https://doi.org/10.3390/vehicles3020018

Academic Editor: Ralf Stetter

Received: 27 March 2021
Accepted: 28 May 2021
Published: 1 June 2021

1. Introduction

Fatigue is a common damage type in the engineering field, which is resulted from cyclic load [1]. The characteristics of fatigue damage are different from static damage, i.e., fatigue failure may occur when the cyclic stress is much smaller than the static strength limit. In railway engineering, the cycle of the braking process can cause serious fatigue damage of the brake disc.

Due to the large tonnage and high speed of a vehicle, a tremendous amount of kinetic energy is transformed into thermal energy during the actions of implementing braking, precision stopping and stabling speeding [2]. This transformation arouses significant temperature increasement and exposes the brake disc to severe thermomechanical loading, resulted from the frictional contact between the brake disc and pad [3]. Because of this, the thermomechanical fatigue damage is considerably severe [4,5] (see Figure 1), and it could be more serious in the long slope line. This problem sets a restriction against the further speed increasement, economic saving and security improvement of railway transportation, and even could cause catastrophic accidents. Therefore, the study of thermomechanical fatigue damage on the brake disc is of great importance from both academic and engineering perspectives.

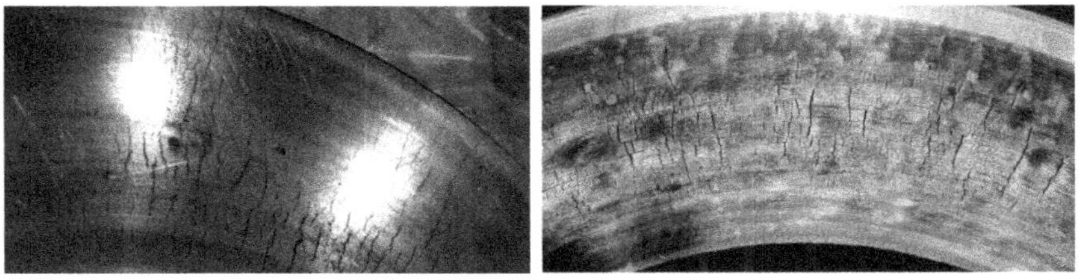

Figure 1. Thermomechanical fatigue of brake disc [6,7].

There are different factors affecting fatigue failure [8,9] and several fatigue damage types can be detected on the brake disc (Figure 1), i.e., heat checking (crackle), radial crack, circumferential crack and hot spot. Kasem et al. [10] analyze the subsurface damage beneath the hot spots, plastic deformation and microcrack propagation are detected under the rubbing surface. Collignon et al. [11] study the damage of brake discs and it is found that residual hoop tensile stress is relevant to radial microcracks. Honner et al. [12] study the thermoelastic instability of the brake system, which is verified to be relevant for non-uniform temperature distribution. Yang et al. [13] analyze the fatigue mechanism on the brake disc, fatigue cracking is observed in the interior and border of hot spots. Panier et al. [14] propose a classification for hot spots and suggest a scenario of their occurrence by experimental methods. Kim et al. [6] find that the maximum thermal stress under a variable pressure distribution is quite relevant with fatigue cracking. Wu et al. [15] study the peak temperature and crack geometry evolution with the help of an inserted semi elliptical surface crack. Adamowica and Grzes [16] study the brake disc temperature distribution, and it is found that the influence of convection cooling is significant in the brake release period. A review of the thermomechanical performance of brake discs with numerical and experimental approaches can be found in the work of Afzal and Mujeebu [17].

According to previous research, the commonly used method for brake disc thermomechanical fatigue investigation is adopting the distribution and amplitude of temperature, stress and strain items. However, due to the rotating movement, the thermomechanical fatigue should belong to multiaxial fatigue damage [18], which is complex with numerous influencing factors [19]. Regrettably, the multiaxial feature of brake disc thermomechanical fatigue is ignored at this moment; this simplification might lead to an underestimation in the thermomechanical fatigue damage and a blur in the thermomechanical fatigue mechanism reveal.

Based on the analysis above, the study of multiaxial thermomechanical fatigue damage on the high-speed railway brake disc is insufficient. In this study, the thermomechanical fatigue damage of the brake disc will be studied with the help of a multiaxial fatigue model instead of using the distribution and amplitude of temperature, stress and strain items as fatigue indicators. A finite element simulation will be used to reproduce the multiaxial material response state under thermomechanical load. Afterward, a multiaxial fatigue model, recently proposed by the first author, will be adopted to analyze the multiaxial thermomechanical fatigue damage and out-of-phase failure status. Finally, the methodology for multiaxial thermomechanical fatigue evaluation will be proposed and the fatigue life prediction will be performed.

2. Finite Element Model

2.1. Theoretical Background

2.1.1. Thermal Formulation

Ignoring the influence of wheel–rail friction and aerodynamic resistance, most vehicle kinetic energy is transferred into heat by friction contact between the brake disc and pad when mechanical braking is implemented. Therefore, the total thermal energy flowed into the braking system can be calculated by friction energy. The corresponding heat flux (q_f) is described in Equation (1):

$$q_f = \frac{\eta \cdot P \cdot \mu}{A} \cdot (v_0 + a \cdot t) \tag{1}$$

where η is the conversion rate between friction energy and heat, P denotes the contact force and μ stands for the friction coefficient, A is the braking contact area, v_0 is the initial relative frictional translation speed between the brake disc and pad, and a and t represent the acceleration and time, respectively.

Apart from the heat input, there are heat exchanges between the braking system and environment through convection heat transfer and heat radiation. The convection heat transfer (q_f) is demonstrated in Equation (2):

$$q_f = h \cdot (T - T_0) \tag{2}$$

where h is the film coefficient, and T and T_0 represent the temperature of braking component and surrounding environment, respectively. The heat radiation (q_h) is calculated by Equation (3):

$$q_h = \text{Ж} \cdot (T - T_*)^4 \tag{3}$$

where Ж is the Stefan-Boltzmann constant, T_* stands for the sink temperature.

2.1.2. Thermal–Mechanical Solution

With regard to the solution technology for the coupled thermal–mechanical problem, the full Newton method is usually implemented with the enonsymmetric Jacobian matrix (see Equation (4)):

$$\begin{bmatrix} K_{dd} & K_{dt} \\ K_{td} & K_{tt} \end{bmatrix} \begin{Bmatrix} \Delta d \\ \Delta t \end{Bmatrix} = \begin{Bmatrix} R_d \\ R_t \end{Bmatrix} \tag{4}$$

where Δd and Δt are increments of displacement and temperature, respectively. K_{ij} is the submatrices of fully coupled Jacobian matrix, R_d and R_t stand for the mechanical and thermal residual vectors. Particularly, with respect to the issues of the coupled thermal–mechanical problem in the brake disc, the coupling effect between the mechanical and thermal solutions is weak. Therefore, the off-diagonal submatrices are assumed to be zero in order to have a less costly solution. The solution equation is described in Equation (5):

$$\begin{bmatrix} K_{dd} & 0 \\ 0 & K_{tt} \end{bmatrix} \begin{Bmatrix} \Delta d \\ \Delta t \end{Bmatrix} = \begin{Bmatrix} R_d \\ R_t \end{Bmatrix} \tag{5}$$

2.2. FE Model Description

There are two common methods for the loading implementation on the brake disc; one is the heat flux functions method and the other one is the mechanical friction method. Particularly, the heat flux functions method can be subdivided into the rotating loading method and the uniform loading method. The advantage of the uniform loading method is the calculation efficiency, but the multiaxial material response state and relevant out-of-phase failure feature during the braking process have to be revealed by the rotating loading method.

The three-dimensional geometry model of the investigated disc brake system is shown in Figure 2a. It can be seen that the disc brake system is complex, comprising the brake disc, brake pad, backplate, pad bracket and brake clamp, etc. In order to improve the calculation efficiency, the brake disc is built with some simplifications and only half of it is established. The finite element model of the brake disc can be seen Figure 2b. The description of the brake disc dimension can be seen in Figure 2c. The red solid area and red dotted area in Figure 2c are the regions for rotating loading and uniform loading, respectively. r_{L1}, r_{L2} and r_M indicate the inner, outer and average friction radii. The geometry dimension can be found in Table 1. A total of 29 elements (denoted as E_{R1}-E_{R29}) are evenly distributed along radial direction (between r_2 and r_3) and 3 elements (represented by E_{A1}-E_{A3}) are uniformly arranged from the contact surface into the brake disc with a distance of 2 mm between each other.

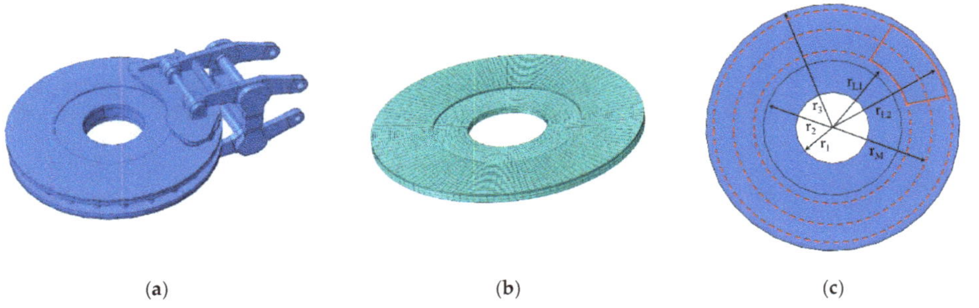

Figure 2. (**a**) Three-dimensional geometry model of the investigated disc brake system, (**b**) simplified finite element model of the brake disc, (**c**) description of brake disc dimension.

Table 1. Geometry dimension of brake disc.

r_1	r_2	r_3	r_{L1}	r_{L2}	r_M
93 mm	175 mm	320 mm	187.5 mm	312.5 mm	250 mm

The finite element model is established in ABAQUS [20] in order to ensure the calculation accuracy and improve the calculation efficiency. An 8-node three-dimensional linear coupled temperature-displacement reduced-integration brick element is used to discretize the model. The material properties are listed in Table 2. The loading (heat flux, normal load and tangential load) is realized by subroutines in ABAQUS [20]. The heat flux can be calculated by Equation (1), which changes with time. The normal load is resulted from the brake force, and the tangential load is resulted from the friction force. The inner ring where the brake disc is in contact with the axle is fully constrained. A global Cartesian coordinate system is employed, and the original point is the center of the brake disc. The z direction is the rotation axis and the rotation speed can be calculated according to the adopted braking condition. The transient coupled temperature-displacement calculation is performed.

Table 2. Material properties of brake disc.

Young's Modulus (MPa)	Poisson's Ratio	Density (kg/m^3)	Thermal Expansion Coefficient	Specific Heat [J/(kg·K)]	Thermal Conductivity [W/(m·K)]
2.02×10^5	0.29	7850	1.04×10^{-5}	457.74	32

The braking operation parameters are listed in Table 3: the initial braking speed is 250 km/h, the wheel radius is 430 mm, the transfer efficiency η is 0.9, the frictional coefficient between brake disc and pad is assumed as 0.4, a uniform acceleration -1.389 m/s^2 is employed under a brake force of 17,000 N, the corresponding brake time is 50 s and the total brake distance is 1736.1 m. The environment temperature is 20 °C.

Table 3. The braking operation parameters.

Wheel Radius	Brake Time	Transfer Efficiency	Acceleration
430 mm	50 s	0.9	-1.389 m/s^2
Brake force	Brake distance	Friction coefficient	Speed
17,000 N	1736.1 m	0.4	250 km/h

3. Multiaxial Fatigue Model

A fatigue model called the multiaxial fatigue space (referred as MFS criterion for convenience), recently proposed by the first author, is adopted to assess the multiaxial thermomechanical fatigue damage of the brake disc. The MFS criterion is validated by the fatigue test data of different materials under various loading conditions and displays a pleasant prediction capability, compared with other commonly used multiaxial fatigue models [21]. A brief introduction is presented, and detailed information can be found in the corresponding reference [21].

The fundamental assumption in the MFS criterion is that the changing of the material response state is the source of the fatigue damage. As shown in Figure 3, subscripts T1 and T2 indicate different time points. σ and τ are normal and shear stresses, subscripts x and y represent the components following the global coordinates, and subscripts 1 and 2 mean the components following the principal directions. $\sigma_1{}^{T1'}$ and $\sigma_2{}^{T1'}$ indicate the response status at moment T2 following the principal directions of T1. C and R are the center and radius of Mohr circle, and A is twice the angle between the loading plane and the principal plane. It can be seen that the changing of the material state can be described by the changing of C, R and A. Therefore, the material response status is described by three parameters (*P1*, *P2* and *P3*), which are the equivalent shear parameter, equivalent normal parameter and out-of-phase parameter, respectively [21]. The changing of them (three fatigue basic units, i.e., $f1(P1)$, $f2(P2)$ and $f3(P3)$) can be used for fatigue damage evaluation, and the multiaxial fatigue space can be established by fatigue basic units (see Figure 4) [21]. The multiaxial fatigue parameter is described in Equation (6):

$$FP = f0(f1(P1), f2(P2), f3(P3)) \tag{6}$$

with

$$P1 = \frac{\sqrt{(N_{C1} - N_{C2})^2 + (S_{C1} - S_{C2})^2}}{2} \tag{7}$$

$$P2 = \frac{N_{C1} + N_{C2}}{2} \tag{8}$$

$$P3 = P1\frac{\theta^2}{\theta_{max}} \tag{9}$$

where *fn* indicates a certain function, *P1* and *P2* are in-phase parameters and *P3* is out-of-phase parameter [21], *N* and *S* indicate the normal and shear components on the critical plane (a couple of two orthogonal planes, noted by subscripts *C1* and *C2*), stress, strain and energy components can be used as the adopted items, $\theta/2$ is the out-of-phase angle, which represents the relative distortion deformation, and $\theta_{max}/2$ is the limit value of the out-of-phase angle [21]. It can be noticed that the MFS criterion belongs to the type of critical plane criterion.

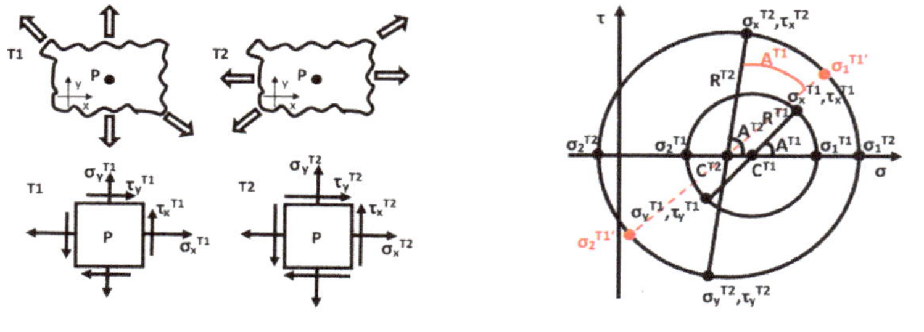

Figure 3. Material response states at different time instants.

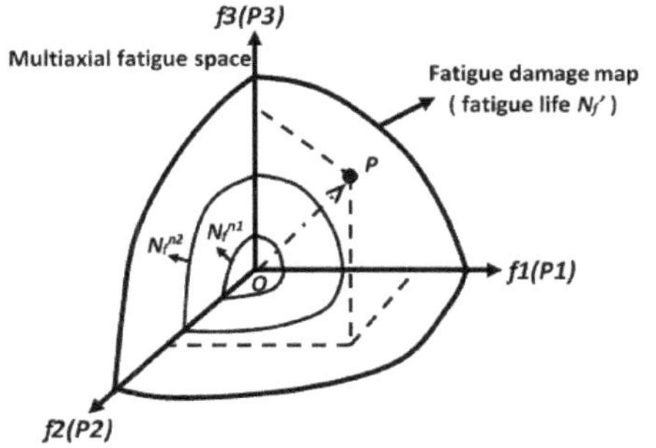

Figure 4. Multiaxial fatigue space.

According to the MFS criterion, the amplitude of parameter variation (changing of *P1*, *P2* and *P3*) is the foundational component in multiaxial fatigue space, which is consistent with the definition of fatigue damage. The influence of the mean scales the vector *OP* in multiaxial fatigue space [21]. Therefore, the multiaxial fatigue parameter in Equation (6) can be expressed by amplitude parameter (FP_a) and mean parameter (FP_m). See Equation (10):

$$FP = FP_a \cdot FP_m \tag{10}$$

with

$$FP_m = (1 + \chi_{P1} P1_m) \cdot (1 + \chi_{P2} P2_m) \cdot (1 + \chi_{P3} P3_m) \tag{11}$$

$$FP_a = \left[\left(FP_a^{in} \right)^{h_{out}} + \left(k_{out} \cdot FP_a^{out} \right)^{h_{out}} \right]^{1/h_{out}} \tag{12}$$

$$FP_a^{in} = \left[(P1_a)^{h_{in}} + (k_{in} \cdot P2_a)^{h_{in}} \right]^{1/h_{in}} \tag{13}$$

$$FP_a^{out} = P3_a \tag{14}$$

where subscripts a and m indicate the amplitude and mean of corresponding parameter, subscripts *in* and *out* denote the in-phase and out-of-phase failure conditions, k and h are material-dependent coefficients, and χ_{Px} is the mean influence coefficient. Afterward, the predicted fatigue life (N_f) can be obtained by Equations (15) or (16), depending on the fatigue curve pattern:

$$FP = \mathrm{A}\left(N_f\right)^B + \mathrm{C}\left(N_f\right)^D \tag{15}$$

$$FP = \mathrm{A}\left(N_f\right)^B \tag{16}$$

where *A*, *B*, *C* and *D* are material-dependent coefficients with solid physical meaning in multiaxial fatigue space [21].

It should be noticed that the original MFS criterion is established and validated in a 2-dimensional condition. With regard to 3-dimensional cases, the general formula is the same; the difference is in the determination of the critical plane. In this work, the directions of the maximum and the second maximum principal strain range are employed for critical plane determination. That is to say, the direction of the maximum principal strain range will be determined firstly. Afterward, the directions perpendicular to the maximum principal strain range will be checked to obtain the direction of the second maximum principal strain range. In this way, the issues of the 3-dimensional condition can be transformed into 2-dimensional cases and the MFS criterion can be used smoothly.

About the material dependent coefficients, normally the fatigue test is performed under mechanical fatigue loading instead of thermomechanical fatigue loading. The thermomechanical fatigue-dependent coefficients of the brake disc are not available at this moment. Thus, it is assumed that they are identical with the fatigue dependent coefficients of hot-rolled 45 steel [21]. In order to present the methodology, this is a common strategy when the fatigue-related information is unavailable [22,23]. The adopted fatigue-dependent coefficients are listed in Table 4. An accurate multiaxial thermomechanical fatigue evaluation of the brake disc can be performed following the proposed approach when the thermomechanical fatigue-dependent coefficients are collected in the future.

Table 4. The adopted material-dependent coefficients.

k_{in}	h_{in}	k_{out}	h_{out}	χ_{P1}	χ_{P2}	χ_{P3}	A	B
1.7	1	1.12	1	0.1	0	0	50.13	−0.4411

4. Results and Discussion

Considering the calculation efficiency, uniform loading is performed first, the calculation time is set as 500 s to have an overall response status, the highest temperature 454 °C occurs at 35 s, and the temperature distribution on the brake disc at four moments before the highest temperature time is demonstrated in Figure 5. As discussed above, the multiaxial characters cannot be revealed with uniform loading. Rotating loading is needed for the multiaxial thermomechanical fatigue evaluation. According to the results with uniform loading, it can be seen that the temperature increases significantly and reaches above 400 °C in the first 20 s. Thus, the calculation time is set as 20 s with rotating loading so as to reduce the simulation time and result file size. The comparison between the results with rotating loading and uniform loading can be seen in Figure 6.

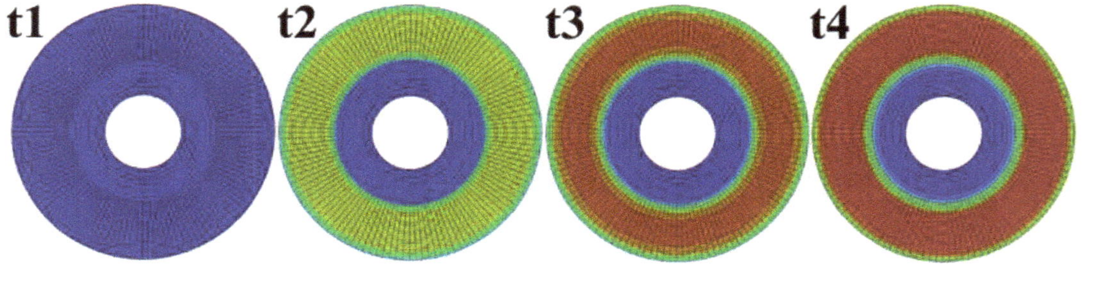

Figure 5. Temperature distribution with uniform loading at different moments (highest temperature 454 °C occurs at t4 = 35 s).

It can be seen in Figure 6 that the results manifestation is reasonable [7,17], which verifies the correctness of the calculation to a certain extent. The difference between the results with uniform and rotating loading is negligible. It can also be noticed that the heating time (50 s) is much shorter than the cooling time (much greater than 450 s) and the results variation during the heating process (see the changing of Huber–Mises results in the first 50 s in Figure 6) is much more violent than that during the cooling process (see the Huber–Mises results from 50 s to 500 s in Figure 6). Thus, as the changing of the material state during the heating process is used for fatigue evaluation, the calculation time is greatly reduced without losing much accuracy. To have a clear view about the multiaxial character of thermomechanical fatigue damage of the brake disc, the evolutions of the strain items with rotating loading can be seen in Figure 7. It can be noticed clearly that the changing of shear strain is non-proportional, which indicates that the principal strain direction is changed during the braking process. It means that the thermomechanical fatigue damage of the brake disc not only belongs to the type of multiaxial fatigue damage, but also the out-of-phase failure status. The out-of-phase failure of the brake disc could be resulted from the mechanical interaction between the brake disc and the brake pad, or could be resulted from the influence of the brake disc structure. The reasons will be discussed in the future.

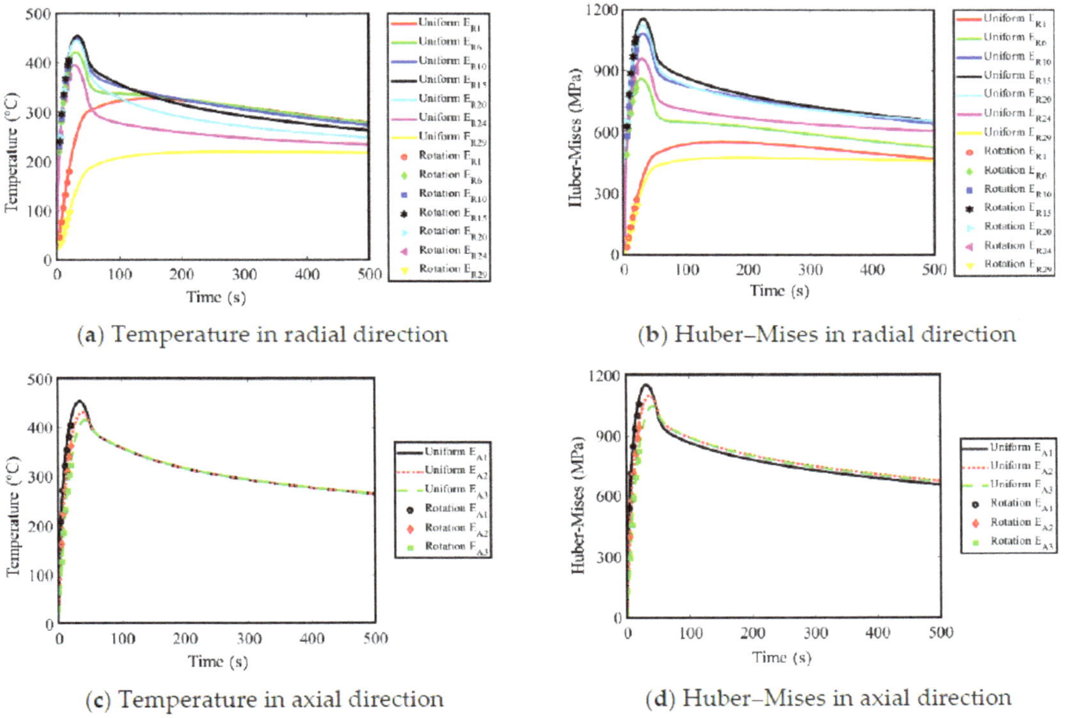

(**a**) Temperature in radial direction (**b**) Huber–Mises in radial direction

(**c**) Temperature in axial direction (**d**) Huber–Mises in axial direction

Figure 6. Comparison between the results with uniform loading and rotating loading.

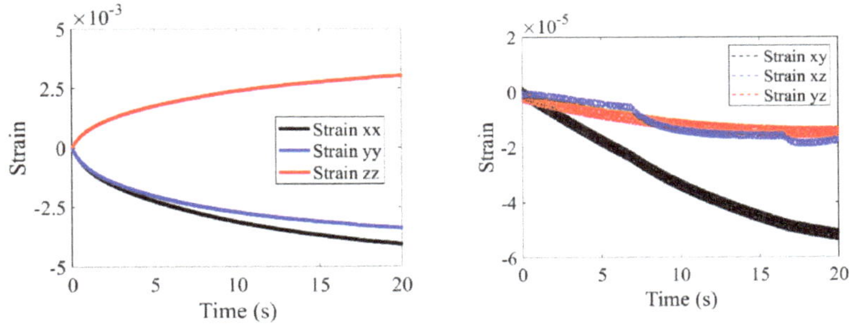

(**a**) (**b**)

Figure 7. (**a**) The changing of strain items following xx, yy and zz directions with rotating loading, (**b**) the changing of strain items following xy, xz and yz directions with rotating loading.

It is known that the changing of the material response state is critical for multiaxial fatigue analysis; the time increment needs to be small enough to capture the details of this changing. This makes the realization with rotating loading unfavorable because of the lengthy calculation time and huge resulting file size. Thus, an equivalent accelerated method (referred to as the equivalent method) is proposed. An amplification factor is used to scale the heat flux in order to reach the target temperature with only one rotation. In order to validate equivalent method, the temperature with rotating loading at 20 s is used as the target temperature (406.73 °C). The accepted tolerance for the obtained temperature by equivalent method is 1%. The temperature distribution with equivalent method at different times in one rotation is shown in Figure 8; the highest temperature is 406.38 °C.

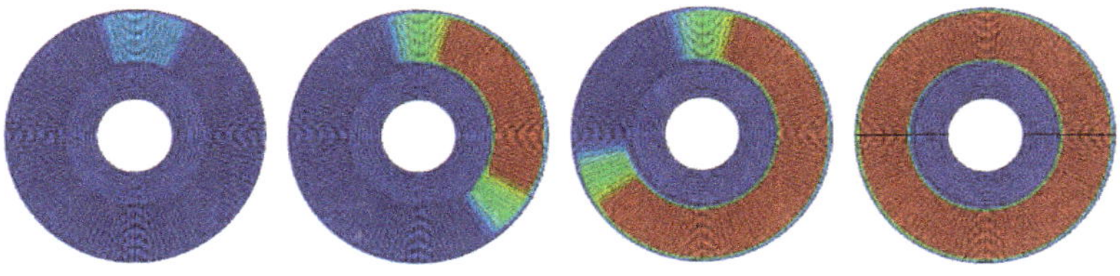

Figure 8. Temperature distribution with equivalent method at different times in one rotation.

It can be seen that the highest temperature by the equivalent method (406.38 °C) is close to that with rotating loading (406.73 °C). The position of the maximum temperature by the equivalent method is located at the average friction radius position (E_{R15}), which is the same as the rotating loading calculation. The comparison of the results among uniform loading, rotating loading and equivalent methods at the maximum temperature position is shown in Table 5. It can be seen that the temperature, thermal strain and Huber–Mises stress are approximately equal with each other. The results with rotating loading is used as reference. The temperature and strain with the equivalent method are quite satisfactory and the relative error is less than 0.1%. The difference of Huber–Mises stress is around 5% because of the history effect and constraint from the extrusion by the surrounding elements. In general, the results with the equivalent method are acceptable and strain items are used for the fatigue evaluation with the MFS criterion.

Table 5. Comparison of the results among uniform, rotating and equivalent methods.

Item	Rotating	Uniform		Equivalent	
		Value	Error %	Value	Error %
Temperature (°C)	406.73	417.82	−2.73	406.38	0.086
Thermal strain (%)	0.40220	0.41373	−2.87	0.40184	0.090
Huber–Mises (MPa)	1063.5	1082.8	−1.81	1117.8	−5.1

Regarding the multiaxial fatigue prediction capability with the equivalent method, the comparison with rotating loading is performed at different temperatures. Because the results of the finite element calculation are described in the global coordinate (coordinate xyz), coordinate transformation is executed with the help of Euler angles to get the material response status in coordinate XYZ (see Figure 9). Here, zXZ rotation order is adopted for the coordinate transformation. Afterward, the MFS criterion can be adopted. The calculated fatigue parameter and relative error (defined in Equation (17)) are demonstrated in Figure 10.

$$E_r = \left(FP_{rotation} - FP_{equivalent}\right)/FP_{rotation} \tag{17}$$

It can be seen that with the increasement of temperature, the relative error between the equivalent method and the rotating loading method is increased slowly, following a linear pattern, approximately. Based on this, the relationship function between the relative error (E_r) and temperature (T) can be written as Equation (18):

$$E_r = 0.00032637 \times T - 0.017 \tag{18}$$

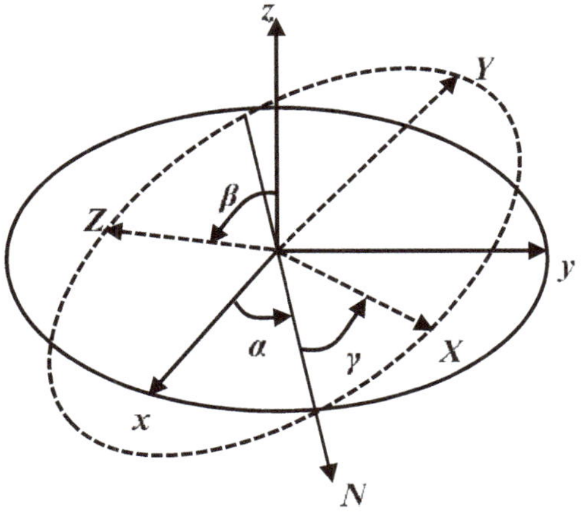

Figure 9. Coordinate transformation with Euler angles.

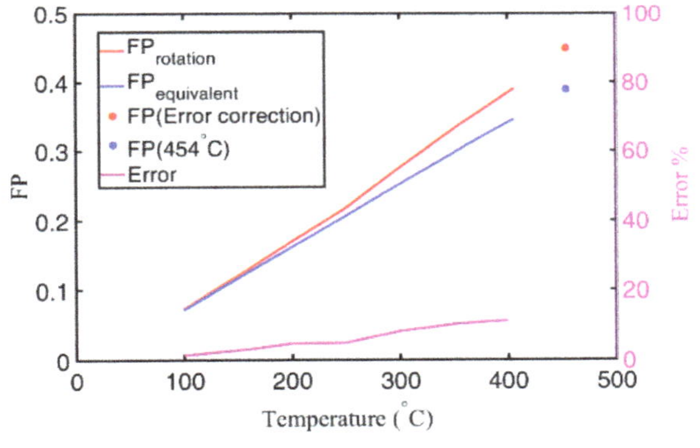

Figure 10. Fatigue parameter comparison between rotating loading and equivalent methods.

Now the multiaxial thermomechanical fatigue of the brake disc under the investigated braking operation condition can be assessed. Due to the fact that the calculation time with rotating loading is quite long, the maximum temperature with uniform loading is used as the target temperature (454 °C, see Figure 6); the value of the fatigue parameter with the equivalent method can be obtained with this target temperature (see Figure 10, blue point). Hereafter, the fatigue parameter value with rotating loading can be obtained with error correction (see Figure 10, red point).

It is discussed above that the amplitudes and distributions of temperature, stress and strain items are not enough for accurate thermomechanical fatigue evaluation of the brake disc. Ignoring the influence of the multiaxial character and out-of-phase feature leads to an underestimation of the thermomechanical fatigue assessment for the brake disc. Based on the result with the rotating loading calculation, the contribution of the equivalent shear parameter, equivalent normal parameter and out-of-phase parameter (represented by *P1*, *P2* and *P3* in the MFS criterion) to the total multiaxial thermomechanical fatigue parameter is demonstrated in Figure 11. It can be seen that just using a uniaxial fatigue parameter causes at least 14% underestimation of the fatigue damage, while employing a multiaxial fatigue parameter without the consideration of the out-of-phase failure leads to an underestimation of 5%. It should be noticed that the relationship between the fatigue parameter and fatigue life is exponential, which means that an insignificant deviation in the fatigue parameter assessment leads to a significant predicted error of fatigue life.

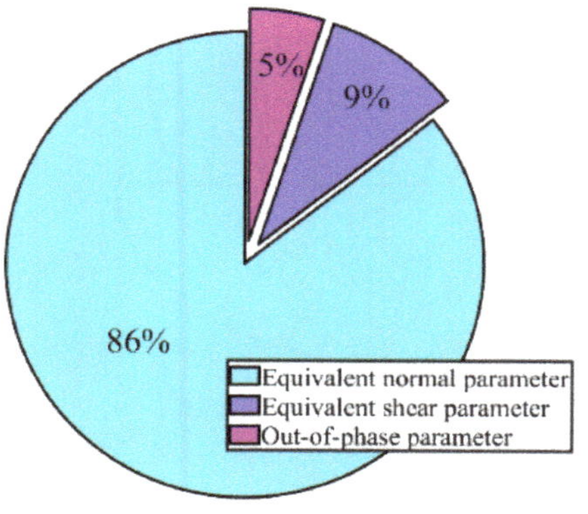

Figure 11. The contribution of different parameters on multiaxial thermomechanical fatigue of the brake disc.

It is known that the material mechanical and fatigue characters change significantly with temperature [24–26]. The material character under high temperature is not available at this moment. As there is an approximate linear relationship between the yield stress and fatigue limit [27], the yield stress decreases exponentially with the increasing of the temperature, approximately (see Figure 12a) [15]. Thus, an exponential relationship between the fatigue limit and temperature is assumed in order to present the methodology in this work (see Figure 12b). The corresponding fatigue curves at different temperatures can be obtained using extrapolation method with the help of the yield stress at different temperatures. The multiaxial thermomechanical fatigue life of the investigated brake disc under the employed braking operation condition can be obtained, which is about 2500 braking cycles. The accurate multiaxial thermomechanical fatigue life prediction can be obtained with the proposed methodology when the temperature related material fatigue characters become available in the future.

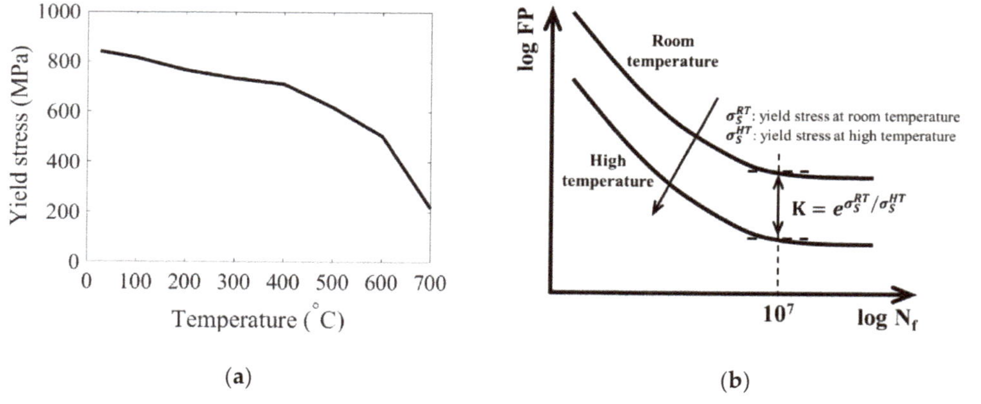

Figure 12. (**a**) The relationship between the yield stress and temperature [15], (**b**) extrapolation of fatigue curve.

5. Conclusions

In this study, the multiaxial characters of thermomechanical fatigue damage on the high-speed railway brake disc is investigated. Through the revealing of the multiaxial response state and the out-of-phase failure feature, it is proved that under mechanical friction braking conditions, the damage of the high-speed railway brake disc belongs to the type of multiaxial thermomechanical fatigue. With the help of a multiaxial fatigue model, the methodology for multiaxial thermomechanical fatigue damage evaluation is proposed and multiaxial thermomechanical fatigue life prediction is implemented. In addition, the contribution of the uniaxial fatigue parameter, multiaxial fatigue parameter and out-of-phase failure parameter to the total multiaxial thermomechanical fatigue damage of the brake disc is discussed. The results indicate that, using the amplitude and distribution of temperature, stress and strain items for fatigue evaluation lead to an underestimation of the brake disc thermomechanical fatigue damage. This work explains the importance of studying the thermomechanical fatigue damage of brake disc from the perspective of multiaxial fatigue.

Author Contributions: Conceptualization, C.L. and J.M.; methodology, C.L. and Z.F.; software, C.L. and Y.W.; validation, C.L. and J.M.; formal analysis, C.L.; investigation, C.L. and R.S.; resources, C.L. and J.M.; data curation, C.L. and Y.W.; writing—original draft preparation, C.L., R.S., Y.W. and Z.F.; writing—review and editing, J.M.; visualization, C.L. and R.S.; supervision, J.M.; project administration, C.L. and J.M.; funding acquisition, C.L. and J.M. All authors have read and agreed to the published version of the manuscript.

Funding: This research was funded by the National Natural Science Foundation of China, grant number 51822508; Sichuan Science and Technology Program, grant number 2020JDTD0012; the Fundamental Research Funds for the Central Universities, grant number 2682021CX028.

Institutional Review Board Statement: Not applicable.

Informed Consent Statement: Not applicable.

Conflicts of Interest: The authors declare no conflict of interest.

References

1. Lu, C.; Melendez, J.; Martínez-Esnaola, J. Fatigue damage prediction in multiaxial loading using a new energy-based parameter. *Int. J. Fatigue* **2017**, *104*, 99–111. [CrossRef]
2. Yevtushenko, A.; Kuciej, M.; Wasilewski, P. Experimental study on the temperature evolution in the railway brake disc. *Theor. Appl. Mech. Lett.* **2019**, *9*, 308–311. [CrossRef]
3. Zhou, S.; Yang, Y.; Xie, J. Analysis of transient temperature and thermal stress distribution on the high-speed strain brake disk by simulatin. *J. Mech. Eng.* **2011**, *47*, 126–131. [CrossRef]
4. Li, Z.; Han, J.; Yang, Z.; Li, W. Analyzing the mechanisms of thermal fatigue and phase change of steel used in brake discs. *Eng. Fail. Anal.* **2015**, *57*, 202–218. [CrossRef]
5. Šamec, B.; Oder, G.; Lerher, T.; Potrč, I. Numerical analysis of railway brake disc. *J. Shanghai Jiaotong Univ.* **2011**, *16*, 149–151. [CrossRef]
6. Kim, D.-J.; Lee, Y.-M.; Park, J.-S.; Seok, C.-S. Thermal stress analysis for a disk brake of railway vehicles with consideration of the pressure distribution on a frictional surface. *Mater. Sci. Eng. A* **2008**, *483–484*, 456–459. [CrossRef]
7. Li, Z.; Han, J.; Yang, Z.; Pan, L. The effect of braking energy on the fatigue crack propagation in railway brake discs. *Eng. Fail. Anal.* **2014**, *44*, 272–284. [CrossRef]
8. Mohtadi-Bonab, M.; Eskandari, M.; Sanayei, M.; Das, S. Microstructural aspects of intergranular and transgranular crack propagation in an API X65 steel pipeline related to fatigue failure. *Eng. Fail. Anal.* **2018**, *94*, 214–225. [CrossRef]
9. Mohtadi-Bonab, M.A.; Eskandari, M.; Ghaednia, H.; Das, S. Effect of Microstructural Parameters on Fatigue Crack Propagation in an API X65 Pipeline Steel. *J. Mater. Eng. Perform.* **2016**, *25*, 4933–4940. [CrossRef]
10. Kasem, H.; Brunel, J.; Dufrénoy, P.; Siroux, M.; Desmet, B. Thermal levels and subsurface damage induced by the occurrence of hot spots during high-energy braking. *Wear* **2011**, *270*, 355–364. [CrossRef]
11. Collignon, M.; Cristol, A.-L.; Dufrénoy, P.; Desplanques, Y.; Balloy, D. Failure of truck brake discs: A coupled numerical–experimental approach to identifying critical thermomechanical loadings. *Tribol. Int.* **2013**, *59*, 114–120. [CrossRef]
12. Honner, M.; Šroub, J.; Švantner, M.; Voldrich, J. Frictionally Excited Thermoelastic Instability and the Suppression of Its Exponential Rise in Disc Brakes. *J. Therm. Stress.* **2010**, *33*, 427–440. [CrossRef]

13. Yang, Z.; Han, J.; Li, W.; Li, Z.; Pan, L.; Shi, X. Analyzing the mechanisms of fatigue crack initiation and propagation in CRH EMU brake discs. *Eng. Fail. Anal.* **2013**, *34*, 121–128. [CrossRef]
14. Panier, S.; Dufrénoy, P.; Weichert, D. An experimental investigation of hot spots in railway disc brakes. *Wear* **2004**, *256*, 764–773. [CrossRef]
15. Wu, S.; Zhang, S.; Xu, Z. Thermal crack growth-based fatigue life prediction due to braking for a high-speed railway brake disc. *Int. J. Fatigue* **2016**, *87*, 359–369. [CrossRef]
16. Adamowicz, A.; Grzes, P. Influence of convective cooling on a disc brake temperature distribution during repetitive braking. *Appl. Therm. Eng.* **2011**, *31*, 2177–2185. [CrossRef]
17. Afzal, A.; Mujeebu, M.A. Thermo-Mechanical and Structural Performances of Automobile Disc Brakes: A Review of Numerical and Experimental Studies. *Arch. Comput. Methods Eng.* **2018**, *26*, 1489–1513. [CrossRef]
18. Lu, C.; Melendez, J.; Martinez-Esnaola, J.M. A universally applicable multiaxial fatigue criterion in 2D cyclic loading. *Int. J. Fatigue* **2018**, *110*, 95–104. [CrossRef]
19. Lu, C.; Melendez, J.; Martínez-Esnaola, J.M. Prediction of crack initiation plane direction in high-cycle multiaxial fatigue with in-phase and out-of-phase loading. *Fatigue Fract. Eng. Mater. Struct.* **2017**, *79*, 65–2007. [CrossRef]
20. D.C.S. Simulia. *ABAQUS Analysis User's Manual*; D.C.S. Simulia: Providence, RI, USA, 2011.
21. Lu, C.; Martínez-Esnaola, J.M. Multiaxial fatigue space: A three-dimensional space constituted of fatigue basic units. *Int. J. Fatigue* **2021**, *143*, 105995. [CrossRef]
22. Liu, Y.; Stratman, B.; Mahadevan, S. Fatigue crack initiation life prediction of railroad wheels. *Int. J. Fatigue* **2006**, *28*, 747–756. [CrossRef]
23. Lu, C.; Meléndez, J.; Martínez-Esnaola, J.M. Stochastic fatigue damage assessment of metro switch blade. *Fatigue Fract. Eng. Mater. Struct.* **2020**, *43*, 137–148. [CrossRef]
24. Han, M.-J.; Lee, C.-H.; Park, T.-W.; Lee, S.-P. Low and high cycle fatigue of automotive brake discs using coupled thermo-mechanical finite element analysis under thermal loading. *J. Mech. Sci. Technol.* **2018**, *32*, 5777–5784. [CrossRef]
25. Pevec, M.; Oder, G.; Potrč, I.; Šraml, M. Elevated temperature low cycle fatigue of grey cast iron used for automotive brake discs. *Eng. Fail. Anal.* **2014**, *42*, 221–230. [CrossRef]
26. Kurek, M.; Lagoda, T.; Walat, K. Variations of Selected Cyclic Properties Depending on Testing Temperature. *Mater. Sci.* **2015**, *50*, 555–563. [CrossRef]
27. Li, J.; Sun, Q.; Zhang, Z.-P.; Li, C.-W.; Qiao, Y.-J. Theoretical estimation to the cyclic yield strength and fatigue limit for alloy steels. *Mech. Res. Commun.* **2009**, *36*, 316–321. [CrossRef]

vehicles

MDPI

Article

Evolution of the Hybrid Aerial Underwater Robotic System (HAUCS) for Aquaculture: Sensor Payload and Extension Development

Casey J. Den Ouden [1,*], Paul S. Wills [1], Lucas Lopes [1], Joshua Sanderson [2] and Bing Ouyang [1]

1 Systems and Imaging Laboratory (SAIL), Harbor Branch Oceanic Institute, Florida Atlantic University, 5600 US 1 North, Fort Pierce, FL 34946, USA; pwills2@fau.edu (P.S.W.); lucaslopez2016@fau.edu (L.L.); bouyang@fau.edu (B.O.)

2 Harriet L. Wilkes Honors College, John D. MacArthur, Florida Atlantic University, 5353 Parkside Dr, Jupiter, FL 33458, USA; jsanderson2020@fau.edu

* Correspondence: cdenouden@fau.edu

Abstract: While robotics have been widely used in many agricultural practices such as harvesting, seeding, cattle monitoring, etc., aquaculture farming is an important, fast-growing sector of agriculture that has not seen significant adoption of advanced technologies such as robotics and the Internet of Things (IoT). In particular, dissolved oxygen (DO) monitoring, a practice in pond aquaculture essential to the health of the fish crops, remains labor-intensive and time-consuming. The Hybrid Aerial Underwater roboticS System (HAUCS) is an IoT framework that aims to bring transformative changes to pond aquaculture. This paper focuses on the latest development in the HAUCS mobile sensing platform and field deployment. To address some shortcomings with the current implementation, the development of a novel rigid Kirigami-based robotic extension subsystem that can expand the functionality of the HAUCS platform is also being discussed.

Keywords: field robotics; hybrid sensing platform; aquaculture; HAUCS; Internet of Things; origami; robotic extension; mobile agent; LoRa

Citation: Den Ouden, C.J.; Wills, P.S.; Lopes, L.; Sanderson, J.; Ouyang, B. Evolution of the Hybrid Aerial Underwater Robotic System (HAUCS) for Aquaculture: Sensor Payload and Extension Development. *Vehicles* **2022**, *4*, 390–408. https://doi.org/10.3390/vehicles4020023

Academic Editors: Ralf Stetter, Udo Pulm, Markus Till and Giulio Reina

Received: 22 February 2022
Accepted: 15 April 2022
Published: 21 April 2022

1. Introduction and Background

Challenges in Aquaculture Farming

Precision agriculture (PA) combines robotic field machines and information technology in agriculture and plays an increasingly important role in farm production. PA-related advanced technologies such as the Internet of things (IoT), robotics, and artificial intelligence (AI) have been an active research topic and have seen robust growth [1–3]. Importantly, research results in CPS have been successfully adopted in many agriculture industry sectors. A BI Intelligence survey expects the adoption of IoT devices in the agriculture industry to reach 75 million in 2020, growing 20% annually [4]. In addition, the global smart agriculture market size is expected to triple by 2025, growing from 5 billion USD in 2016 to over 15 billion USD [4].

However, a critical agriculture sector that has been left behind is aquaculture. Aquaculture is farming in an aquatic environment. As an agricultural practice, aquaculture is characterized by considerable diversity in habitats, methods, and species. The species range from "livestock" (e.g., fish, mollusks, and crustaceans) to plants (e.g., microalgae, macroalgae, and vascular plants). The systems employed include earthen ponds, tanks, or open water (nearshore or offshore), depending on the habitat where production occurs. Pond and tank systems are generally land-based, and net pens or bottom cultures are in open water. Underpinning all of these are the energy systems powering the farm operations.

Worldwide aquaculture plays an essential role in food security in the seafood sector, filling the need and demand gap due to stagnant capture fisheries output. The transition from fisheries to aquaculture has been growing at an average rate of >6% annually. Since 2014, more farmed seafood than wild-caught seafood has been consumed globally, with more than half of all seafood coming from farms [5]. It is also important to emphasize that compared with other farmed proteins (e.g., chicken and beef), seafood has the highest resource efficiency, lowest feed conversion ratio (i.e., most efficient in converting feed to animal proteins) (Figure 1a), aquaculture produces lower greenhouse gas emissions than other types of farming [6]. Farmed fish are less resource-intensive overall than other common animal-based protein products and consume less water for production than pork and beef in many cases [7,8]

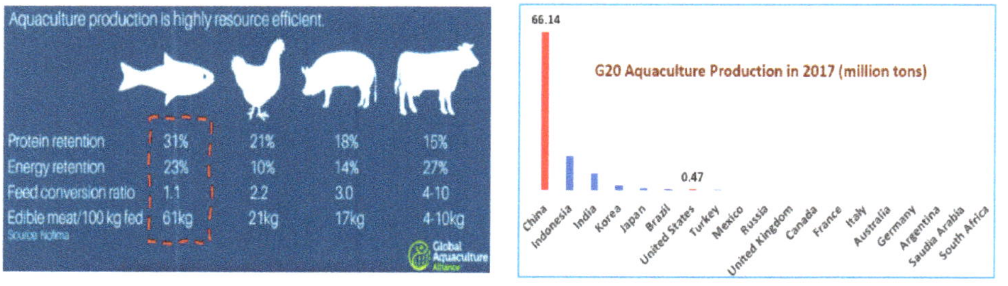

(a) (b)

Figure 1. (a) The high resource efficiency of fish compared with other farmed proteins [6]; (b) aquaculture production in 2017 among the G20 countries. China (66.14 Mts) and the US (0.47 Mts) are highlighted in red [9].

Management of water quality, particularly dissolved oxygen (DO), is critically important for successful operation in aquaculture fish farming. DO depletion is a leading cause of fish mortality on farms [10]. Catastrophic loss can occur within hours if ponds are not appropriately managed. The current management practice on pond-based farms uses human operators who drive trucks or other all-terrain vehicles throughout the day, especially at night, to sample and monitor the DO in each pond (Figure 2a). The associated labor and equipment costs limit the scope and frequency of such sampling since each sensor-equipped truck must manage dozens of ponds. To obtain the required monitoring frequency for proper management, large farms require multiple drivers and sampling instruments. The level of resolution that this approach can achieve on any pond is generally restricted to a single nearshore measurement at a point on the pond with a well-maintained roadbed. On large ponds (e.g., 2 to 8 hectares), this may result in a failure to promptly identify localized water quality problems that can ultimately affect a large percentage of the stock. Even though readings should be taken hourly on each pond, large farms (>400 hectares) with hundreds of ponds may only be outfitted to take readings at much lower frequencies due to the high labor and equipment costs of operating large fleets of monitoring vehicles. Measurements of additional water quality parameters cannot be performed due to the demanding schedules required of drivers to achieve the minimum frequency for DO management. The current monitoring practice relies on human operators to drive a truck around the farm to measure the DO concentration in each pond. An example of a typical DO measurement truck is shown in Figure 2a. Such practice has deficiency on multiple fronts: (1) human data collection may be difficult to interpret or recorded incorrectly; (2) the sensor is mounted onto a boom that is operated from within the cab of the vehicle, meaning the operator is unaware if they breach the air/water interface for acceptable readings to occur; (3) weather variability may cause the water levels of the ponds to decrease, this prevents the operators from receiving acceptable data as the sensor may not even reach the current water level and (4) fuel cost may become prohibitively high on large farms that may require a few dozen of such trucks. Furthermore, with the current practice, operators have a very limited window of time (e.g., less than an hour in the middle of the night) to react to potential oxygen depletion, increasing the likelihood of catastrophic events. The response (e.g., putting emergency aeration equipment in a pond) takes time away from achieving DO measurement frequencies.

Figure 2. (**a**) The typical truck-based boom senor system. (**b**) The "top-of-the-line" in situ buoy system from In-Situ Inc [11].

The industry has attempted to remedy the deficits of present truck-based aquaculture monitoring through in situ pond buoy monitoring systems (Figure 2b) [11]. However, these platforms are a temporary fix as they suffer negative attributes. In an aquaculture farm system, there is always a plethora of bacteria and other organics in the ponds. This will inevitably result in biofouling. Since the aquaculture ponds are generally very productive, the growth of organic material such as algae, Bivalvia, etc., on a sensor surface due to biofouling would quickly render the platform useless.

Similarly, maintenance on these platforms is labor-intensive. It requires a laborer to enter the pond to retrieve the device or an elaborate pully system that would cause stress to the stock within the pond. The farm technicians need to remove the sensor suite from the pond before starting stock extraction when harvesting. Many farms have employed the buoy-based systems and subsequently abandoned them due to long-term difficulties in application.

The Hybrid Aerial/Underwater robotiC System (HAUCS) framework aims to mitigate the aforementioned issues to provide automated, high-density monitoring of key environmental metrics of each aquaculture pond on a farm using relatively inexpensive robotic sensing platforms.

A critical aspect of the HAUCS sensing platform is a novel deployable robotic extension. This subsystem is essential to reduce the drag of the HAUCS platform during the flight and improve the platform stability during the sensing operation. Stability during sensing is even more critical to expand our sensor payload beyond DO monitoring to include underwater cameras or sonars to support other farm operations.

This design relies on Origami principles to realize rigid structures that retain their shape even under transverse loading. We will therefore provide a brief discussion on Origami. Origami is the ancient art of folding, generally, the word conjures images of intricate paper structures [12,13]. Origami has been adopted for many engineering applications where a package with limited initial storage space is required to morph into a structure with a significantly larger span. This is accomplished by employing the governing geometric calculations at the heart of Origami to create more complex structures. One of the well-known applications is the NASA Solar Panel Array [14]. Additional applications include circuit board design [15], shelters [16], and biomedical devices [17]. Peraza-Hernandez et al. provide an excellent review in [18].

At the fundamental level, the deployable robotic extension under consideration is a cylindrical tube that can be folded in the axial direction to a flat state on demand. One Origami pattern of interest is the twist buckling pattern or the Kresling Origami pattern [19–22], formed by the buckling of a thin cylindrical shell under torsional loading. The properties of the Kresling pattern have been studied extensively [23–26].

In particular, the Kresling pattern is preferable to the axially loaded Yoshimura or diamond origami [26] in that the Yoshimura pattern is not continuously foldable, and significant stretching occurs during the re-stabilization stage [24]. The Kresling pattern was demonstrated experimentally in [24]. In the experiments, a thin-walled sheet is wrapped around two coaxial mandrels. When the mandrels are twisted, a highly regular self-organized folding pattern appears across the gap, formed by inclined and elongated parallelograms (mountain-folds), divided on their long diagonal by a valley-fold [24,27].

Origami-based deployable robotic extension for drones has been previously studied by Kim et al. [28]. The extension is designed to "fold-flat" whilst flying and deploy into the rigid structure while hovering at a given point. The end effector was designed to be plug-and-play so they could easily switch between a gripper mechanism and a camera. In their design, the axially loaded Origami pattern was adopted. In this regard, the pattern can be considered a variant of the Yoshimura Origami. The folding and unfolding are achieved via an indirect tendon system—a wire runs the length of the structure and is controlled by a motor attached to the drone body. The team designed a locker system to counteract transverse force loadings on the expanded structure. These lockers are moved in conjunction with the wire deployment and retraction. Additionally, they are strengthened with magnets.

In the context of the HAUCS platform, there are several deficiencies in the aforementioned approach. The first constraint in utilizing this design is the mechanical restrictions. The structure must fully expand or retract as the system is operated by indirect force. The arm cannot be halted partway through the deployment or retrieval. This is intrinsic to the Yoshimura Origami [29].

Additionally, due to the complexity of the locker system, while the design succeeds in its objective of rigidity, it severely limits the flexibility of the robotic extension. As such, the current design will not be able to achieve more complex configurations. Thirdly, the folded footprint will be twice as large, which may not be desirable for field robots in aquaculture applications.

With the main focus of our current endeavors being to realize the HAUCS framework, the paper structure is as follows. In Section 2, we review the HAUCS concept and previous development efforts. Section 3 presents our current sensor, platform development, and field experimental results. We discuss our effort to develop a versatile robotic extension in Sections 4 and 5 and conclude in Section 6.

2. Overview of the HAUCS Framework

The overarching goal of HAUCS is to relieve human operators from the most labor-intensive, time-consuming, and expensive tasks in aquaculture farming operations through a group of cooperative robotic sensing and actuator platforms. With support from the National Institute of Food and Agriculture (NIFA), USDA, through the Ubiquitous Collaborative Robots (NRI-2.0) program, the project was launched in the Spring of 2019.

HAUCS is a framework that aims to achieve collaborative monitoring and decision-making on aquaculture farms of varying scales. The HAUCS framework consists of three basic modules: aero-amphibious robotic sensing platforms integrated with underwater sensors, land-based infrastructure, and backend modeling and processing infrastructure, particularly an ML-based water quality prediction model. Each HAUCS autonomous unmanned platform (AUP) consists of an unmanned robotic vehicle and submerged underwater sensors. Data from the underwater sensors attached to the AUP, such as DO and temperature sensors, will be sent to the farm control center via a radio link during sensing operations. Sensor data from all the ponds on the farm and the associated weather data will be used to train an ML-based prediction model. The model prediction can, in turn, guide other instruments to mitigate an emergency situation (e.g., turning on a fixed aerator or instructing human operators to move mobile emergency aeration equipment into place in a pond). The overall concept of operations is illustrated in Figure 3 [30].

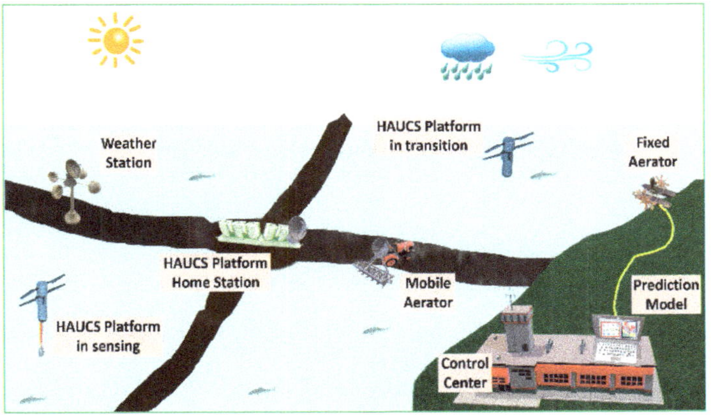

Figure 3. A basic HAUCS concept of operations [30].

The agility of the HAUCS platform supports a level of precision that is unobtainable by the truck-based monitoring systems. The HAUCS platform expands the sampling capabilities by allowing the farmers to sample multiple points within a pond. In contrast, truck-based monitoring is limited to a single-point sampling near the edge of the pond. Various novel sensing schemes can be investigated to optimize spatial and temporal regions being studied by utilizing the platform's mobility. One potential sampling scheme sees the platform enter a new pond where the platform will hover for approximately 30 s so the DO sensor can record an accurate reading. The platform will then move through the pond toward a neighboring pond, the sensor maintaining its submergence below the water surface to collect a stream of water quality data (Figure 4). Through sampling routes such as this, it acquires significantly more data samples and captures the spatial variability of water quality in any given pond.

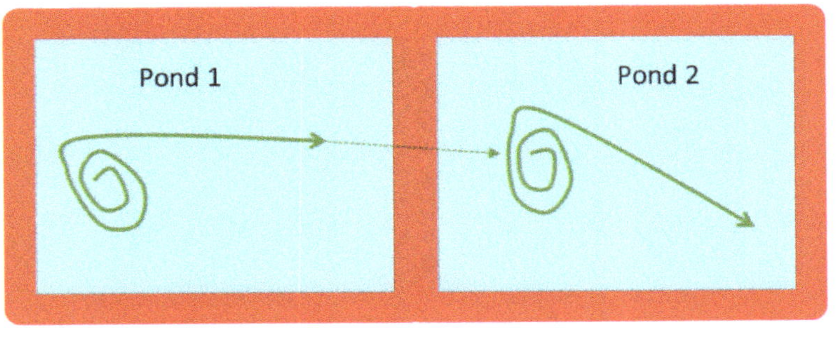

Figure 4. Illustration of the mobile platform-based sensing [30].

This highly scalable framework will convert aquaculture farm operations to an "Internet of Aquaculture." In particular, the DO sensing platform will transition from a truck-based manual measurement practice to an intelligent, automated scheme relying on aerial drones integrated with underwater sensors. Compared with the state-of-the-art solutions, the advantages of HAUCS include:

- Improved scalability—compared with the sensor buoys, HAUCS design can be easily adapted to farms of varying scales.
- Broad spatial coverage—capable of realizing novel sensing patterns to cover different areas on a large pond, hence providing more accurate reporting of pond conditions.
- Mitigated biofouling—avoiding sensors in bio-productive water.

- Supporting novel sensing schemes to cover extended spatial regions and generate more robust readings than the traditional truck-based data acquisition.

A stop-gap truck-based automated sensor system was developed for three reasons: (1) to provide an automated DO data acquisition system compatible with the current monitoring practice on the farm that will transition seamlessly to the drone-based platform; (2) to evaluate the infrastructure need for future deployment of HAUCS platforms on the farm, and (3) to collect high-quality data to support prediction model development. This prototype sensor system consists of a mobile data acquisition system (DAQ) and a central server. The mobile DAQ was installed on the farm sensing truck, side by side with the DO sensor used in the farm operation at the Logan Hollow Fish Farm (Figure 5a). The central server was located inside the farm manager's office.

(a) (b)
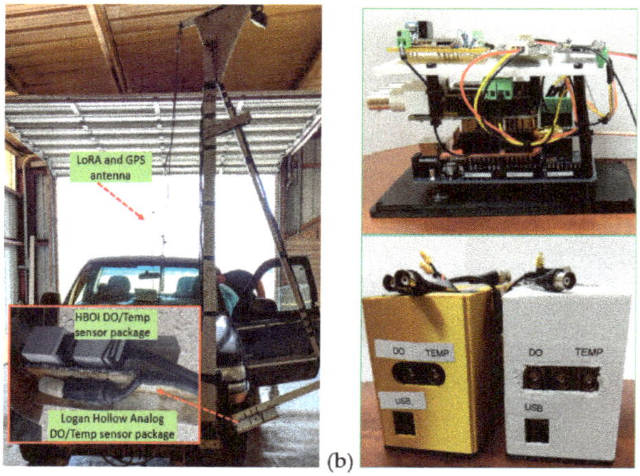

Figure 5. DO sensing truck with HBOI sensor head co-mounted with the original DO sensor. (**a**) HBOI sensors mounted side-by-side with Logan Hollow sensors. (**b**) In-truck control unit.

As shown in Figure 5b, the sensor control system is an Arduino Mega integrated with Atlas Scientific™ temperature and DO shields to simplify the integration of the sensor with the backend controller. In addition to the sensor shields, a GPS shield was used to identify the pond being sampled. For communication, a Dragino LoRa [31] shield was used to stream DO and temperature sensor data to the server in the farm manager's office. The LoRa antenna was mounted on a pole at the back of the truck (Figure 5a).

The system was deployed on Logan Hollow Farm between September and October 2019. Logan Hollow Fish Farm is located in southwest Illinois in the bottomland of the Mississippi River. This is a comparatively small farm with 70 ponds. For DO monitoring, the farm employed two technicians to drive the trucks and sample each pond during the night. A total of 15 days of data were collected using the automated system on the farm during this period. The systems were again deployed at another collaborating farm-Flowers Fish Farm, in the summer of 2021 to collect additional data.

These datasets are critical to improving the prediction model. The model development effort will be presented in future work. More importantly, these deployments helped us gain our farm collaborators' trust.

As evidence of the success of the system deployment in reducing the barriers to the acceptance of robotic technology by the fish farmers, Mr. Pete Reiff, Logan Hollow's Owner/Operator, requested the HBOI sensing system to be installed on the second truck of his farm [32]. One immediate tangible benefit that the farm can gain from the automated system is to allow the farm to collect high-quality DO data consistently. The current paper-based operation is crude and fraught with potential decision-making problems (e.g., transcription issues of the data collected), as shown in Figure 6.

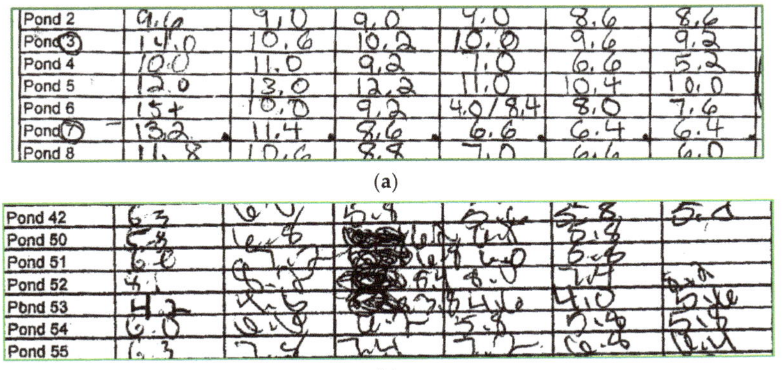

Figure 6. Hand-written sheets of DO data collected on the Logan Hollow farm by different farm technicians. The quality of the sheet ranges from good in (**a**) "good" quality DO datasheet to almost illegible in (**b**) "bad" quality DO datasheet.

3. Development and Field Demonstration of Drone-Based HAUCS Sensing Payload

3.1. Development of Drone-Based HAUCS Sensing

For this effort, a low-cost Swellpro Splashdrone 3+ was adopted for payload integration (https://www.swellpro.com/waterproof-splash-drone.html, accessed on 11 March 2021). Splashdrone™ is a waterproof drone that is surface buoyant. The Splashdrone 3+ employs the conventional quadcopter design, sports an axis diameter of 450 mm, a flight time of 20–23 min, a weight of 2 kg, and a max payload capacity of 1 kg. Such payload capability is ideal for HAUCS platforms. Another important consideration is that this drone can operate in heavy rain with winds up to 18 mph and gusts of up to 31 mph. In addition, the vendor will provide API to support the integration of external logic with the drone in the near future. The programming capability will be desirable for the future implementation of environmental adaptive path planning algorithms [33]. However, one important insight the team arrived at in designing for the project is that instead of being locked into specific platform designs, the HAUCS sensing payload should be platform-neutral. Therefore, while the new sensing payload targets an aerial drone, it can be easily adapted to other platforms, such as unmanned surface vehicles (USVs) or ground vehicles.

The sensing system has two subsystems: a payload module containing the sensors and microcontroller to support data acquisition and transmission to the topside module and a topside module that interfaces with the platform (aerial drone in this case) forward data to the control center via a long-range communication link. The two modules are connected with a winch that will be released to allow the sensors to go underwater during the sensing operations (Figure 7).

The topside is the gateway between the sensing module and the control center. The topside engages the winch to release and retrieve the sensing module during the sensing operation. The engagement can be triggered by signals from the sensing platform (which is the current implementation) or by GPS-driven waypoint programming (i.e., positioned over a pond and at proper altitude). To perform these tasks, the topside module contains a micro-controller, a GPS unit, and a servo that controls the winch (Figure 8a). A LoRA link is adopted to support the communication between the sensing platform and the control center [30]. However, other options may also be considered. For example, the SIYI 2.4 G Datalink can support video links up to 15 km range, albeit at increased weight and power than LoRA links [34]. To support the LoRA link, the ESP32 LoRA controller is ideal due to its rich connectivity options: LoRA, WiFi, and Bluetooth [35]. The payload module handles the sensor data acquisitions and quality control (QC). Temperature and DO sensors are included in the module. Here, the controller is an ESP32S [36], which consumes less power than ESP32 LoRA while supporting multiple analog to digital converter (ADC) channels and WiFi links. The communications between the topside and the sensing modules are achieved using the ESP-NOW link [37]. ESP-NOW is a connectionless communication protocol developed by Espressif to support short packet transmission (up to 250 bytes) between ESP32 boards. This protocol enables multiple devices to talk to each other using the 2.4GHz RF circuits on the ESP32x boards without the WiFi protocol. Therefore, this protocol is ideal for linking topside and sensing modules. The lightweight ESP-NOW link helps eliminate the need for a physical electrical connection between topside and sensing modules and simplifies the winch system. As a result, the winch essentially consists of a servo that controls the release and retrial of the sensing module using a metal chain (Figure 9).

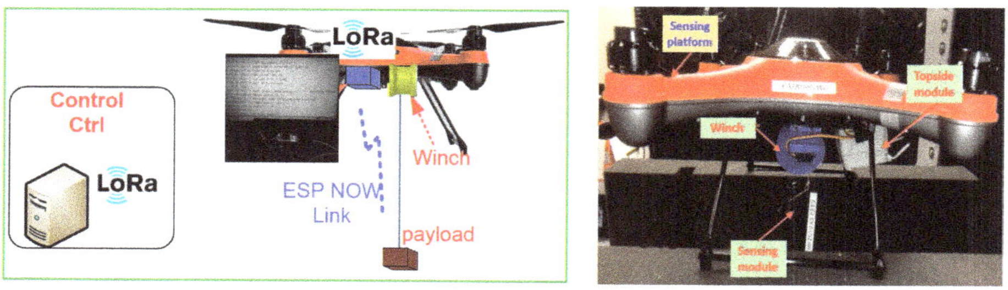

(**a**) (**b**)

Figure 7. Illustration of the platform-independent sensing system design. (**a**) A flow chart describes the components in the sensing system. The sensor data will be sent via the ESP ONE wireless link from the topside module. The topside will relay the data to the control center via long-range communication links such as LoRa. (**b**) Showing the integration of the sensing system with the Splashdrone.

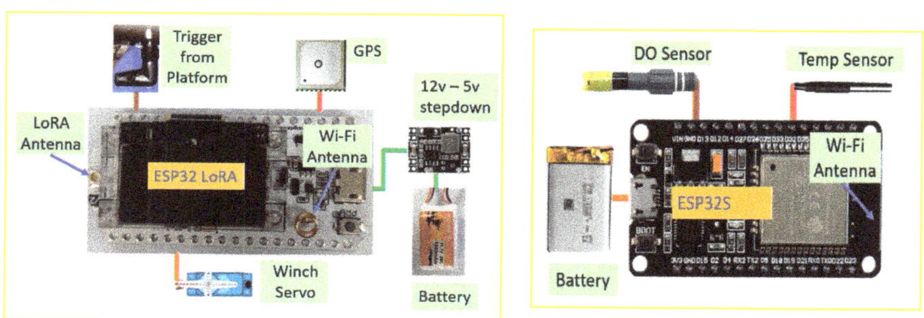

(**a**) (**b**)

Figure 8. Flow charts describe the topside and payload modules. (**a**) The flow chart illustrates the topside electronic. (**b**) A flow chart shows the sensing module electronic.

The drone will be programmed to reach a pre-defined location over the pond and transition to hovering mode during the sensing operation. Once the drone is on location, the sensing module release will be triggered (either via a signal from the sensing platform or a pre-determined waypoint). The topside will, in turn, engage the winch to lower the payload module into the water. The payload module loses communication with the topside module when fully submerged in water. The payload module will start acquiring DO and temperature data for a pre-determined time (i.e., 20 s) and store the data onboard the controller. The topside will then retrieve the winch after the data acquisition period. Once the payload module is fully retrieved, the data stored on the payload module controller will be sent via the ESP-NOW link to the topside. The repackaged data (i.e., adding the GPS coordinates and timestamps) will be forwarded to the control center for processing.

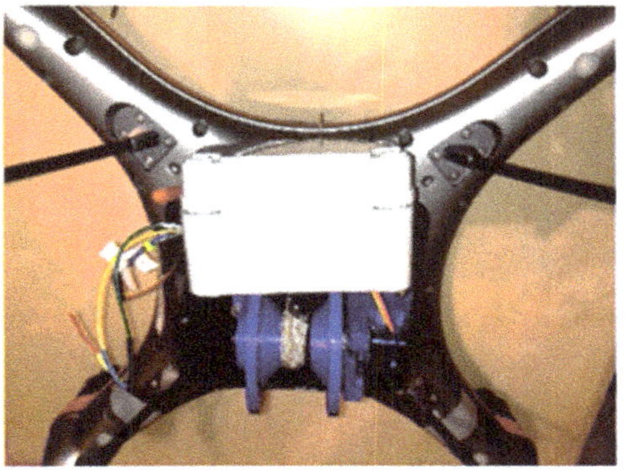

Figure 9. Winch to release/retrieve the sensing module.

While drones are the primary platform, the platform-neutral nature of the sensing system design ensures that the payload is compatible with other platforms such as all-terrain vehicles (ATVs) or unmanned surface vehicles (USVs).

3.2. Lab and Field Deployment

The initial test was a static test inside the System and Imaging Laboratory (SAIL) lab. During the test, the payload is mounted on the drone. However, the drone was fixed to the lab ceiling and kept idle. A trigger signal was sent from the drone transmitter to the drone during the test. The signal triggered the sensing operation: the topside module engaged the winch to release the payload; the payload then idled for 20 s to simulate the data acquisition period; the topside retrieved the payload (Figure 10). All these were done in an automated fashion. The communication links from the sensing module to the topside (through ESP-NOW) and from the topside to the control center (via LoRA) were validated.

Figure 10. Lab test for the payload release and retrieve.

Following the lab test, the system was further tested over the ground at the HBOI campus to validate the feasibility of the sensing operation. Here, the objective was to deploy the payload into a plastic tub with a 1.5 m diameter. Video frames that illustrate the whole process are in Figure 11.

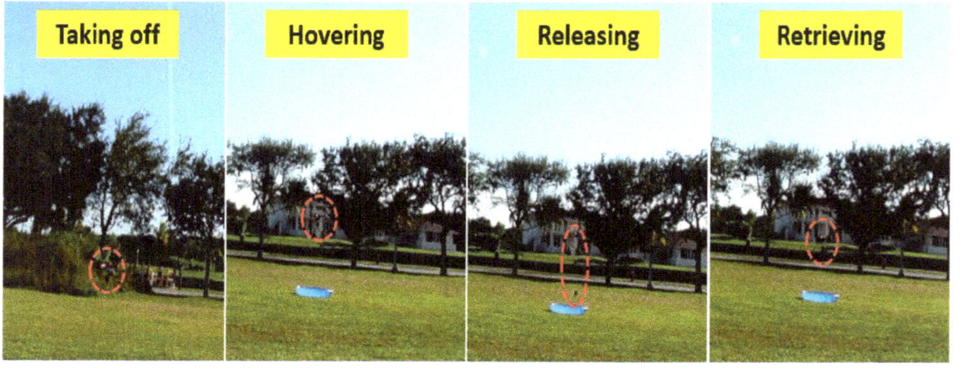

Figure 11. Field validation of the sensing operation.

The system was set up at our research collaborator Flowers Fish Farm in October 2021 for the initial field test of the drone-based sensing operation. With the consent of the farm owner, Ms. Kelly, and manager, Mr. Frampton, the HBOI team (Dr. Paul S. Wills) piloted the drone during the tests (Figure 12).

(**a**) (**b**)

Figure 12. (**a**) Photos were taken during the preparation of the field test at Flowers Fish Farm. From left to right: Ms. Kelly Flowers, the Flowers Fish Farm owner. Mr. Mike Frampton, Farm Manager, and Dr. Paul Wills. (**b**) The Splashdrone integrated with the sensing system during the flight at Flowers Fish Farm in Dexter, MO.

Since the primary goal of the experiment was to validate the functionalities of the sensing system and identify any potential issues, the tests were conducted during the daytime when the farmworkers were present. During the tests, Dr. Wills acted as the pilot to command the drone takeoff and landing, and flew the drone to the desired location in the pond for a simulated sensing operation (Figure 13a).

(**a**)

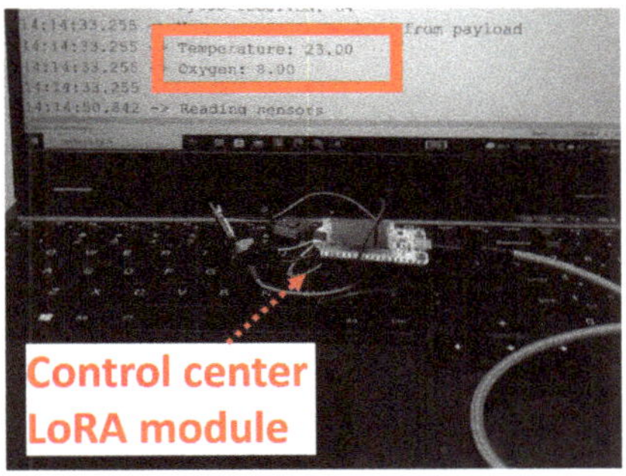

(**b**)

Figure 13. (**a**) Frames from the video clip taken during the field test to validate the functionality of the drone-based sensing system. The frames illustrate the drone taking off, deploying payload, retrieving the payload, and flying back to base. (**b**) Photo shows the sensor data acquired during the test received at a laptop equipped with a LoRA module at the control center.

While the field tests validated the basic concept, one issue we identified during the tests was that since the payload was connected to the drone body with a string, it is more susceptible to ambient environmental conditions. For example, strong side wind could induce a pendulum effect on the payload, impacting the drone's flight stability.

To address this issue, we aim to develop a deployable robotic extension that can provide better controllability and expand the functionality of the sensing payload. In this regard, we are exploring an Origami-based robotic extension. We will discuss the initial feasibility study of the Kresling Kirigami drone robotic extension design through both CAD modeling and laboratory testing. A more focused study of the Kresling Kimigami robotic extension will be presented in a future manuscript.

4. Feasibility Study of a Kresling Kirigami Robotic Extension Design

The design choice is motivated by several factors. First, the robotic extension for the HAUCS will need the flexibility of support with varying extension lengths. Secondly, the folding and unfolding should be confined to the same horizontal footprint to avoid interference with other sensors to be integrated into the drone, such as cameras and other environmental sensors. Thirdly, the actuation needs to be supported via the drone flight controller. For this reason, we investigated a Kresling buckling pattern-based design.

Furthermore, in our implementation, we opted for Kirigami instead of Origami. In addition to folding, Kirigami also involves cutting. There are more than simple semantic differences between Kirigami and Origami. For example, Li et al. adopted Kirigami enhancement to prevent wrinkling during continuous folding/unfolding and improve structure reliability [15]. Yasuda uses a similar methodology where he makes small, precise cuts and inserts small holes in areas where continual stresses may lead to creasing and deformation in the mesh [38]. Many others use Kirigami to construct self-folding structures [39,40]. For the HAUCS robotic extension, we also envision the need for continuous folding/unfolding during the sensing operations. This, therefore, motivated us to adopt a Kirigami-based design.

4.1. Kresling Kirigami Prototype Design

The proposed robotic extension consists of multiple Kresling Kirigami sections. Each section consists of two plates connected with hinges (i.e., Kirigami creases). The torsional loading is realized via the plate rotating at the center axis, driven by a gear system at one end and a ball bearing at the other. The hinges then rotate in conjunction with the rotation of the main body, allowing the structure to fold and unfold. The gear and bearing will be mounted on a set of collapsible tapered rods.

A Solidworks model of this design was developed to support laboratory evaluations. In the prototype design, the full extension of one Kirigami section is 82.8 mm, and a folded section has a height of 20.2 mm. The diameter of the structure is 85 mm (Figure 14).

Figure 14. The three images from left to right show the folding of a single Kirigami section.

The central axis is a telescopic rod that can collapse in the current design. The structure is composed of multiple Kirigami sections. Each Kresling Kirigami section can collapse independently to realize a variable-length robotic extension (Figure 15a).

One key component in supporting the structure folding and unfolding is the uniquely designed hinge. This Kirigami design requires each hinge to rotate 180 degrees. An action that is not supported by traditional straight hinge pins. Instead, the devised hinge pin sports an 'L' shaped attachment joint (Figure 15b). This hinge design accomplishes two essential requirements. The first, as previously mentioned, is the ability to rotate 180 degrees. The second is a locker system on one side to maintain rigidity, and compression only occurs in one direction.

Another critical design consideration is to use a gear system to realize the torsional loading. The gear system can be driven by a servo, worm drive, or stepper motor (Figure 16), With this design, each Kirigami section can be actuated independently from the drone flight controller (Figure 16).

Figure 15. (**a**) The CAD models of a single section and two-section Kresling Kirigami incorporate gear-driven actuation. (**b**) The 'L' shaped hinges support 180-degree rotation.

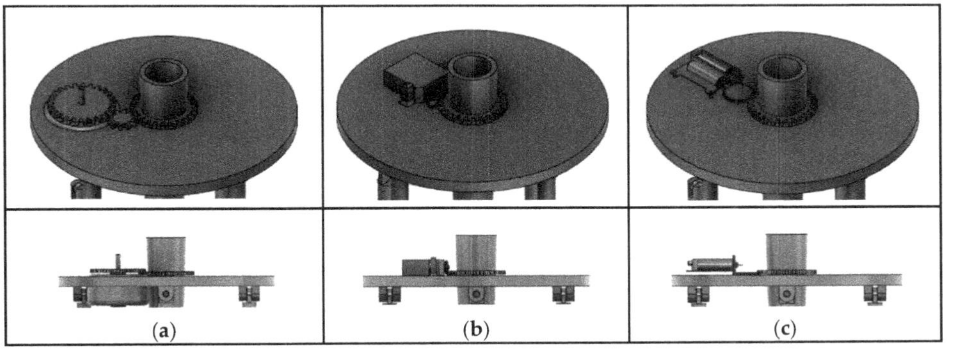

Figure 16. Three different actuation options to drive the utilizing: (**a**) a step motor, (**b**) micro servo, and (**c**) a linear actuator.

4.2. Laboratory Validations

Small-scale models were constructed and subjected to dynamic force to determine efficacy. During the CAD modeling, "mockups" were first created by utilizing the most simplistic materials, foam board, and bendable straws to provide physical proof of concept. Once CAD models were established, the components were printed on 3D printers and assembled. Figure 17 illustrates the folding of a sing-section structure and a two-section structure. Figure 17a–c illustrates the single-section structure begins in the locked position, with hinges at full extension. Through manual actuation, the structure compresses along the *z*-axis until it is fully folded. The same process is conducted for a two-section unit shown in Figure 17d–g.

Load tests were performed on the single unit model. The tests were conducted by adding water to various containers placed on top of the unit. This test determined that the model would suffer structural failure at 1802 g of applied weight (Table 1). The test measured structural integrity to the applied load. A green indicator value was given when the structure maintained its integrity with little to no deflection, similar to the elastic region. Yellow was given when the structure exhibited signs of deflection consistent with the plastic region. Red was given for the point of failure or fracture. Figure 18 illustrates the corresponding behavior of the structure under different loads.

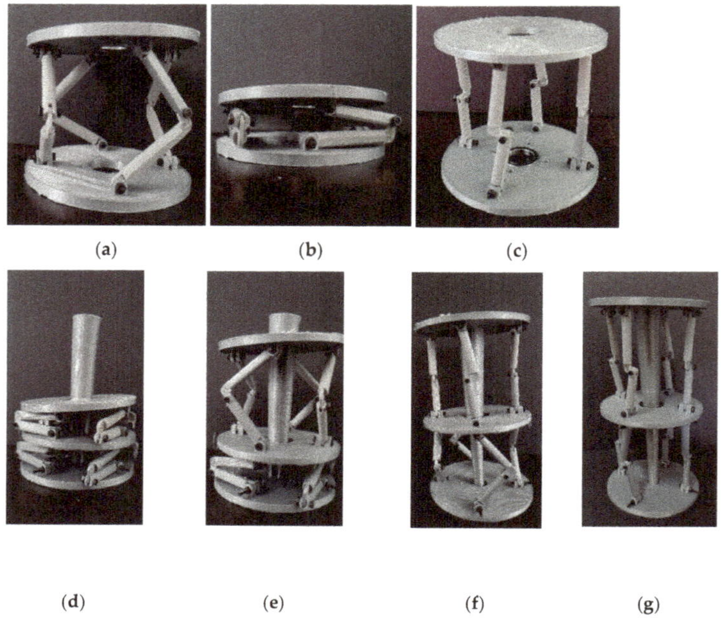

Figure 17. Revolute body frame in expansionsequence, in both single unit (**a**–**c**) and double-stacked arrangements (**d**–**g**).

Table 1. Results of load failure test of a single-section structure. Structure support denotes any deformations suffered during loading: green means little to no deformation, yellow is significant deformation, and red marks the point of failure.

Weight (g)	401	550	700	1512	1605	1727	1802
Structure	green	green	green	green	yellow	yellow	red

Figure 18. Weight of load applied to the revolute structure and resulting structural deformations. The photos in the top row show the actual load and those in the bottom row show the responses to the structure deformations.

Similar tests were conducted against a two-section structure. Here, two different cases were evaluated. In the first case, the top section was initially fully extended. The results are summarized in Table 2 and the actual deformations are illustrated in Figure 19. Here, the top section experienced failure at a lower loading than the bottom layer, which behaved similarly to the single-section case.

Table 2. Results of load failure test of a two-section structure. The structural support is ranked the same as Table 1, with only two points of failure as there are two individual units.

Weight (g)	401	701	1000	1251	1654	1800	1900	2050
Structure 1								
Structure 2								

Figure 19. Weight of load applied to a fully extended two-section structure.

Further investigation may be needed to explain the cause fully. Still, one likely reason was the misalignment between the two sections—the parts were printed in a low-quality setting to support rapid prototyping. In the second test, the top section was half folded. As results in Table 3 and Figure 20 show, the behavior of the bottom section is similar to that of a single section as expected. However, the top section was able to withstand less loading. We did not integrate any electronic actuator into the prototype.

Table 3. Results of load failure test of a fully extended two-section structure with the top section half folded initially.

Weight (g)	18	100	306	506	1350	1852	1900
Structure 1							
Structure 2							

Upon completion of these experiments, the model was examined. While the prototype was structurally weakened from the collapse, it was not irreparably damaged. The prototype was still able to fold and unfold properly. The prototype could also withstand axial loading, although with decidedly lighter weights than previously observed.

The next step in laboratory testing will be to evaluate the actuation, in particular, comparing the effectiveness of step motor, micro servo, and worm drive to actuate the Kirigami structure through the gear system. This is a critical and necessary step before further mechanical design optimizations. Unfortunately, we are hindered by the short supply of the electronic components needed for this work. However, we intend to provide a thorough study of the Kirigami robotic extension design in a separate manuscript.

Figure 20. Weight of load applied to a two-section structure with the top section half folded initially.

5. Conclusions

A significant expense in most pond aquaculture farm operations is the labor and vehicle fuel cost related to the DO sensing practice. The overarching goal of the HAUCS framework is a transformative robotic solution for aquaculture farms of varying scales to achieve collaborative monitoring and decision-making. Once fully developed, the HAUCS sensing platform can overcome the shortcomings in many existing aquaculture farm monitoring solutions such as the static pond buoy type sensors. These sensors may face the challenges of being susceptible to biofouling, high deployment costs, and maintenance difficulty.

An essential aspect of our work is the experience we gained on how to move forward in any field of robotic development in an industry that is generally risk-averse regarding new technology due to the high cost for initial adopters and the low profit margins such as the aquaculture farms. As we realized from early on, one of the most critical components of the HAUCS project is the close engagement with the farms and buying-in from the farmers. In this regard, the team achieved great success, demonstrated by the support from our collaborative farms throughout the development and testing of HAUCS infrastructure and platforms. Our success in this aspect can benefit many similar field robotic projects.

One of the crucial achievements in technology development is the HAUCS AUP—the mobile environmental sensing system. The successful deployment of the prototype system on an operational fish farm validated the effectiveness of the design and will be the foundation to build upon for future efforts. It is worth noting that our work is the first known attempt at drone-based mobile DO monitoring system for the aquaculture farm setting. The HAUCS system is a transformative technology that can revolutionize how pond aquaculture farms operate. In addition to significant savings in labor and time through the automation of the DO monitoring of the ponds on the farm, HAUCS can provide unprecedented insight into the spatial variations of each pond that none of the existing state-of-the-art techniques are capable of. At the same time, the HAUCS system avoids many drawbacks that plague these techniques, such as biofouling and maintenance difficulties.

While the initial motivation for developing the Kirigami-based robotic extension is to improve the drone's stability during the sensing operation, this extension can significantly enhance the AUP's functionalities, such as supporting/stabilizing underwater cameras or sonars or collecting water samples. These capabilities will help expand the HAUCS framework's applicability to support other environmental monitoring efforts, such as surveillance of coastal zone harmful algae bloom (HAB) [41]. While this development is still in the early stages, the laboratory tests of the mechanical models verified that the design is sound.

It is worth noting that while the primary objective is to support drone-based sensing, one unique aspect of the system is that it is platform-neutral. As a result, the system can be easily adapted to other platforms, such as unmanned surface vehicles (USV).

Going forward, one main focus of the project will be the operational deployment of the HAUCS framework. Here, multiple HAUCS platforms will be deployed on our collaborative farms to conduct farm environmental monitoring to complement the current DO sensing practices during farm operations. In addition to improving the reliability and reducing the power consumption of the sensor payload, one critical component to support such deployment is implementing the hybrid path planning algorithm [33] in the HAUCS framework.

The initial modeling of the Kirigami-based robotic extension validated the basic design concept. With further development, the Kirigami-based extension will be able to perform the desired HAUCS tasks. It can be adapted for more advanced operations (e.g., attaching a gripper or tube for physical collection). Going forward, several topics need to be further explored. Different drive options to actuate the Kirigami structure (i.e., servo, worm drive, and step motor) will be investigated through laboratory experiments to test the integration with the drone flight controller while avoiding a significant weight increase. The mechanical design will be further improved to reduce the volume of the structure in the compressed form. Following the same trajectory of our payload development, we aim to integrate the robotic extension with the drone and field test in the implementation on the collaborative farm.

6. Patents

Bing Ouyang, Paul Wills, Casey Den Ouden and Lucas Lopes, "Platform-Independent Mobile Environmental Sensing System," US Patent Application 63/311,937, 02/18/2022.

This invention was made with government support under Contract No. 2019-67022-29204, awarded by the National Institute of Food and Agriculture/United States Department of Agriculture.

Author Contributions: Conceptualization, B.O., P.S.W. and C.J.D.O.; methodology, C.J.D.O., B.O. and P.S.W.; software, L.L.; validation, P.S.W., B.O., L.L. and J.S.; resources, B.O. and P.S.W.; data curation, C.J.D.O.; Kirigami design and testing, C.J.D.O.; writing—original draft preparation, C.J.D.O., B.O., L.L., P.S.W. and J.S.; writing—review and editing, C.J.D.O. and B.O.; visualization, C.J.D.O.; supervision, B.O.; funding acquisition, B.O. and P.S.W. All authors have read and agreed to the published version of the manuscript.

Funding: This research was funded with support under Contract No. 2019-67022-29204, awarded by the National Institute of Food and Agriculture/United States Department of Agriculture.

Acknowledgments: The authors want to thank Yanjun Li and Dennis Estrada for helping during the laboratory tests. Ben Metzger for the discussions on the mechanical design of the Kirigami structure.

Conflicts of Interest: The authors declare no conflict of interest. The funders had no role in the design of the study, in the collection, analyses, or interpretation of data, in the writing of the manuscript, or in the decision to publish the results.

References

1. Aravind, K.R.; Raja, P.; Pérez-Ruiz, M. Task-based agricultural mobile robots in arable farming: A review. *Span. J. Agric. Res.* **2017**, *15*, e02R01. [CrossRef]
2. Bhandari, S.; Raheja, A.; Green, R.L.; Do, D. Towards collaboration between unmanned aerial and ground vehicles for precision agriculture. In *Autonomous Air and Ground Sensing Systems for Agricultural Optimization and Phenotyping II*; SPIE: Orlando, FL, USA, 2017; Volume 10218.
3. Schimmelpfennig, D. Precision Agriculture Technologies and Factors Affecting Their Adoption. 12 May 2016. Available online: https://www.ers.usda.gov/amber-waves/2016/december/precision-agriculture-technologies-and-factors-affecting-their-adoption/ (accessed on 1 February 2018).
4. Meola, A. IOT-101: The Essential Guide to the Internet of Things. April 2018. Available online: https://artillry.co/wp-content/uploads/2019/08/Business-Insider-IOT-101.pdf (accessed on 11 August 2020).
5. *The State of World Fisheries and Aquaculture 2020*; Food and Agriculture Organization of the United Nations: Rome, Italy, 2020.
6. What Is Aquaculture and Why Do We Need It? *Global Aquaculture Alliance*. Available online: https://www.aquaculturealliance.org/blog/what-is-aquaculture-why-do-we-need-it/ (accessed on 24 November 2020).
7. Yuan, Q.; Song, G.; Fullana-I-Palmer, P.; Semakula, H.; Mekonnen, M.; Zhang, S. Water footprint of feed required by farmed fish in China based on a Monte Carlo-supported von Bertalanffy growth model: A policy implication. *J. Clean. Prod.* **2017**, *153*, 41–50. [CrossRef]
8. Hall, S.; Delaporte, A.; Phillips, M.; Beveridge, M.; O'Keefe, M. *Blue Frontiers: Managing the Environmental Costs of Aquaculture*; The WorldFish Center: Penang, Malaysia, 2011.
9. Aquaculture Production (Indicator). OECD. 2020. Available online: https://doi.org/10.1787/d00923d8-en (accessed on 24 November 2020).
10. Dureisseix, D. An Overview of Mechanisms and Patterns with Origami. *Int. J. Space Struct.* **2012**, *27*, 1–14. [CrossRef]
11. Lang, R.J. Origami: Complexity increases (again). *Eng. Sci.* **2004**, *67*, 5–19.
12. Zirbel, S.; Magleby, S.; Howell, L.; Lang, R.; Thomson, M. Accommodating thickness in origami-based deployable arrays. *J. Mech. Des.* **2013**, *135*, 111005. [CrossRef]
13. Li, Y.; Liu, W.; Deng, Y.; Hong, W.; Yu, H. Miura-ori enabled stretchable circuit boards. *Nat. Partn. J.-Flex. Electron.* **2021**, *5*, 1–9. [CrossRef]
14. Thrall, A.P.; Quaglia, C.P. Accordion shelters: A historical review of origami-like deployable shelters developed by the US military. *Eng. Struct.* **2013**, *59*, 686–692. [CrossRef]
15. Randall, C.L.; Gultepe, E.; Gracias, D.H. Self-folding devices and materials for biomedical applications. *Trends Biotechnol.* **2012**, *30*, 138–146. [CrossRef]
16. Peraza-Hernandez, E.A.; Hartl, D.J.; Malak, R.J., Jr.; Lagoudas, D.C. Origami-inspired active structures: A synthesis and review. *Smart Mater. Struct.* **2014**, *23*, 094001. [CrossRef]
17. Kresling, B. Plant Design: Mechanical Simulations of Growth Patterns and Bionics. *Biomimetics* **1995**, *3*, 105–222.
18. Guest, S.D.; Pellegrino, S. The Folding of Triangulated Cylinders Part I: Geometric Considerations. *ASME J. Appl. Mech.* **1994**, *61*, 773–777. [CrossRef]
19. Guest, S.D.; Pellegrino, S. The Folding of Triangulated Cylinders Part II: The Folding Process. *ASME J. Appl. Mech.* **1994**, *61*, 778–783. [CrossRef]
20. Guest, S.D.; Pellegrino, S. The Folding of Triangulated Cylinders Part III: Experiments. *ASME J. Appl. Mech.* **1996**, *63*, 77–83. [CrossRef]
21. Kresling, B. Folded Tubes as Compared to Kikko ("Tortoiseshell") Bamboo. In *Origami*; AK Peters: Natick, MA, USA, 2001; pp. 197–207.
22. Hunt, G.W.; Ario, I. Twist buckling and the foldable cylinder: An exercise in origami. *Int. J. Non-Linear Mech.* **2005**, *40*, 833–843. [CrossRef]
23. Nishiyama, Y. Miura Folding: Applying Origami to Space Exploration. *Int. J. Pure Appl. Math.* **2012**, *79*, 269–279.
24. Yoshimura, Y. *On the Mechanism of Buckling of a Circular Cylindrical Shell under Axial Compression*; NACA: Washington, DC, USA, 1955.
25. Kresling, B. Origami-structures in natures: Lessons in designing 'smart' materials. In *Symposium OO—Multiscale Mechanics of Hierarchical Materials*; Cambridge University Press: Cambridge, UK, 2012.
26. Kim, S.J.; Lee, D.Y.; Jung, G.P.; Cho, K.J. An origami-inspired, self-locking robotic arm that can be folded flat. *Sci. Robot.* **2018**, *3*, 1–10. [CrossRef]
27. Miura, K.; Tachi, T. Synthesis of rigid-foldable cylindrical polyhedra. In Gmuend, Austria. *Symmetry Art Sci.* **2010**, 204–213.
28. Ouyang, B.; Wills, P.S.; Tang, Y.; Hallstrom, J.O.; Su, T.C.; Namuduri, K.; Mukherjee, S.; Rodriguez-Labra, J.I.; Li, Y.; Ouden, C.J. Initial Development of the Hybrid Aerial Underwater Robotic System (HAUCS): Internet of Things (IoT) for Aquaculture Farms. *IEEE Internet Things J.* **2021**, *8*, 14013–14027. [CrossRef]
29. LoRa Alliance. A technical overview of LoRa®and LoRaWAN™. May 2015. Available online: https://lora-alliance.org/sites/default/files/2018-04/what-is-lorawan.pdf (accessed on 25 June 2018).

30. Ouyang, B.; Wills, S.P.; Li, Y.; den Ouden, C. Hybrid Aerial Underwater Robotic System (HAUCS): The Initial Instruments Development and Deployment. In Proceedings of the 2020 IEEE 92nd Vehicular Technology Conference (VTC2020-Fall), Victoria, BC, Canada, 18 November–16 December 2020.
31. SIYI. 2.4G Datalink User Manual. Available online: https://discuss.ardupilot.org/uploads/short-url/sgfUtaqa7P5KZ8r4jRcTUtVpzoW.pdf (accessed on 9 July 2021).
32. Heltec. WiFi LoRa 32 (V2). Available online: https://resource.heltec.cn/download/WiFi_LoRa_32/WIFI_LoRa_32_V2.pdf (accessed on 5 April 2021).
33. Shenzhen Anxinke Technology. ESP-32SDatasheet. 3 October 2016. Available online: https://www.es.co.th/Schemetic/PDF/ESP32.PDF (accessed on 1 June 2021).
34. espressif. ESP-NOW. Available online: https://docs.espressif.com/projects/esp-idf/en/latest/esp32/api-reference/network/esp_now.html (accessed on 1 June 2021).
35. Van Manen, T.; Janbaz, S.; Ganjian, M.; Zadpoor, A.A. Kirigami-enabled self-folding origami. *Mater. Today* **2020**, *32*, 59–67. [CrossRef]
36. Reid, A.; Lechenault, F.; Rica, S.; Adda-Bedia, M. Geometry and design of origami bellows with tunable response. *Phys. Rev. E* **2016**, *95*, 013002. [CrossRef]
37. Saito, K.; Tsukahara, A.; Okabe, Y. Designing of self-deploying origami structures using geometrically misaligned crease patterns. *R. Soc.* **2016**, *472*, 20150235. [CrossRef] [PubMed]
38. Sellner, K.G.; Doucette, G.J.; Kirkpatrick, G.J. Harmful algal blooms: Causes, impacts and detection. *J. Ind. Microbiol. Biotechnol.* **2003**, *30*, 383–406. [CrossRef] [PubMed]
39. Mukherjee, S.; Ouyang, B.; Kamesh, N.; Wills, P.S. Multi-Agent Systems (MAS) related data analytics in the Hybrid Aerial Underwater Robotic System (HAUCS). In *Big Data III: Learning, Analytics, and Applications*; SPIE: Washington, DC, USA, 2021; Volume 11730.
40. Kresling, B. Natural twist buckling in shells: From the hawkmoth's bellows to the deployable Kresling-pattern and cylindrical Miura-ori. In Proceedings of the 6th International Conference on Computation of Shell and Spatial Structures, Ithaca, NY, USA, 28–31 May 2008.
41. Srinvasan, A. *Handbook of Precision Agriculture: Principles and Applications*; CRC Press: Boca Raton, FL, USA, 2006.

vehicles

Article

Modernization of Fire Vehicles with New Technologies and Chemicals

Cagri Un * and Kadir Aydın

Automotive Engineering Department, Çukurova University, Adana 01330, Turkey; kdraydin@cu.edu.tr
* Correspondence: cagri_un@hotmail.com

Abstract: Fire is a stable exothermic chain reaction of flammable materials brought together with oxygen or other oxidizing substances under certain conditions, occurring uncontrollably. Fire vehicles interfere with many types of fire, such as wildfires, factory fires, building fires, etc. During this intervention, fire vehicles generally use water or foam. In this study, new effective fire suppression applications are investigated. Thermal camera applications in fire trucks and also new extinguishing agents—boron-based chemicals—were tested in forest fire simulations. In these experiments, it was observed that the thermal camera detected the fire as soon as it occurred. It seemed appropriate to use thermal cameras for all types of fire vehicles (foam trucks, water tankers, rescue trucks, etc.). It was seen that the thermal camera application could detect and monitor the fire during the fire-extinguishing work of the firefighters. The boron-based fire suppressant had a better extinguishing and cooling effect than water in the experiments. Compared to the water used as a traditional method, the liquid boron-based extinguisher provided 22% faster—while the solid boron-based extinguisher provided 42% faster—suppression and cooling. With three separate experiments, it is predicted that thermal camera applications and the use of boron-based extinguishers in fire vehicles can lead to an effective and positive transformation in the coming years.

Keywords: vehicle technology; energy; fire suppression; boron

Citation: Un, C.; Aydın, K. Modernization of Fire Vehicles with New Technologies and Chemicals. *Vehicles* **2023**, *5*, 682–697. https://doi.org/10.3390/vehicles5020037

Academic Editors: Ralf Stetter and Mohammed Chadli

Received: 13 March 2023
Revised: 29 May 2023
Accepted: 2 June 2023
Published: 4 June 2023

1. Introduction

Fire is a flammable and uncontrolled process in which substances react with oxygen or other oxidizing agents under certain conditions. It is a stable, exothermic chain reaction.

In addition to causing material damage in the geographical area in which they occur, fires also have great negative effects on living creatures and ecological balance.

Concomitant heat generation is also achieved by the fire. This mechanism includes the atoms and molecules reacting with the oxygen in the air. Combustion is the term used to describe the event that starts the fire.

In order to start combustion, the temperature must reach the combustible material's ignition temperature. The ignition temperature that can ignite the fuel in air without a separate ignition source can be defined as the lowest temperature required.

The ignition temperatures of some solid-form combustible materials are shown in Table 1 [1].

Fire vehicles are one of the most important symbols of emergency response. In this respect, the history of fire trucks and how these have evolved from the earliest days of steed-drawn carriages into today's massive vehicles are below.

18th century

In the early 18th century, Englishman Richard Newsham filed two patents for his fire machine design. In 1721, an invention was depicted that could direct a sluice of water toward a fire. The alternate patent, issued in 1725, detailed an outfit operated through a hand pump for which two people worked a large switch while others directed the sock.

Revolutionary in design, these original constructions paved the way for ultramodern firefighting outfits.

Table 1. Ignition temperatures of some solid combustible materials (°C).

Solid Combustible Materials	Ignition Temperature (°C)
Cotton	400 °C
Wool	600 °C
Cotton fabric (Raw cloth)	225 °C
Nylon (6.66)	425 °C
Wood	240–270 °C
Pine tree	260 °C
Polyester	450–485 °C

19th century

Horse-drawn brume pumpers were still the most popular system for putting out fires in the early 19th century. As time passed, nags proved ineffective for larger extremities. After the American Civil War, small levy companies began to combine into larger ones with larger coffers and the power to attack the increasingly frequent and destructive fires sweeping through metropolises. In 1885, the first electric fire machine system patent, designed by Schuyler Wheeler, entered into force in the United States. These great leaps forward from former styles of pumping water allowed firefighters to receive the help they demanded to extinguish fires much more briskly and with added safety.

20th century

Firefighting technology advanced at an inconceivable pace throughout the 20th century. In the early 1900s, motorized pumps replaced gasoline-powered pumps. In the 1960s, water pumps, graduations, and cherry selectors surfaced as commonplace tools for firefighters. This ultramodern equipment allowed firefighters to target areas that were initially inaccessible. The introduction of elevating thunderclaps, enchantresses, and other inventions helped shape the ultramodern firefighting outfit.

21st century

The fire truck of the 21st century is a high-tech, high-powered machine that has been modified over centuries to meet the requirements of the ultramodern firefighting community. In addition to ergonomics, some departments now concentrate on terrain and community configurations when designing their lines. Currently, numerous fire departments focus on the aspects below:

- Safety;
- Optimized storehouses;
- Vehicle-to-vehicle communication;
- Advanced norms [2].

Currently, fire trucks are equipped with motorized machines with sophisticated after-treatment bias and automatic transmissions. Although the changes related to the machine were backed by immigration regulations set forth by the government, the use of motorized technology has brought an edge to the machine and the vehicle. The machine can now tell you when something is wrong with fairly accurate and specific information.

As vehicles become more motorized, the traditional red fire truck may evolve into a red fire truck that can talk to other vehicles, have some automated systems that do not involve human intervention, and have innovative ways to acclimate safety systems ahead of nonsupervisory changes. It will be intriguing to see how these effects manifest themselves to shape the fire truck of the future [3].

Forest fires can develop very quickly and cause great damage in very little time. Special applications should always be considered for forest fires. Thermal camera applications, which are one type of such applications, provide important information about the formation and progression of the fire by detecting temperature increases in places where there is a risk of fire or burning. These efficient and solution-oriented devices can produce results very quickly and can be used in fire-prone areas.

Thermal camera systems can be used to observe heat and how it spreads. These systems work by sensing heat and displaying the data from this heat as a heat map. High-tech thermal camera systems are also preferred for multiple detections of objects.

Many applications have been implemented in fire-extinguishing systems over the years. Fire trucks with water suppressant are considered to be the first generation of firefighter trucks and emerged more than 100 years ago [4].

Second-generation firefighter trucks appeared during the First World War. In these vehicles, there are powder and foam extinguishers. Second-generation firefighter trucks have been successful in fighting fires involving petrochemical-based products in B and C fire classes because these types of fires cannot be extinguished with water.

Increasing industrial developments in recent years have made fires more complex and multifaceted. In this respect, new technologies need to be applied and researched [5].

Foams and synthetic or detergent additives are known as traditional firefighting methods. Such applications may be insufficient against fires reaching or exceeding 700 °C. When fire interventions are examined in general, it is seen that the expectation of firefighters is to control the fire. Compared to traditional fire suppression systems, water mist has shown many benefits for sustainable fire suppression The extinguishing performance of a water mist system under various gas-spray momentum ratios were evaluated, and the effect of fire extinguishing was observed using laser-based measurement [6]. At the same time, Liu et al. [7] showed intervention studies with water mist and demonstrated the possible success of water mist systems in natural gas leakage fires.

Grant et al. [8] stated that water has become the most commonly used fire extinguishing agent. The thermal properties of water make it an ideal extinguishing medium for most types of fires, whether used to extract heat directly from flames, hot combustion products, or fuel surfaces.

Researchers such as Shukhman et al. [5] examined the emission values that occur during fire suppression with different extinguishing products. As a result of the chromatographic analysis of exhaust gases, hazardous formations have been observed for human health and engine inner surfaces. Additionally, early studies showed that Halon 1301 contains the poisonous products of its thermal destruction and chemical responses as HF, HBr, CF_3H, and CF_2O. These species are dangerous to human health and may be anticipated to be present in exhaust fumes while operating in a fire area. These may damage the machine internal shells. It is important to research the extinguishing chemicals that are the least harmful to humans and the environment [5].

Thermal imaging cameras (TIC) are becoming a significant device for many firefighters and other users. Increasing use of these cameras against fires in the future will provide a more effective application against fires. TIC can provide first responders with valuable information to measure track fire growth, a fire incident, allowing them to determine the location of victims [9].

TIC is used extensively in the development of imaging technology for firefighting operations. Firefighters may encounter high temperatures, open flames, sprays of water, etc. Thus, it is important that TIC is capable of seeing in these obstructive conditions with minimal hindrance from the surrounding terrain. Fire service thermal imagers are generally designed to detect radiant thermal energy in the 8 μm–14 μm spectral range. This energy is radiated from solid shells and from certain fumes that radiate.

The traditional method of using water or foam as extinguishing agents is widely used. However, this study investigates new effective fire suppression applications, such as thermal camera applications in fire trucks and the use of boron-based chemicals as extinguishing agents in forest fire simulations. The study found that the thermal camera application in fire trucks can detect fire flames as soon as they occur, making it suitable for all types of fire vehicles. The use of boron-based chemicals had better extinguishing and cooling effects than water in the experiments, indicating their potential to be an effective alternative to traditional extinguishing agents. The study predicts that the combination of thermal camera applications and the use of boron-based extinguishers in fire vehicles can lead to an effective and positive transformation in the coming years. The thermal camera application can detect and monitor the fire at the right time and place during firefighting work, allowing firefighters to take timely and appropriate measures in order to suppress fires. Overall, this highlights the importance of exploring new effective fire suppression applications and methods to improve the safety and effectiveness of firefighting work. The use of boron-based extinguishers and thermal camera applications in fire vehicles can potentially lead to a positive transformation in the firefighting industry, ultimately enhancing public safety.

2. Materials

For this study, water and boron-based solid and liquid fire extinguishers were used for combustion and extinguishing experiments. In addition, a thermal imaging camera (TIC) was used to detect the initial flame. This camera was also used to monitor the extinguishing effects and temperature progressions of the fire extinguishers. The properties of the materials used in the experiments are given below.

2.1. Solid and Liquid Boron-Based Fire Suppressant

In the Exp.SB and Exp.LB experiments, boron-based fire suppressants were used. Boron is used in many different fields. Boron production from boron minerals in the form of ores requires special processes. Boron is generally found in nature as borate salts or boric acid [10].

It is highly resistant to combustion reactions. For this reason, it is used as a fire and flame retardant. Additionally, it can be mixed with other substances with different properties. It prolongs the rate of absorption of the smoke from the fire in the form of embers quickly. Boric acid and borates provide resistance to cellulosic materials and fire. Before reaching the ignition temperature, they remove the water molecules in the cellulose and prevent further combustion by covering the surface of the coal. The use of cellulosic insulating materials such as durable material boric acid for fires, has led to increased demand [11].

Boron-based fire extinguishing chemicals are non-carcinogenic. These also have a long shelf life, superior properties in cooling, and narrow the environment for fire progression and ignition. With these features, they act by suffocating the fire.

Boron shows oxide properties. The melting temperature of the boron chemical is around 2300 °C. Due to this feature, it is very advantageous to use in combustion applications. Its use as a flame retardant or as an innovative step in fire extinguishing applications is very important. Boron is a chemical in flame retardants; it is used as a component in compounds such as borax, borate, or ammonium fluoroborate. The structure of the boron chemical provides a cooling process and prevents ignition [12].

In these studies, the properties of the liquid boron-based fire extinguisher used in Exp.LB are shown in Table 2. Generally, this product has a melting point above 600 °C. In addition, this product has water-repellent and odorless properties.

Table 2. The properties of boron-based liquid fire suppressant components [13].

Component Name	CAS No	Concentration
$Na_2B_4O_7\,5H_2O$	12179-04-3	10–20%
SiO_2	112926-00-8	<1%
Sodium bicarbonate	144-55-8	15–30%

For Exp.SB, the solid boron-based fire suppressant was used. In this solid extinguisher, borax pentahydrate is predominantly involved. Borax pentahydrate is obtained from tincal ore in crystal or powder form. The chemical formula of tincal is shown as $Na_2B_4O_7{\cdot}10H_2O$.

In the production stages of this boron chemical component have melting, filtration, centrifugation, crystallization and drying [12]. The molecular weight of borax pentahydrate is 291.35 g/mol. Regarding physical properties, it can be found in powder, white, crystal, and odorless forms. Boron-based solid fire extinguisher was used in the Exp.SB experiments. This chemical component also has water-repellent properties. Moisture content is 0.25% at maximum, and its bulk density is 1.13 ± 0.007 g/cm^3 at 20 °C. In the tables below, the properties of the boron chemical used in the Exp.SB experiment are indicated. The properties of the liquid boron-based fire extinguisher used in Exp.SB are shown in Tables 3–5.

Table 3. Physical properties of $Na_2B_4O_7$▪$5H_2O$ [14].

Features	Value
Bulk Density	0.966 g/cm^3 (granule)
Boiling Point	1575 °C
Heat Capacity	7.6 J/g °C
Thermal Conductivity	0.647 W/mK
Specific Weight	1.815 g/cm^3
Specific Surface Area	<1 m^2/g
Diffusion Coefficient	1.0 × 10^{-5} cm^2/s
Surface Tension	67.19 mN/m (1.0 wt% aqueous solution)
Melting Point	741 °C
Color Measurement Test	91.92 (avg. L value)
Molecular Weight	291.35 g/mole

Table 4. The properties of boron-based solid fire suppressant component [15].

Component Name	CAS No	Concentration
Boraxpentahydrate ($Na_2B_4O_7$▪H_2O)	12179-04-3	10–20%
Silicate (SiO_2)	112926-00-8	<2%

Table 5. Chemical properties of $Na_2B_4O_7$ [14].

Component	Content	
	Granular	Powder
B_2O_3	48.00–49.35%	47.80–49.00%
Na_2O	21.37–21.95%	21.36–21.81%
SO_4	mxm. 135 ppm	mxm. 200 ppm
Cl	mxm. 70 ppm	mxm. 70 ppm
Fe	mxm. 3 ppm	mxm. 3 ppm
Water Insoluble	mxm. 150 ppm	mxm. 150 ppm

Borax pentahydrate transforms into borax dihydrate at 160 °C–170 °C, into borax monohydrate at 290 °C–299 °C, and into anhydrous borax at 400 °C–450 °C. These situations are expressed by the equations below [16].

$$Na_2B_4O_7.5H_2O \rightarrow Na_2B_4O_7.2H_2O \text{ (160 °C–170 °C)}$$

$$Na_2B_4O_7.5H_2O \rightarrow Na_2B4O_7.H_2O \text{ (290 °C–299 °C)}$$

$$Na_2B_4O_7.5H_2O \rightarrow Na_2B_4O_7 \text{ (400 °C–450 °C)}$$

In these studies, an innovative extinguishing technology using boron-based fire extinguishers was tried. Boron chemicals have properties that prevent the progression and re-ignition of the fire. These chemicals form a film layer in the areas to which they are applied and thus provide a suffocating effect in the fire surround. In addition to all these, with their radiation dispersion feature, they provide and facilitate breathing during intervention and ensure the continuity of the intervention. Boron-based suppressants are environmentally friendly in the fire sector. They do not cause a carcinogenic effect on humans and do not cause an explosion. They are safe with fast extinguishing and cooling features [16].

2.2. *Thermal Imaging Cameras (TIC)*

Thermal imaging cameras (TIC) have become efficient for firefighters in recent years. As this is a new field for firefighters; in order to use TICs correctly and effectively during firefighting applications, performance trials, and standard application, procedures must be put into use [9].

TICs contain many features. They are very successful in detecting and tracking people and objects affected by fire, especially in low-visibility conditions. When the working principles of thermal cameras are examined in the electromagnetic spectrum, special optics and imaging sensors appear to be applied to create a digital image of the infrared range. Such applications are superior in low-visibility conditions, locating heat sources, detecting the possible existence of living things, and temperature measurement [17].

TICs can be used to very quickly detect bad contact of an electrical device and overheated mechanical parts so as to prevent serious fires and short circuits. For those parts that cannot be seen directly due to shielding, the thermal hidden danger can be found according to heat conduction to the external parts.

For fire trucks, the vehicle rescue equipment is the necessary equipment to ensure operational safety.

Compared to the main firefighting vehicle used in fire detection vehicle infrared cameras, the vehicle-mounted thermal camera is safe, simple, easy to operate, and easy to use. The entire vehicle thermal imaging camera system is based on an advanced, excellent vehicle thermal camera with a simple control system and a control terminal set in the cab. The vehicle-mounted thermal camera is controlled by the system motor and can have a remote connection through a tablet computer from the cab remote control.

It is suitable for all kinds of fire trucks (command vehicle, city main fighting vehicle, water tanker, foam truck, rescue truck, etc.). With large petrochemical fires; tire factories, paper mills, and other large factory fires; large residential fires; vehicle accident rescue and relief, etc., it can be a great help.

Thermal camera systems are a technology that makes it possible to see thermal radiation from objects that cannot be seen with the human eye (Figure 1). Thermal radiation is the energy transmitted by an object through its temperature and the wavelengths emitted by the electromagnetic spectrum. This technique is the result of accelerated charges emitting radiation in accordance with the laws of electrodynamics. Thermal cameras generally show heat density information in the range specified on the color scale (red = warm, blue = cold) [18].

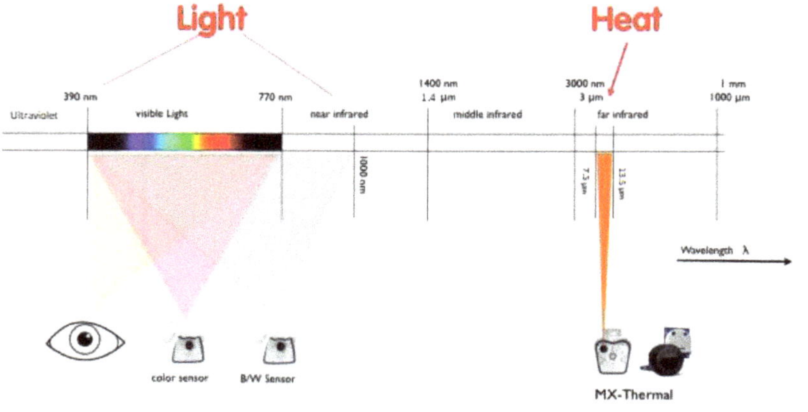

Figure 1. Spectral visible range [18].

Thermal cameras offer a distinctive feature compared to cameras with optical image sensors: the ability to show the slightest temperature differences with different colors. Thermal cameras can show upward temperature changes from a very low temperature value, such as 0.05 °C. In addition, it can indicate and signal the slightest temperature and heat change thanks to its features, such as the time of fire and set value. Thermal cameras have been used in many places where there is a risk of fire. In recent years, they have also been located in observation towers, especially during forest fires. The use of thermal cameras in fire trucks is also important in terms of responding to fires very quickly and efficiently.

The main characteristics of the thermal camera which was used in the experiments are given below and shown in Figures 2 and 3.

- ✓ Integrated speaker and microphone;
- ✓ Mx6 system platform (2nd generation) compatible with MxBus, H.264 and ONVIF;
- ✓ Thermal field of view; 45°, 25°, or 17°;
- ✓ Possibility of recording to an internal MicroSD card (4 GB as standard);
- ✓ Can be used in total darkness with MxActivitySensor feature;
- ✓ TR: calibrated thermal image sensor for measurable added value;
- ✓ Thermal camera POE with power consumption < 10 W;
- ✓ Fixed premium thermal image sensor with 50 mK.

The advantages of TIC are given below:

- ✓ High sensitivity and resolution in temperature and heat measurements;
- ✓ High temperature alarm detector feature;
- ✓ Provides image sharpening with auto focus feature;
- ✓ High temperature digital alarm, special alarm algorithm against forest fires is available;
- ✓ False alarm reduction can be achieved thanks to the monitoring protection area feature;
- ✓ Enlarged and finalized range of information for fire occurrence confirmation;
- ✓ 3D positioning function;
- ✓ This type of camera searches for fire and flame at a wide range. In fire formations, they can be observed even from a short distance.

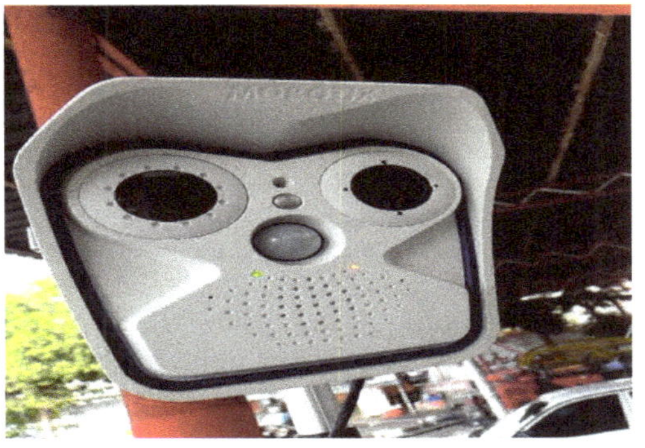

Figure 2. Thermal imaging camera which was used in the experiment.

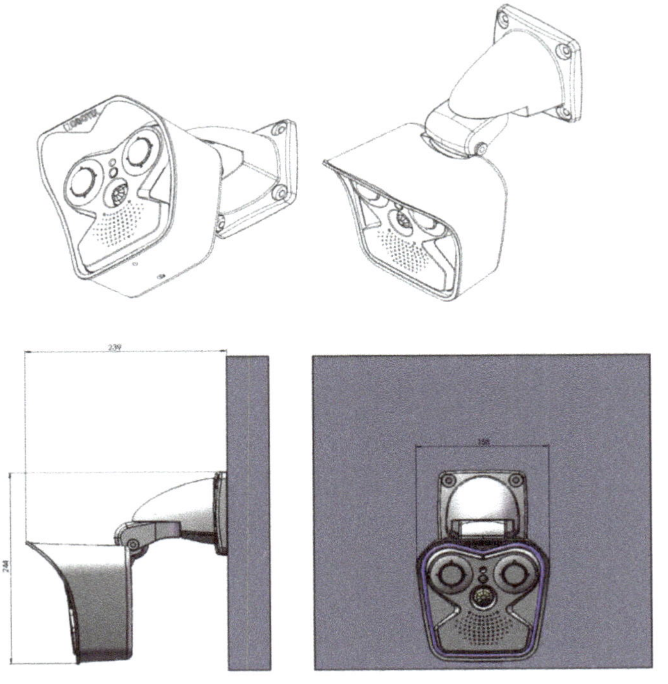

Figure 3. TICs technical drawings.

3. Experimental Setup

To observe the benefits of using thermal cameras and boron-based fire suppressants for fire trucks, a well-designed experimental setup is required. First, a fleet of fire trucks should be selected as participants in the experiment. These fire trucks should be equipped with thermal cameras and boron-based fire suppressants, respectively. Next, a controlled fire should be set up to simulate a real-world fire scenario. This could involve creating a fire in a controlled environment, such as a large field or empty lot. The fire should be allowed to grow and spread naturally, giving the fire trucks ample opportunity to respond and test the effectiveness of their equipment.

During the fire, data should be collected on several key metrics. First, thermal imaging footage should be captured to analyze the effectiveness of the thermal cameras in detecting hotspots and tracking the spread of the fire. Secondly, visual observations should be made to assess the effectiveness of the boron-based fire suppressants in extinguishing the flames. Additional data should also be gathered on factors such as response time, water usage, and the overall efficiency of the fire trucks equipped with thermal cameras and boron-based fire suppressants. This data can then be analyzed to determine the effectiveness of these technologies in real world fire scenarios. Overall, a well-designed experimental setup that carefully measures key metrics is essential to accurately assessing the benefits of thermal cameras and boron-based fire suppressants for fire trucks. By conducting such an experiment, researchers can better understand the potential of these technologies to improve the safety and effectiveness of firefighting efforts.

In the experimental setup, which was prepared by utilizing the steps made by Li et al. [19]. In the prepared experimental setup for this study, examinations of the first flame formation in the fire, the follow-up of the fire, and its extinction at every stage were carried out with a thermal camera.

For this research, three experimental setups were prepared and classified as Exp.W, Exp.SB, and Exp.LB. For the fire suppression operations, water was used in Exp.W, boron-based solid fire suppressant was used in Exp.SB, and boron-based liquid fire suppressant was used in Exp.LB. In all three experiments, the fire studies appeared to be similar to forest fires. For this, pinecones and branches were prepared in equal amounts and placed in the burning pan. Red pine, which is frequently found in the Mediterranean region, was used. Additionally, in this study, thermal imaging cameras (TIC) were used to detect the initial exacerbation of the generated fire pattern and to monitor the extinguishing effect and temperature progression of the fire extinguishers. All of the experiments were conducted with Çukurova University and Adana Metropolitan Municipality Department of Fire and Rescue officers.

As can be seen in Figure 4, the distance between the fire environment and the thermal camera is 15 m. The thermal camera is located at a height of 3 m. The experiment was carried out at ambient temperatures. Different extinguishing agents were manufactured by firefighters.

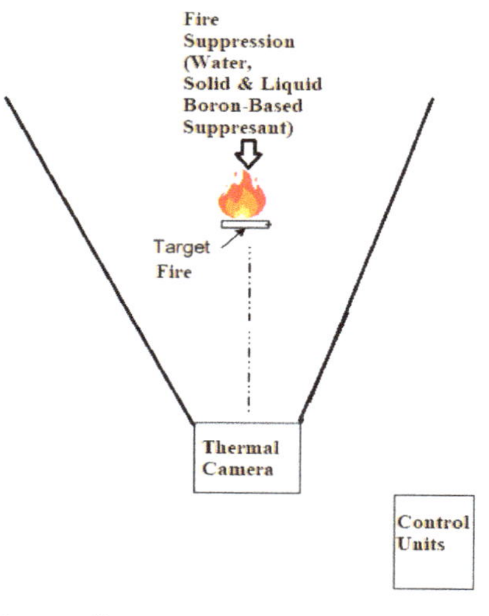

Figure 4. Experimental setup.

For this study, a forest fire model was prepared. For this model, pine tree branches and cones were prepared as flammable materials in three different experiments. (Exp.W, Exp.SB, Exp.LB).

The extinguishing and cooling times of these three extinguishers were tested by using water, boron-based liquids, and solid fire suppressants. The weight of the burning brushwood in each trial was 15 kg. Totals of 50 L of water, 5 L of boron-based liquid suppressant, and 5 kg of boron-based solid suppressant were used in the experiments (Table 6, Figures 5 and 6). In addition, a thermal camera system was used in all three trials, and thermal analysis was investigated for all three trials with a thermal camera. All trials were conducted outdoors and under the supervision of the firefighters. The fire start temperature was assumed to be ambient. The fire environment did not intervene until the temperature reached 700 °C. After reaching 700 °C, the fire extinguishing process started. The fire temperature was allowed to drop to 80 °C, and it was determined how long the extinguishers took to bring the fire under control and extinguish it.

Table 6. The experimental setups.

Experiment Name	Combustible Materials	Extinguishing Agent
Exp.W	pine cones and branches (15 kg.)	water (50 lt.)
Exp.SB	pine cones and branches (15 kg.)	boron-based solid fire suppressants (5 kg.)
Exp.LB	pine cones and branches (15 kg.)	boron-based liquid fire suppressants (5 lt.)

Figure 5. Brushwood fire setup.

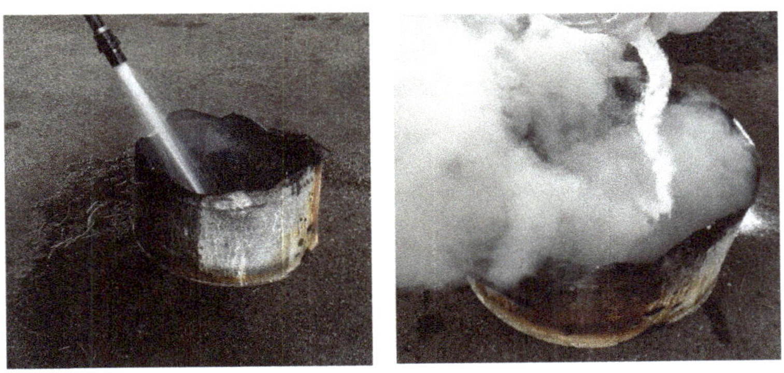

Figure 6. Brushwood fire suppressions with water and solid boron-based fire suppressant.

4. Results and Discussion

The thermal energy that radiates onto the TIC lens is concentrated on the detector. The signals generated by the detector are also manipulated into a video signal that is transferred to the display. Some TICs also have a video camera that can be used to record the signal via a hardwired connection or wireless transmitter. The signal that goes to the video display is not generally exactly the same as the signal that goes to the TIC display. The overall mileage of the TIC is therefore truly dependent on the capability of the observer to interpret the image displayed on the screen well enough to perform a task or operation successfully.

The thermal imaging camera is located in a place that sees the fire point from above for testing. In all three trials, it started to signal immediately after manual exacerbation. Afterwards, the thermal camera recorded the temperature and heat conditions. With these images, the heat dissipation and extinction conditions between the three experiments were observed (Figure 7). Experimental changes for the experiments, according to the types of chemical used in Exp.W, Exp.SB, and Exp.LB, are shown in the figures (Figures 8–10) and tables (Tables 7–9) below.

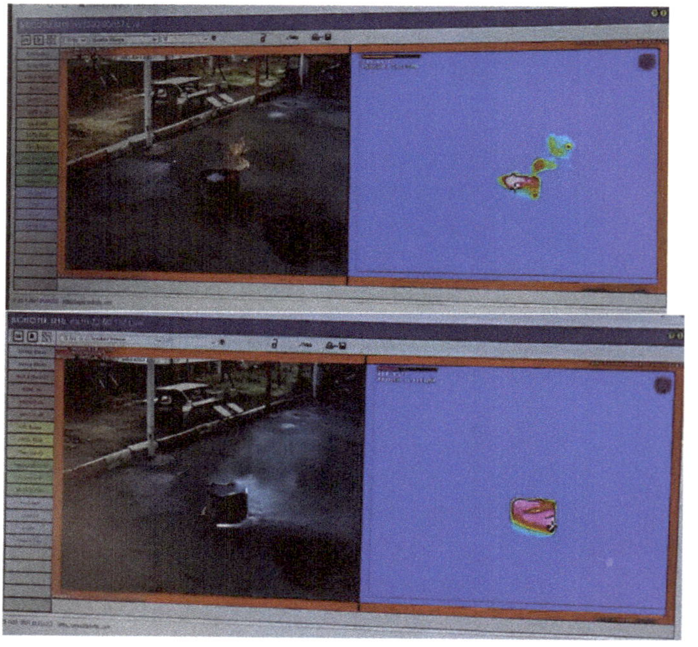

Figure 7. Thermal imaging camera images.

Table 7. Exp.W results.

Steps	Temperature and Time
Starting the fire	33 °C, 14:12:38
Start of suppression with water	700 °C, 14:16:20
End of the fire	80 °C, 14:23:34

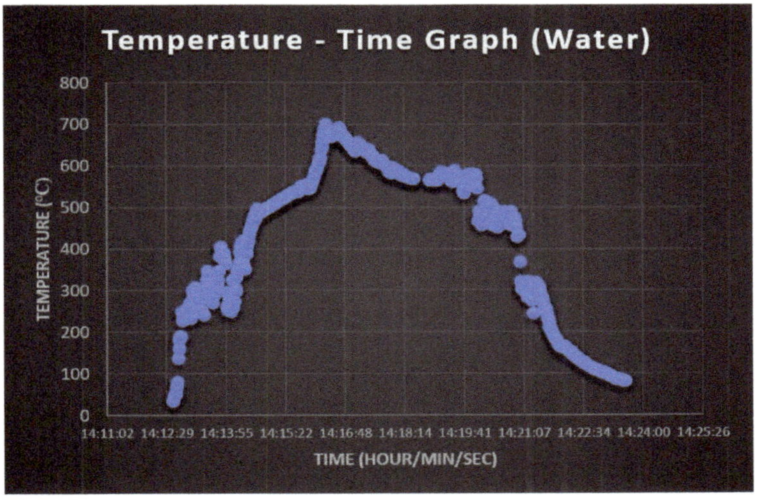

Figure 8. Exp.W results (temperature/time).

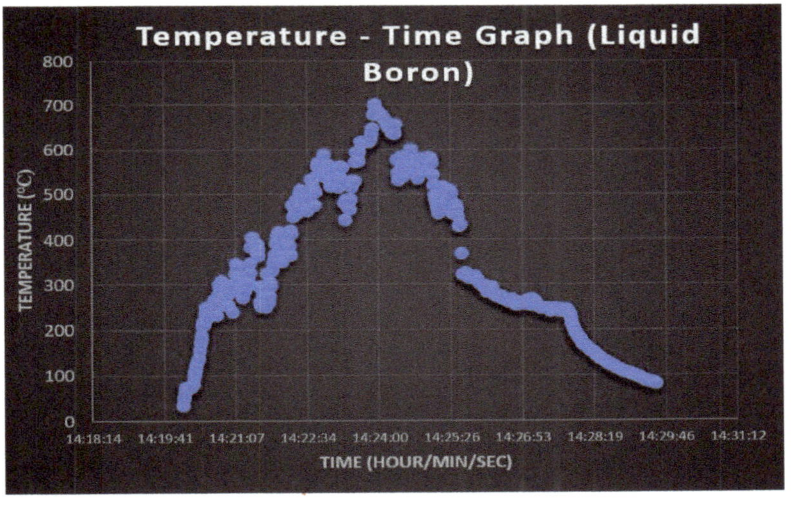

Figure 9. Exp.LB results (temperature/time).

Table 8. Exp.LB results.

Steps	Temperature and Time
Starting the fire	33 °C, 14:20:03
Start of suppression with liquid boron suppressant	700 °C, 14:23:56
End of the fire	80 °C 14:29:34

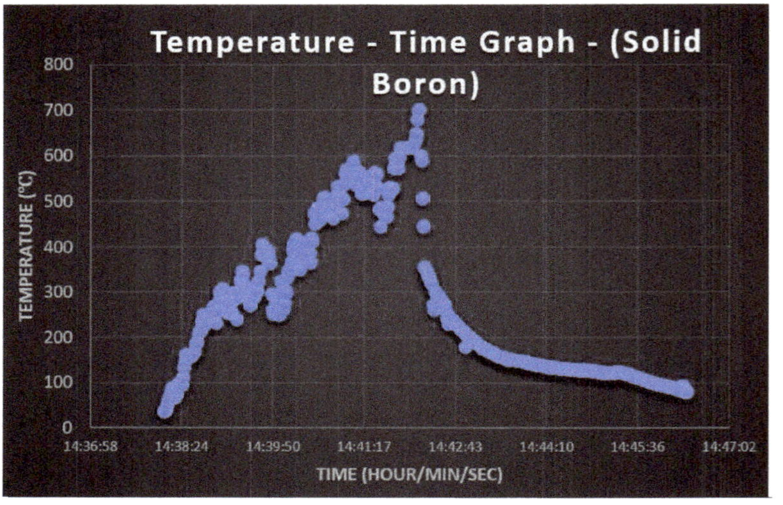

Figure 10. Exp.SB results (temperature/time graph).

Table 9. Exp.SB results.

Steps	Temperature and Time
Starting the fire	35 °C, 14:38:07
Start of suppression with solid boron suppressant	700 °C, 14:42:09
End of the fire	80 °C, 14:46:23

In the experiments, it has been determined that the solid boron-based fire extinguisher is more successful than the liquid boron-based fire extinguisher and the water extinguisher. It has been observed that the solid boron-based extinguisher provides a faster temperature drop and controls heat dissipation better. It is important to use different extinguishers instead of water in fires belonging to different types of fire classes.

It has been observed that, instead of water, boron-based extinguishers extinguish the fire much faster and provide cooling.

A boron report specifies that boron-based fire suppressants work by forming an insulating layer of borate glass. Boron, which the mineral contains in its makeup, reacts with oxygen and moisture to produce this fire-fighting protective barrier when heated up or contacted with water vapor or misty air. Its composition includes four elements: sodium, oxygen, hydrogen, and boric oxide, which we call "borate," as it provides protection against wildfires through the chemical reactions described above during exposure to heat or fire conditions. Boron is a chemical element that does not leave harmful residues, and its high cost makes it less desirable for widespread use in consumer products. Still, this property has made it the most effective material to utilize when combating large-scale fires from petroleum or coal processing facilities. Working with boron-based fire suppressants allows firefighters time to reach locations where they are needed before there is any risk from flames. After this compound starts burning, the smoke released is white and non-toxic [20].

As a result of the experiments, it was seen as an appropriate step to install the TIC systems on the stairs of the fire vehicles. Observations can be made from high points with the activation of the ladder and the gain of height. Additionally, the rotary head feature can provide even better observation capabilities. The figure below provides a visual of where to install TICs on a fire truck (Figure 11).

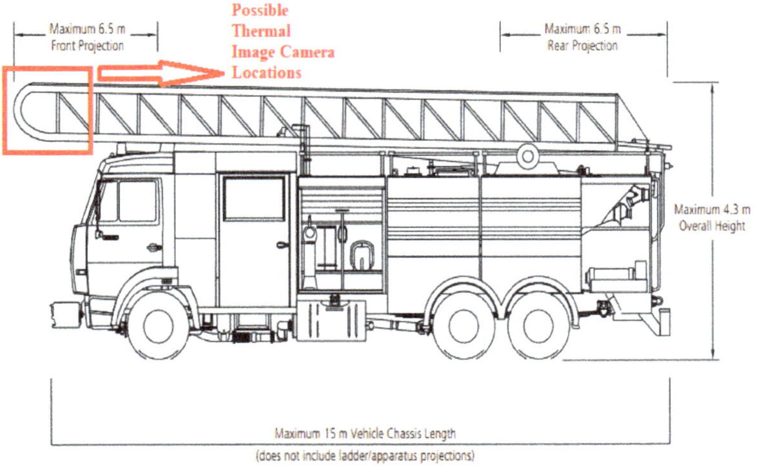

Figure 11. Possible TIC location for fire trucks.

5. Conclusions

For this study, three experiments were carried out with fire extinguishers with boron properties and with water. Suppression experiments were investigated with water and liquid and solid boron-based suppressants.

In each of these trials, plus the initial stage of flare-ups with a thermal camera, the heat dissipation view of the fires formed was given an overview.

In all trials, the extinguishing process was carried out by burning combustible material (pine tree twigs) under the same conditions. In this study, it was seen that boron-based suppressants displayed better performance for restraining the heat release rate than water. If water is used in fire suppression applications, a large amount of water should be used in general.

Boron-based fire extinguishers are very effective for extinguishing and cooling high-temperature fires. In the future, the usage of boron-based fire extinguishers should be further increased. Boron based extinguishers to be used during fire extinguishing works should be applied especially in forest fires. Boron-based extinguishers do not harm either nature or firefighters while extinguishing and cooling fires. With this feature, this type of extinguisher can be considered an environmentally friendly product and can also contribute to the rehabilitation of forests. The existence of boron-based extinguishers in the air and long-term performance effects on the engines of fire trucks must be investigated in the future. The boron-based fire suppressant had better extinguishing and cooling effects than water in the experiments. Compared to the water used as a traditional method, the liquid boron-based extinguisher provided 22% faster—and the solid boron-based extinguisher provided 42% faster—suppression and cooling.

Heat release rate (HRR) is considered the single most important fire response feature, as the heat released by a burning material can contribute to the growth and spread of fire.

At this point, the fight against HRR in large-scale fires includes a difficult process. These studies should be carried out in the future.

Thermal cameras, which have been applied in areas with fire risk in recent years, have the ability to quickly catch the onset of flare-ups. With laser illumination, infrared thermal imaging, night vision, image acquisition, and network transmission, thermal cameras can quickly detect fires from their onset.

Thermal cameras generally detect abnormal flame, smoke, and fire in the fire monitoring area in the desired time interval. They analyze the fire and make a quick comment about the characteristic situation of the fire. With these features, they help firefighters to deal with fires. They can greatly reduce the occurrence of false alarm situations. One of the most important features of these cameras is direct command and dispatch fire extinguishing according to the intuitive picture based on a real-time view of the scene. These cameras can be installed in places where there is a fire risk as well as by attaching them to the rising stairs of fire trucks. In the event of a fire, they can quickly analyze the fire area. Thus, they can give information to firefighters about the existence of living things in the area, the progress of the fire, and its extinguishing properties.

In addition, thermal cameras also assist firefighters in navigating in low-visibility conditions to find heat sources and to measure the temperature. These cameras are suitable for use in observation towers in forests where there is a risk of fire, in areas such as factories and landfills, as well as on the head of the rotating and rising ladder parts of fire trucks.

Firefighters can benefit from the use of thermal imaging cameras. The cameras have sensors and lenses that provide them with access to the IR spectrum. A firefighter can conduct successful search and rescues thanks to these features.

Author Contributions: Conceptualization, C.U. and K.A.; methodology, C.U. and K.A.; software, C.U. and K.A.; validation, C.U. and K.A.; formal analysis, C.U. and K.A.; investigation, C.U. and K.A.; resources, C.U. and K.A. data curation, C.U. and K.A. writing—original draft preparation, C.U. and K.A.; writing—review and editing, C.U. and K.A.; visualization, C.U. and K.A.; supervision, C.U. and K.A.; project administration, C.U. and K.A.; funding acquisition, C.U. and K.A. All authors have read and agreed to the published version of the manuscript.

Funding: This research was funded by the Çukurova University Scientific Research Projects Unit, grant number FDK-2020/12753 "Development of New Generation Automatic Fire Suppression Systems for Vehicles".

Institutional Review Board Statement: Not applicable.

Informed Consent Statement: Not applicable.

Data Availability Statement: Most of the data and the results are reported in the paper. Additional data can be requested from the corresponding author.

Conflicts of Interest: The authors declare no conflict of interest.

References

1. IBITEM. *Fire and Accident Fighting Training Book*; Istanbul Metropolitan Municipality Publications (IBB): İstanbul, Turkey, 2009. Available online: https://www.itfaiye.ibb.gov.tr/img/111657432020__7851983905.pdf (accessed on 12 December 2022).
2. A Brief History of Fire Trucks. Available online: https://www.fentonfire.com/blog/fire-truck-history/ (accessed on 28 May 2023).
3. Fire Truck of the Future. Available online: https://www.fama.org/forum_articles/fire-truck-future/ (accessed on 28 May 2023).
4. Wang, G.; Ji, Y.-X.; Shen, Y.-Z. The 3rd Generation Fire Truck and its Spraying Technique. *Procedia Eng.* **2011**, *11*, 424–430.
5. Shukhman, Y.; Baibikov, V.; Marmur, A.; Veinblat, M.; Tartakovsky, L. Internal Combustion Engine Response to Presence of Combustion Inhibitors in Ambient Air. *SAE Int. J. Eng.* **2013**, *6*, 1138–1144. [CrossRef]
6. Liu, Y.; Shen, J.; Ma, J.; Li, G.; Zhao, Z.; Ni, X.; Wang, X. Laser-based measurement and numerical simulation of methane-air jet flame suppression with water mist. *Process Saf. Environ. Prot.* **2021**, *148*, 1033–1047. [CrossRef]
7. Liu, Y.P.; Wang, X.S.; Zhu, P.; Li, G.C.; Ni, X.M.; Zhang, J. Experimental study on gas jet suppressed by water mist: A clean control technique in natural gas leakage incidents. *J. Clean. Prod.* **2019**, *223*, 163–175. [CrossRef]
8. Grant, G.; Brenton, J.; Drysdale, D. Fire suppression by water sprays. *Prog. Energy Combust. Sci.* **2000**, *269*, 79–130. [CrossRef]
9. Francine, A.; Nelson, B.; Andrew, L.; Anthony, H. *Performance Metrics for Fire Fighting Thermal Imaging Cameras—Small- and Full-Scale Experiments. NIST Technical Note 1499*; Fire Research Division Building and Fire Research Laboratory, National Institute of Standards and Technology: Gaithersburg, MD, USA, 2008.
10. Yonglan, X. Technologies for Boron Removal. *Ind. Eng. Chem. Res.* **2008**, *47*, 16–24.
11. Durgun, Z.G. Synthesis of Various Calcium Borates, Characterization and Investigation of Flame Retardant Efficiencies. Master Thesis, Ankara University, Ankara, Turkey, 2010.
12. Un, C.; Aydın, K. Thermal Runaway and Fire Suppression Applications for Different Types of Lithium Ion Batteries. *Vehicles.* **2021**, *3*, 480–497. [CrossRef]
13. Fireborex Boron Based Liquid Fire Extinguisher/Coolant Material Safety Information Form, Demircan Fire. Available online: http://www.novayangin.com/pdf/FIREBOREX%20%C3%87ok%20Ama%C3%A7l%C4%B1%20Jel%20MSDS.pdf (accessed on 12 December 2022).

14. ETI Mining Sodium Tetraborate Pentahydrate Product Technical Data Sheet. Available online: https://www.etimaden.gov.tr/rafine-urunler (accessed on 12 December 2022).
15. Fireborex Boron Based Solid Fire Extinguisher/Coolant Material Safety Information Form, Demircan Fire. Available online: http://www.novayangin.com/pdf/FIREBOREX%20ABC%20Dry%20Chemical%20MSDS.pdf (accessed on 12 December 2022).
16. Demir, C. The Evaluation of Boron Minerals as Energy Sources. Master Thesis, Ankara University, Ankara, Turkey, 2006.
17. Thermal Imaging for Aircraft Rescue and Fire Fighting Applications-Final Report. Available online: http://www.tc.faa.gov/its/worldpac/techrpt/tc17-27.pdf (accessed on 12 December 2022).
18. Mobotix Thermal Solutions, Technical Features and Areas of Application, Compact Guide. Available online: https://www.symquest.com/wp-content/uploads/mobotix-thermal-solutions-guide.pdf (accessed on 12 December 2022).
19. Li, Z.; Zhu, H.; Qiu, F.; Zhao, J.; Tian, C. Experiment Study on the Effectiveness of Various and Mixed Kinds of Low Expansion Foam of 120# Gasoline Pool Fire Suppression. *Fire* **2023**, *6*, 121. [CrossRef]
20. Boron: Fire Retardant, BENEFITS, Heat Resistant. Available online: https://borates.today/boron-fire-retardant/ (accessed on 12 March 2023).

MDPI
St. Alban-Anlage 66
4052 Basel
Switzerland
www.mdpi.com

Vehicles Editorial Office
E-mail: vehicles@mdpi.com
www.mdpi.com/journal/vehicles

www.ingramcontent.com/pod-product-compliance
Lightning Source LLC
Chambersburg PA
CBHW061502210926
51850CB00008B/1795
9783036586656